# Multimedia Signals and Systems

Srdjan Stanković • Irena Orović • Ervin Sejdić

# Multimedia Signals and Systems

## Basic and Advanced Algorithms for Signal Processing

Second Edition

Springer

Srdjan Stanković
University of Montenegro
Podgorica, Montenegro

Irena Orović
University of Montenegro
Podgorica, Montenegro

Ervin Sejdić
University of Pittsburgh
Pittsburgh, USA

ISBN 978-3-319-23948-4 ISBN 978-3-319-23950-7 (eBook)
DOI 10.1007/978-3-319-23950-7

Library of Congress Control Number: 2015954627

Springer Cham Heidelberg New York Dordrecht London


Printed on acid-free paper

Springer International Publishing AG Switzerland is part of Springer Science+Business Media (www.springer.com)

# Preface to the 2nd Edition

Encouraged by a very positive response to the first edition of the book, we prepared the second edition. It is a modified version which intends to bring slightly different and deeper insight into certain areas of multimedia signals. In the first part of this new edition, special attention is given to the most relevant mathematical transformations used in multimedia signal processing. Some advanced robust signal processing concepts are included, with the aim to serve as an incentive for research in this area. Also, a unique relationship between different transformations is established, opening new perspectives for defining novel transforms in certain applications. Therefore, we consider some additional transformations that could be exploited to further improve the techniques for multimedia data processing. Another major modification is made in the area of compressive sensing for multimedia signals. Besides the standard reconstruction algorithms, several new approaches are presented in this edition providing efficient applications to multimedia data. Moreover, the connection between the compressive sensing and robust estimation theory is considered. The chapter "Multimedia Communications" is not included because it did not harmonize with the rest of the content in this edition and will be a subject of a stand-alone publication. In order to enable a comprehensive analysis of images, audio, and video data, more extensive and detailed descriptions of some filtering and compression algorithms are provided compared to the first edition.

This second edition of the book is composed of eight chapters: Chapter 1—Mathematical transforms, Chapter 2—Digital audio, Chapter 3—Digital data storage and compression, Chapter 4—Digital image, Chapter 5—Digital video, Chapter 6—Compressive sensing, Chapter 7—Digital watermarking, and Chapter 8—Telemedicine. As described above, the chapter entitled "Mathematical transforms" (Chap. 1) and the chapter entitled "Compressive sensing" (Chap. 6) have been significantly modified and supplemented by advanced approaches and algorithms. In order to facilitate the understanding of the concepts and algorithms, the authors have put in efforts to additionally enrich information in other chapters as well.

Each chapter ends with a section of examples and solved problems that may be useful for additional mastering and clarification of the presented material. Also, these examples are used to draw attention to certain interesting applications. Besides the examples from the previous editions, the second edition contains some advanced problems as a complement to the extended theoretical concepts. A considerable number of Matlab codes are included in the examples, so that the reader can easily reconstruct most of the presented techniques.

Regardless of the efforts that the authors made to correct errors and ambiguities from the first edition, the authors are aware that certain errors may appear in this second edition as well, since the content was changed and extended. Therefore, we appreciate any and all comments made by the readers.

Further, the authors gratefully acknowledge the constructive help of our colleagues during the preparation of this second edition, particularly to the help of Prof. Dr. Ljubiša Stanković and Dr. Milica Orlandić. Also, we are thankful to the Ph.D. students Miloš Brajović, Andjela Draganić, Stefan Vujović, and Maja Lakičević.

Finally, we would like to extend our gratitude to Prof. Dr. Moeness Amin whose help was instrumental together with the help of Prof. Dr. Sridhar Krishnan to publish the first edition of this book. Prof. Dr. Zdravko Uskoković and Prof. Dr. Victor Sucic also contributed to the success of the first edition.

| | |
|---|---|
| Podgorica, Montenegro | Srdjan Stanković |
| Podgorica, Montenegro | Irena Orović |
| Pittsburgh, USA | Ervin Sejdić |
| July 2015 | |

# Introduction

Nowadays, there is an intention to merge different types of data into a single vivid presentation. By combining text, audio, images, video, graphics, and animations, we may achieve a more comprehensive description and better insight into areas, objects, and events. In the past, different types of multimedia data were produced and presented by using a separate device. Consequently, integrating different data types was a demanding project by itself. The process of digitalization brings new perspectives and the possibility to make a universal data representation in binary (digital) format. Furthermore, this creates the possibility of computer-based multimedia data processing, and now we may observe computer as a multimedia device which is a basis of modern multimedia systems.

Thus, ***Multimedia*** is a frequently used word during the last decade and it is mainly related to the representation and processing of combined data types/media into a single package by using the computer technologies. Nevertheless, one should differentiate between the term multimedia used within certain creative art disciplines (assuming a combination of different data for the purpose of efficient presentation) and the engineering aspect of multimedia, where the focus is towards the algorithms for merging, processing, and transmission of such complex data structures.

When considering the word etymology, we may say that the term multimedia is derived from the Latin word *multus*, meaning numerous (or several), and *medium*, which means the middle or the center.

The fundamentals of multimedia systems imply creating, processing, compression, storing, and transmission of multimedia data. Hence, the multimedia systems are multidisciplinary (they include certain parts from different fields, especially digital signal processing, hardware design, telecommunications and computer networking, etc.).

The fact that the multimedia data can be either time-dependent (audio, video, and animations) or space-dependent (image, text, and graphics) provides additional challenges in the analysis of multimedia signals.

Most of the algorithms in multimedia systems have been derived from the general signal processing algorithms. Hence, a significant attention should be paid to the signal processing theory and methods which are the key issues in further enhancing of multimedia applications. Finally, to keep up with the modern technologies, the multimedia systems should include advanced techniques related to digital data protection, compressive sensing, signal reconstruction, etc.

Since the multimedia systems are founded on the assumption of integrating the digital signals represented in the binary form, the process of digitalization and its effect on the signal quality will be briefly reviewed next.

## Analog to Digital Signal Conversion

The process of converting analog to digital signals is called digitalization. It can be illustrated by using the following scheme:

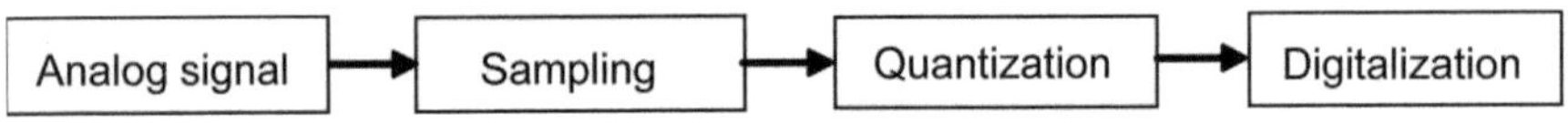


The sampling of an analog signal is performed by using the sampling theorem which ensures the exact signal reconstruction from its digital samples. The Shannon-Nyquist sampling theorem defines the maximal sampling interval (the interval between successive samples) as follows:

$$T \leq \frac{1}{2f_{\max}},$$

where $f_{max}$ represents the maximal signal frequency. According to the analog signal nature, the discrete signal samples may have any value from the set of real numbers. It means that, in order to represent the samples with high precision in the digital form, a large number of bits are required. Obviously, this is difficult to realize in practice, since the limited number of bits is available for representing signal samples. The number of bits per sample defines the number of quantization intervals, which further determines a set of possible values for digital samples. Hence, if the value of the sample is between two quantization levels, it is rounded to the closer quantization level. The original values of samples are changed and the changes are modeled as a quantization noise. The signal, represented by $n$ bits, will have $2^n$ quantization levels. As illustrations, let us observe the examples of 8-bit and 16-bit format. In the first case the signal is represented by 256 quantization levels, while in the second case 65536 levels are available.

Working with digital signals brings several advantages. For instance, due to the same digital format, different types of data can be stored in the same storage media, transmitted using the same communication channels, and processed and displayed by the same devices, which is inapplicable in the case of an analog data format. Also, an important property is robustness to noise. Namely, the digital values "0" and "1" are associated with the low (e.g., 0 V) and high voltages (e.g., 5V). Usually the threshold between the values 0 and 1 is set to the average between their corresponding voltage levels. During transmission, a digital signal can be corrupted by noise, but it does not affect the signal as long as the digital values are preserved, i.e., as long as the level of "1" does not become the level of "0" and vice versa.

However, the certain limitations and drawbacks of the digital format should be mentioned as well, such as quantization noise and significant memory requirements, which further requires the development of sophisticated masking models and data compression algorithms.

In order to provide a better insight into the memory requirements of multimedia data, we can mention that text requires 1.28 Kb per line (80 characters per line, 2 bytes per character), stereo audio signal sampled at 44100 Hz with 16 bits per sample requires 1.41 Mb, and a color image of size $1024 \times 768$ requires 18.8 Mb (24 bits per pixel are used), while a video signal with the TV resolution requires 248.8 Mb (resolution $720 \times 576$, 24 bits per pixel, 25 frames per second).

# Contents

# List of Figures

# List of Tables

# Chapter 1
# Mathematical Transforms Used for Multimedia Signal Processing

Various mathematical transformations are used for multimedia signal processing due to the diverse nature of these signals. Specifically, multimedia signals can be time-dependent, i.e., the content changes over time (audio, video) or time-independent media (text, images). In addition to the Fourier analysis, the time-frequency and wavelet transforms are often used. In some cases, other advanced methods (e.g., the Hermite projection method) may be of interest as well. In this chapter, we consider the fundamentals of the commonly used signal transformations together with some advanced signal processing approaches.

## 1.1 Fourier Transform

The Fourier transform is one of the basic mathematical transformations used for multimedia signal processing. Moreover, many other mathematical transformations are based on the Fourier transform.

To understand the Fourier transform, let us consider a simple example involving a sinusoidal signal, $f(t) = \cos(\omega_1 t)$, as shown in Fig. 1.1a.

The considered signal is completely defined by its frequency, initial phase, and amplitude. The frequency and the amplitude can be obtained by using the Fourier transform as depicted in Fig. 1.1b. Also, we may observe from Fig. 1.1b that a sinusoid is represented by two peaks in the frequency domain. This occurs due to the nature of the Fourier transform. Namely, symmetrical components at negative frequencies appear for real signals. Hence, the signal is often transformed into its analytical form before processing.

Consider the signal in Fig. 1.2a. It is more beneficial to represent the signal in the frequency domain, since the signal consists of two sine waves of different frequencies and amplitudes (Fig. 1.2b).

 
S. Stanković et al., *Multimedia Signals and Systems*,
DOI 10.1007/978-3-319-23950-7_1

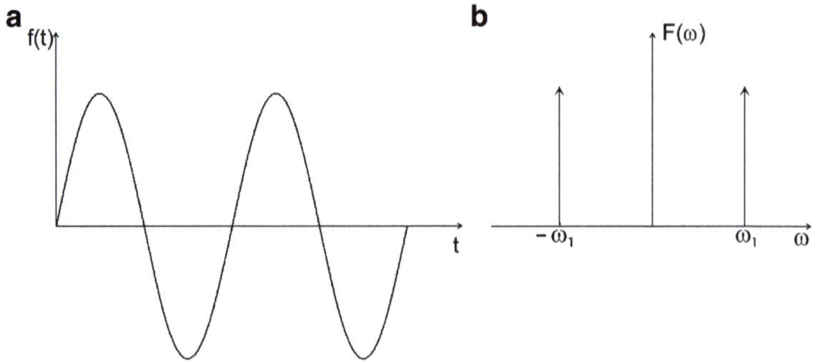


**Fig. 1.1** Signal representations in: (**a**) the time domain; (**b**) the frequency domain

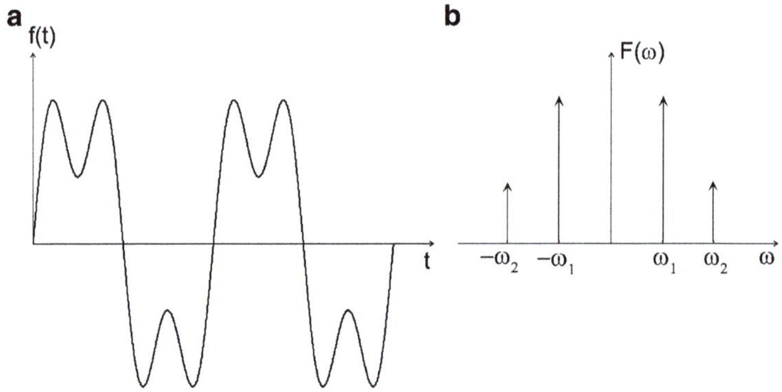


**Fig. 1.2** Representations of a multicomponent signal: (**a**) the time domain representation, (**b**) the frequency domain representation

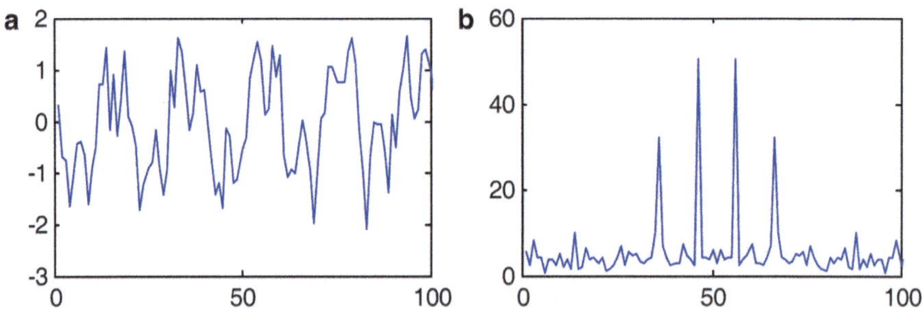


**Fig. 1.3** Signal representation (obtained using Matlab): (**a**) the time domain; (**b**) the frequency domain

The time domain representation can be especially difficult to interpret for signals corrupted by noise (e.g., white Gaussian noise as shown in Fig. 1.3a). As shown in Fig. 1.3b, it is easier to interpret the signal parameters in the frequency domain representation. Specifically, the energy of certain type of noise can be scattered across all frequencies, while the signal is concentrated at the frequencies of the sinusoidal components.

Let us introduce the mathematical definition of the Fourier transform for a signal $f(t)$:

$$F(\omega) = \int_{-\infty}^{\infty} f(t)e^{-j\omega t}dt. \tag{1.1}$$

The inverse Fourier transform is used to obtain the time domain representation of the signal:

$$f(t) = \frac{1}{2\pi}\int_{-\infty}^{\infty} F(\omega)e^{j\omega t}d\omega. \tag{1.2}$$

Next, we briefly review some of the Fourier transform properties.

*Linearity*: The Fourier transform of a linear combination of signals is equal to the linear combination of their Fourier transforms:

$$\begin{aligned}\int_{-\infty}^{\infty} (\alpha f(t) + \beta g(t))e^{-j\omega t}dt &= \alpha\int_{-\infty}^{\infty} f(t)e^{-j\omega t}dt + \beta\int_{-\infty}^{\infty} g(t)e^{-j\omega t}dt \\ &= \alpha F(\omega) + \beta G(\omega).\end{aligned} \tag{1.3}$$

In other words, $FT\{\alpha f(t) + \beta g(t)\} = \alpha FT\{f(t)\} + \beta FT\{g(t)\}$, where $FT$ denotes the Fourier transform.

*Time shift*: Shifting the signal $f(t)$ by $t_0$ in the time domain results in multiplying the Fourier transform with a phase factor:

$$\int_{-\infty}^{\infty} f(t-t_0)e^{-j\omega t}dt = e^{-j\omega t_0}F(\omega). \tag{1.4}$$

*Frequency shift*: Modulating the signal with a complex exponential function shifts the Fourier transform $F(\omega)$ along the frequency axis:

$$\int_{-\infty}^{\infty} \left(e^{j\omega_0 t}f(t)\right)e^{-j\omega t}dt = F(\omega-\omega_0). \tag{1.5}$$

*Convolution*: The Fourier transform of convolution of two functions $f(t)$ and $g(t)$ is equal to the product of the Fourier transforms of the individual signals:

$$FT\{f(t)*g(t)\} = FT\left\{\int_{-\infty}^{\infty} f(\tau)g(t-\tau)d\tau\right\} = F(\omega)G(\omega). \tag{1.6}$$

On the other hand, the Fourier transform of the product of two signals equals to convolution of their Fourier transforms:

$$FT\{f(t)\cdot g(t)\} = F(\omega)*_{\omega}G(\omega), \tag{1.7}$$

where $*_{\omega}$ denotes the convolution in frequency domain.

### 1.1.1 Discrete Fourier Transform

Given that discrete signals are mainly used in applications, it is necessary to introduce the Fourier transform in its discrete form. Specifically, for a limited duration discrete signal (with $N$ samples) illustrated in Fig. 1.4, the discrete Fourier transform is given by:

$$DFT(k) = \sum_{n=0}^{N-1} f(n)e^{-j\frac{2\pi}{N}nk}. \tag{1.8}$$

The inverse discrete Fourier transform is defined as:

$$f(n) = \frac{1}{N}\sum_{k=0}^{N-1} DFT(k)e^{j\frac{2\pi}{N}nk}. \tag{1.9}$$

It should be mentioned that computationally efficient algorithms known as the Fast Fourier Transform (FFT) algorithms have been derived to calculate discrete Fourier transform and its inverse.

To become more familiar with the Fourier transform, we recommend considering problems found at the end of this chapter.

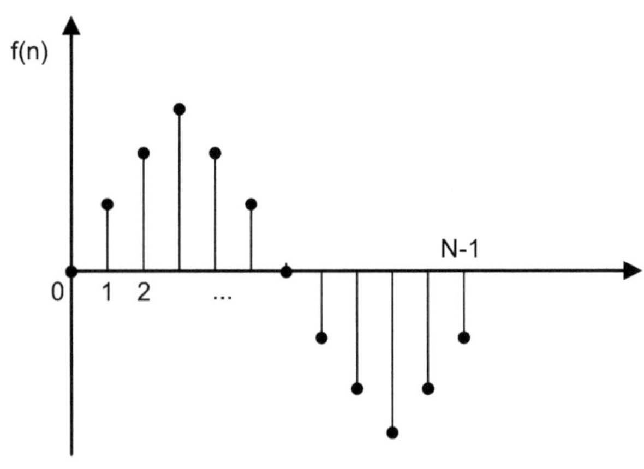


**Fig. 1.4** Finite duration discrete signal

### 1.1.2 Discrete Cosine Transform

Beside the Fourier transform, the discrete cosine transform (DCT) is used in many applications with real signals. The DCT is real-valued transform and represents the positive part of the spectrum. The most commonly used form of DCT (DCT-II which is usually referred simply as DCT) is defined:

$$DCT(k) = c(k)\sum_{n=0}^{N-1} f(n)\cos\frac{(2n+1)k\pi}{2N},\ k = 0, \ldots, N-1, \tag{1.10}$$

where the normalization coefficient $c(k)$ is:

$$c(k) = \begin{cases} \sqrt{1/N}, & k = 0 \\ \sqrt{2/N}, & k = 1,..,N-1, \end{cases}$$

and $f(n)$ represents a signal with $N$ being its length.

The inverse DCT is given by:

$$f(n) = \sum_{k=0}^{N-1} c(k)DCT(k)\cos\frac{(2n+1)k\pi}{2N},\ n = 0, \ldots, N-1. \tag{1.11}$$

## 1.2 Filtering in the Frequency Domain

The frequency domain representation of signals is suitable for signal filtering, which can be done by using low-pass, high-pass, and/or band-pass filters. The ideal forms of these filters are defined as follows (Fig. 1.5):

Low-pass filter:

$$H(\omega) = \begin{cases} 1, & \text{for } |\omega| < \omega_L, \\ 0, & \text{otherwise.} \end{cases} \tag{1.12}$$

High-pass filter:

$$H(\omega) = \begin{cases} 1, & \text{for } |\omega| > \omega_H, \\ 0, & \text{otherwise.} \end{cases} \tag{1.13}$$

Band-pass filter:

$$H(\omega) = \begin{cases} 1, & \text{for } \omega_L < |\omega| < \omega_H, \\ 0, & \text{otherwise.} \end{cases} \tag{1.14}$$

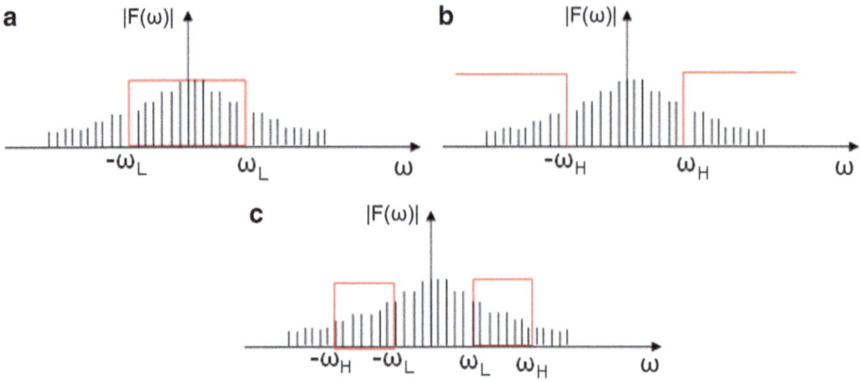


**Fig. 1.5** Frequency domain representation of filters: (**a**) a low-pass filter, (**b**) a high-pass filter, (**c**) a band-pass filter

Filtering in the frequency domain is simply performed by multiplying the Fourier transform of the signal with the filter transfer function. Then, the time domain representation of the filtered signal ($g(t)$) can be obtained by the inverse Fourier transform of their product:

$$\begin{aligned} G(\omega) &= F(\omega)H(\omega), \\ g(t) &= \frac{1}{2\pi}\int_{-\infty}^{\infty} G(\omega)e^{j\omega t}d\omega. \end{aligned} \tag{1.15}$$

## 1.3 Time-Frequency Signal Analysis

Time-frequency analysis is used to represent signals with a time-varying spectral content, since the Fourier transform does not provide sufficient information about these signals. Specifically, the Fourier transform provides information about the frequency content of the signal, but there is no information about the time instants when spectral components appear. For example, using the Fourier transform to analyze a speech signal, we obtain the spectral content of different sounds, but not their timing.

Using a simple example, let us illustrate the advantages of using the time-frequency analysis in comparison to the Fourier transform of the signal. For example, Fig. 1.6 depicts that the amplitude spectra (the amplitude of the Fourier transforms) of two different signals can be almost the same.

Hence, to obtain more information about these signals, it is necessary to use a representation from which one can follow temporal changes of the spectrum. Such a representation can be obtained by using the time-frequency analysis, as illustrated in Fig. 1.7 (the temporal changes of the spectrum are depicted, but not the energy distribution of spectral components).

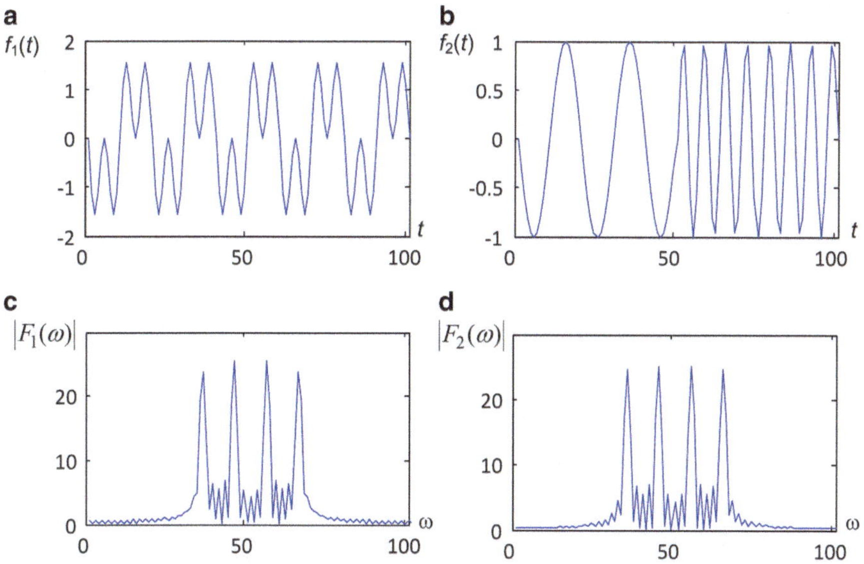


**Fig. 1.6** Comparing frequency domain representations of two different signals: (**a**) a sum of two sinusoids with equal duration, (**b**) a composition of two sinusoids appearing at different time instants, (**c**) the Fourier transform for the first signal, (**d**) the Fourier transform for the second signal

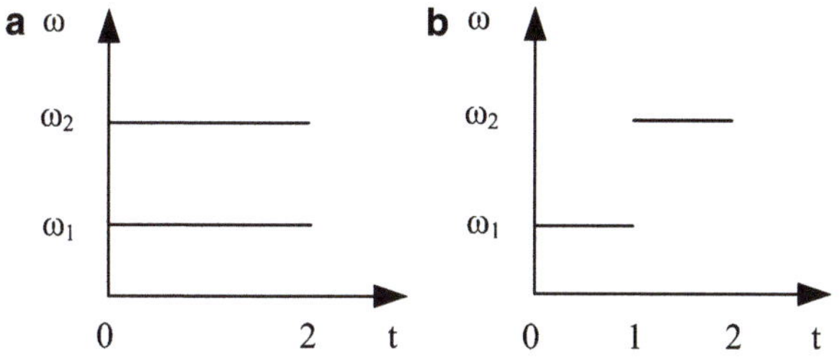


**Fig. 1.7** (**a**) The ideal time-frequency representation of the sum of sinusoids from Fig. 1.6a, (**b**) the ideal time-frequency representation of the time-shifted sinusoids from Fig. 1.6b

Time-frequency distributions also provide information about the energy distribution around the instantaneous frequency. It is important to note that there is no single time-frequency distribution that is optimal for all types of signals. That is, different time-frequency distributions are used depending on the application and on the signal type. The most commonly used distributions are the spectrogram and the Wigner distribution. The spectrogram is the squared module of the short-time Fourier transform, while the Wigner distribution is a quadratic distribution and exhibits significant drawbacks when applied to multicomponent signals that are often found in practical applications.

## 1.4 Ideal Time-Frequency Representation

Before we consider various time-frequency representations, let us introduce an ideal time-frequency representation. Consider a signal defined as:

$$f(t) = Ae^{j\phi(t)}, \tag{1.16}$$

where $A$ is the amplitude, and $\phi(t)$ is the phase of the signal. Note that the first phase derivative has the physical meaning and represents the instantaneous frequency, i.e., $\omega = \phi'(t)$. Therefore, the ideal time-frequency representation should concentrate energy along the instantaneous frequency of the signal and can be written in the form:

$$ITF(t,\omega) = 2\pi A^2 \delta(\omega - \phi'(t)). \tag{1.17}$$

## 1.5 Short-Time Fourier Transform

The Short-Time Fourier transform (STFT) of a signal $f(t)$ is defined as:

$$STFT(t,\omega) = \int_{-\infty}^{\infty} w(\tau) f(t+\tau) e^{-j\omega\tau} d\tau, \tag{1.18}$$

where $w(t)$ is a window function. It provides the time-frequency representation by sliding the window and calculating the local spectrum for each windowed part of the signal as illustrated in Fig. 1.8.

The STFT is a linear transform. In other words, the STFT of a multicomponent signal: $f(t) = \sum_{m=1}^{M} f_m(t)$, is equal to the sum of the STFTs of the individual components:

$$STFT(t,\omega) = \sum_{m=1}^{M} STFT_{f_m}(t,\omega). \tag{1.19}$$

This is an important feature of STFT, since many practical signals are the multicomponent ones. As previously mentioned, the spectrogram is the squared module of the STFT:

$$SPEC(t,\omega) = |STFT(t,\omega)|^2. \tag{1.20}$$

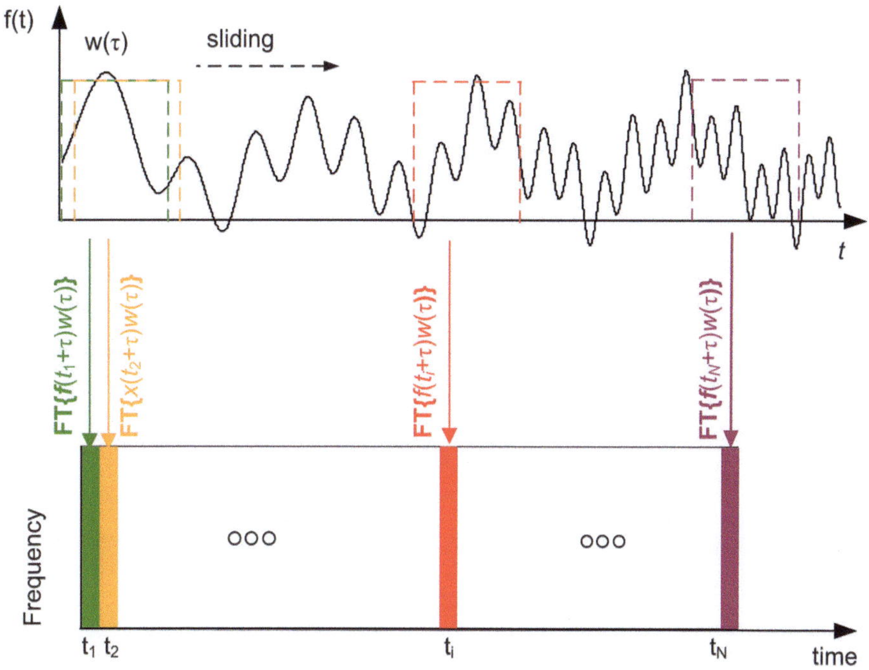


**Fig. 1.8** An illustration of the STFT calculations

In practical application, the discrete version of the STFT is used:

$$STFT(n,k) = \sum_{m=-N/2}^{N/2-1} w(m)x(n+m)e^{-j2\pi mk/N}, \tag{1.21}$$

where $N$ is the length of the signal, $n$ and $k$ are discrete time and frequency parameters, respectively.

Unlike the STFT, the spectrogram is a real-valued function. The main drawback of STFT (and the spectrogram) is the fact that the time-frequency resolution highly depends on the window width. Specifically, we obtain good time resolution (and poor frequency resolution) using a narrow window. On the other hand, a wider window enhances the frequency resolution, but decreases the time resolution. To illustrate this trade-off between time and frequency resolutions, let us consider the following example:

$$f(t) = \delta(t-t_1) + \delta(t-t_2) + e^{j\omega_1 t} + e^{j\omega_2 t}. \tag{1.22}$$

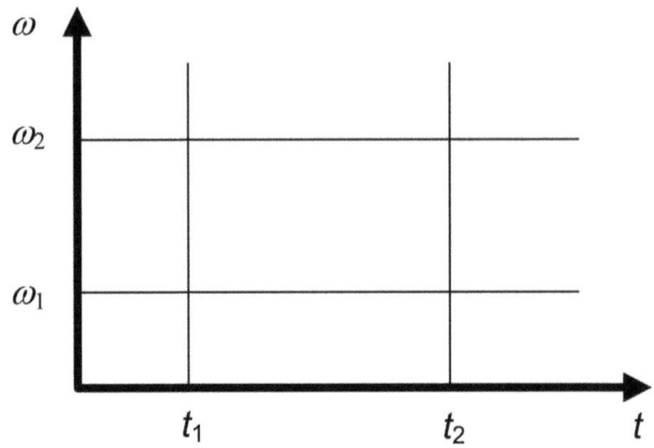


**Fig. 1.9** Ideal time-frequency representation of signal *f*(*t*)

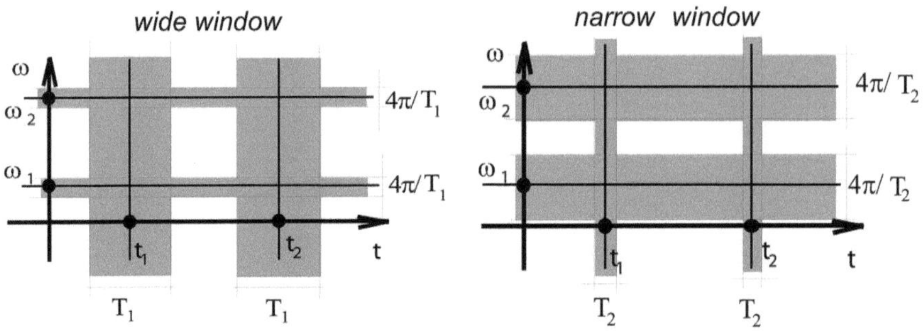


**Fig. 1.10** Illustration of the uncertainty principle ($T_1$ and $T_2$ are window widths in the time domain)

The ideal time-frequency representation of *f*(*t*) is illustrated in Fig. 1.9.

Using the definition of the STFT, we obtain the following time-frequency representation:

$$\begin{aligned} STFT(t,\omega) &= w(t_1-t)e^{-j\omega(t_1-t)} + w(t_2-t)e^{-j\omega(t_2-t)} \\ &\quad + W(\omega-\omega_1)e^{j\omega_1 t} + W(\omega-\omega_2)e^{j\omega_2 t}, \end{aligned} \tag{1.23}$$

where $W(\omega)$ is the Fourier transform of the window function. Figure 1.10 clearly shows how the time-frequency representation depends of the window function (length and type). When using the rectangular window, the product of time and frequency resolutions for the considered example is $D \cdot d = 4\pi$, where $d$ is the window width in the time domain, while $D$ is the window width in the frequency domain. Hence, increasing the resolution in one domain decreases the resolution in other domain.

According to the uncertainty principle, it is not possible to arbitrarily concentrate a signal in time and frequency domain. The more concentrated the signal is in the time domain, the wider band it occupies in the frequency domain.

Generally, the uncertainty principle in signal processing states that for any function satisfying $w(t)\sqrt{t} \to 0 \; as \; t \to \pm\infty$, the product of measures of duration in time and frequency is:

$$M_T M_W \geq \frac{1}{2}, \tag{1.24}$$

where:

$$M_T = \frac{\int_{-\infty}^{\infty} \tau^2 |w(\tau)|^2 d\tau}{\int_{-\infty}^{\infty} |w(\tau)|^2 d\tau}, \qquad M_w = \frac{\int_{-\infty}^{\infty} \omega^2 |W(\omega)|^2 d\omega}{\int_{-\infty}^{\infty} |W(\omega)|^2 d\omega}.$$

The lowest product is obtained for the Gaussian window function: $M_T M_W = \frac{1}{2}$. In addition, Fig. 1.11 demonstrates time-frequency representations using wide and narrow windows of two multicomponent signals: the first one consists of two sinusoids and a linear-frequency modulated signal, i.e., chirp (Fig. 1.11a), while the second consists of three sinusoids of short duration and a chirp (Fig. 1.11b). The ideal time-frequency representations are presented as well.

Using a narrow window, we achieve a good time resolution for sinusoidal signal components as shown in the second column of Fig. 1.11. Using a wide window, a good frequency resolution of these components is achieved, but the time resolution is significantly decreased. Notice that the time-frequency resolution of the chirp component is poor in both cases.

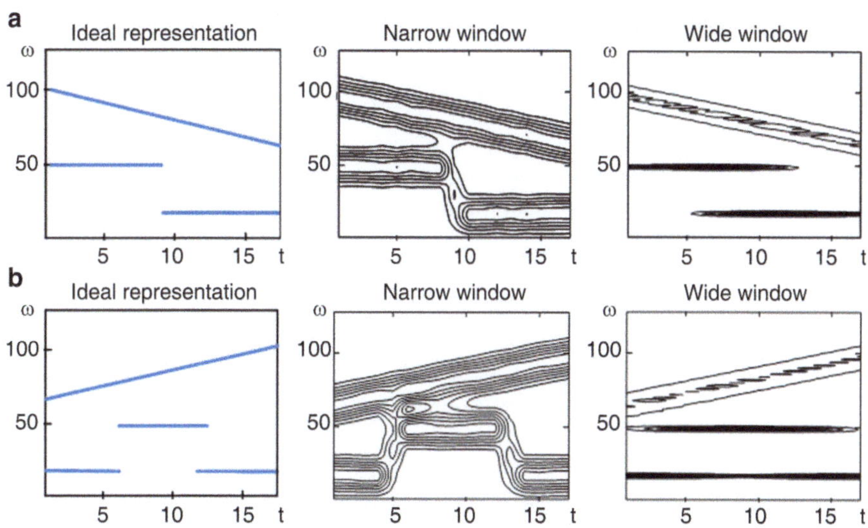


**Fig. 1.11** Spectrograms of multicomponent signals: (**a**) two sinusoids and chirp, (**b**) three sinusoids and chirp

### 1.5.1 Window Functions

The window functions are important for localization of signals in the time-frequency domain. Some of the most commonly used windows are: Rectangular, Gaussian, Hann(ing), Hamming and Blackman, defined as follows:

1. *Rectangular window*: $w(\tau) = \begin{cases} 1, \text{ for } |\tau| < T \\ 0, \text{ otherwise} \end{cases}$
2. *Gaussian window*: $w(\tau) = e^{-\tau^2/\alpha^2}$ (it localizes signal in time although it is not time-limited).
3. *Hann*(ing) *window*: $w(\tau) = \begin{cases} 0.5(1 + \cos(\pi\tau/T)), \text{ for } |\tau| < T \\ 0, \text{ otherwise} \end{cases}$
4. *Hamming window*: $w(\tau) = \begin{cases} 0.54 + 0.46\cos(\pi\tau/T)), \text{ for } |\tau| < T \\ 0, \text{ otherwise} \end{cases}$
5. *Blackman window*:

$$w(\tau) = \begin{cases} 0.42 + 0.5\cos(\pi\tau/T) + 0.08\cos(2\pi\tau/T)), \text{ for } |\tau| < T \\ 0, \text{ otherwise} \end{cases}$$

## 1.6 Wigner Distribution

In order to improve the localization of a signal in the time-frequency domain, a number of quadratic distributions are introduced. A common requirement is that they meet the marginal conditions, which will be discussed in the sequel.

Given a time-frequency representation $P(t,\omega)$, the signal energy within the region $[(t,\ t + \Delta t),\ (\omega\ ,\ \omega + \Delta\omega)]$ is equal to:

$$P(t,\omega)\frac{\Delta\omega}{2\pi}\Delta t. \tag{1.25}$$

Projections of the distribution on time and frequency axes provide the spectral energy density and the instantaneous power of the signal, respectively:

$$\int_{-\infty}^{\infty} P(t,\omega)dt = |F(\omega)|^2,$$
$$\frac{1}{2\pi}\int_{-\infty}^{\infty} P(t,\omega)d\omega = |f(t)|^2. \tag{1.26}$$

These conditions are known as marginal conditions (Fig. 1.12).

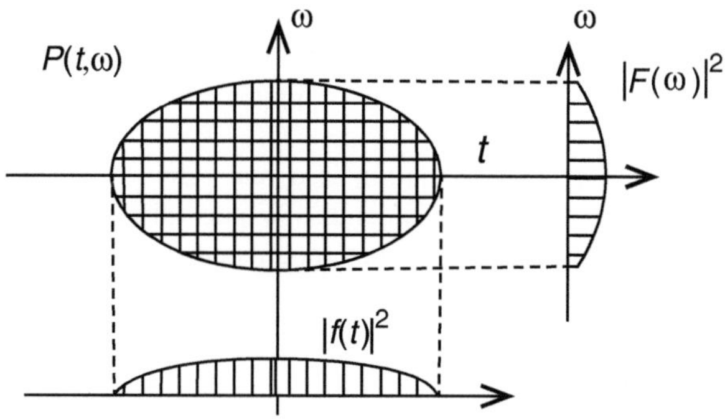


**Fig. 1.12** Calculation of marginal conditions

The signal energy can be obtained as:

$$\frac{1}{2\pi}\int_{-\infty}^{\infty}\int_{-\infty}^{\infty}P(t,\omega)d\omega dt=\int_{-\infty}^{\infty}|f(t)|^2dt=E_x. \tag{1.27}$$

One of the distributions that satisfy marginal conditions is the Wigner distribution, which originated from the quantum mechanics. The distribution is defined as:

$$WD(t,\omega)=\int_{-\infty}^{\infty}R(t,\tau)e^{-j\omega\tau}d\tau=\int_{-\infty}^{\infty}f\left(t+\frac{\tau}{2}\right)f^*\left(t-\frac{\tau}{2}\right)e^{-j\omega\tau}d\tau, \tag{1.28}$$

where:

$$R(t,\tau)=f\left(t+\frac{\tau}{2}\right)f^*\left(t-\frac{\tau}{2}\right),$$

is the local auto-correlation function. In real applications, we use a windowed version of the Wigner distribution:

$$PWD(t,\omega)=\int_{-\infty}^{\infty}w\left(\frac{\tau}{2}\right)w^*\left(-\frac{\tau}{2}\right)f\left(t+\frac{\tau}{2}\right)f^*\left(t-\frac{\tau}{2}\right)e^{-j\omega\tau}d\tau, \tag{1.29}$$

which is often referred as the pseudo Wigner distribution (PWD). A window function does not play a significant role for PWD as for the STFT (or spectrogram). For example, it is possible to use a wider window and to keep good time resolution. A discrete version of the PWD can be defined in the form:

$$PWD(n,k)=\sum_{m=-N/2}^{N/2-1}w(m)w^*(-m)x(n+m)x^*(n-m)e^{-j4\pi mk/N}, \tag{1.30}$$

where $n$ and $k$ are discrete time and frequency parameters, respectively. Note that for the realization of the discrete PWD, the signal should be oversampled by the factor of 2 (sampled at a twice higher sampling rate than required by the sampling theorem).

The Wigner distribution is a real-valued function and satisfies the marginal conditions. Let us consider the Wigner distribution of delta pulse, a sinusoidal signal and a linear frequency modulated signal. For the delta pulse:

$$f(t) = A\delta(t - t_1), \tag{1.31}$$

the Wigner distribution equals to:

$$WD(t,\omega) = 2\pi A^2\delta(t - t_1). \tag{1.32}$$

For sinusoidal and chirp signals, we still obtain the ideal representation by using the Wigner distribution:

$$\begin{array}{ll} f(t) = Ae^{j\omega_1 t} & f(t) = Ae^{jat^2/2} \\ WD(t,\omega) = 2\pi A^2\delta(\omega - \omega_1), & WD(t,\omega) = 2\pi A^2\delta(\omega - at). \end{array}$$

However, for a multicomponent signal $f(t) = \sum_{m=1}^{M} f_m(t)$, the Wigner distribution is:

$$WD(t,\omega) = \sum_{m=1}^{M} WD_{f_m f_m}(t,\omega) + \sum_{m=1}^{M}\sum_{\substack{n=1 \\ m\neq n}}^{M} WD_{f_m f_n}(t,\omega). \tag{1.33}$$

Therefore, the Wigner-distribution of a multicomponent signal equals to the sum of Wigner distributions of all signal components (auto-terms) and of quadratic terms obtained by multiplying different signal components ($f_m$ and $f_n$, $m\neq n$), called cross-terms. Hence, the Wigner distribution can be useless for the time-frequency representation of multicomponent signals, since it can yield the time-frequency components that do not exist in the analyzed signal. For example, the Wigner distribution of a multicomponent signal, whose spectrogram is shown in Fig. 1.11a, is shown in Fig. 1.13a. The presence of strong cross-terms is obvious and they diminish the accuracy of the representation. However, if the cross-terms are removed, a concentrated representation is obtained as shown in Fig. 1.13b.

In order to reduce or completely eliminate the cross-terms, many distributions have been defined over the years. One such distribution is the S-method (SM),

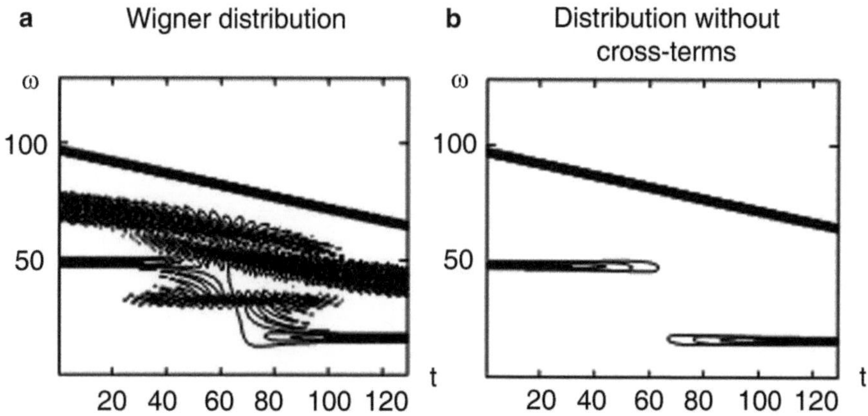


**Fig. 1.13** (**a**) The Wigner distribution of a multicomponent signal, (**b**) auto-terms of the Wigner distribution

which combines good properties of the spectrogram and of the Wigner distribution. The SM is defined as:

$$SM(t,\omega) = \frac{1}{\pi}\int_{-\infty}^{\infty} P(\theta)STFT(t,\omega+\theta)STFT^{*}(t,\omega-\theta)d\theta, \tag{1.34}$$

where $P(\theta)$ represents a finite frequency domain window function. A discrete version of the S-method is given by:

$$\begin{aligned} SM(n,k) &= \sum_{i=-L_d}^{L_d} P(i)STFT(n,k+i)STFT^{*}(n,k-i) \\ &= |STFT(n,k)|^2 + 2\cdot \mathrm{Re}\left\{\sum_{i=1}^{L_d} STFT(n,k+i)STFT^{*}(n,k-i)\right\}, \end{aligned} \tag{1.35}$$

where the parameter $L_d$ determines the frequency window length. As we increase the value of $L_d$ (starting from $L_d = 0$ which corresponds to the spectrogram), we gradually approach toward the Wigner distribution. In order to avoid the presence of cross-terms, the value of $L_d$ should be less than half of the distance between two auto-terms. Note that the SM is suitable for hardware implementation, and does not require oversampling of the signal.

Let us consider the previous example and its time-frequency representations obtained with the SM, for various values of $L_d$ (Fig. 1.14).

In many real applications, $L_d = \{3, 4 \text{ or } 5\}$ can provide satisfactory results, since it eliminates the cross-terms and provides good concentration of the auto-terms, which almost equals the concentration achieved by the Wigner distribution.

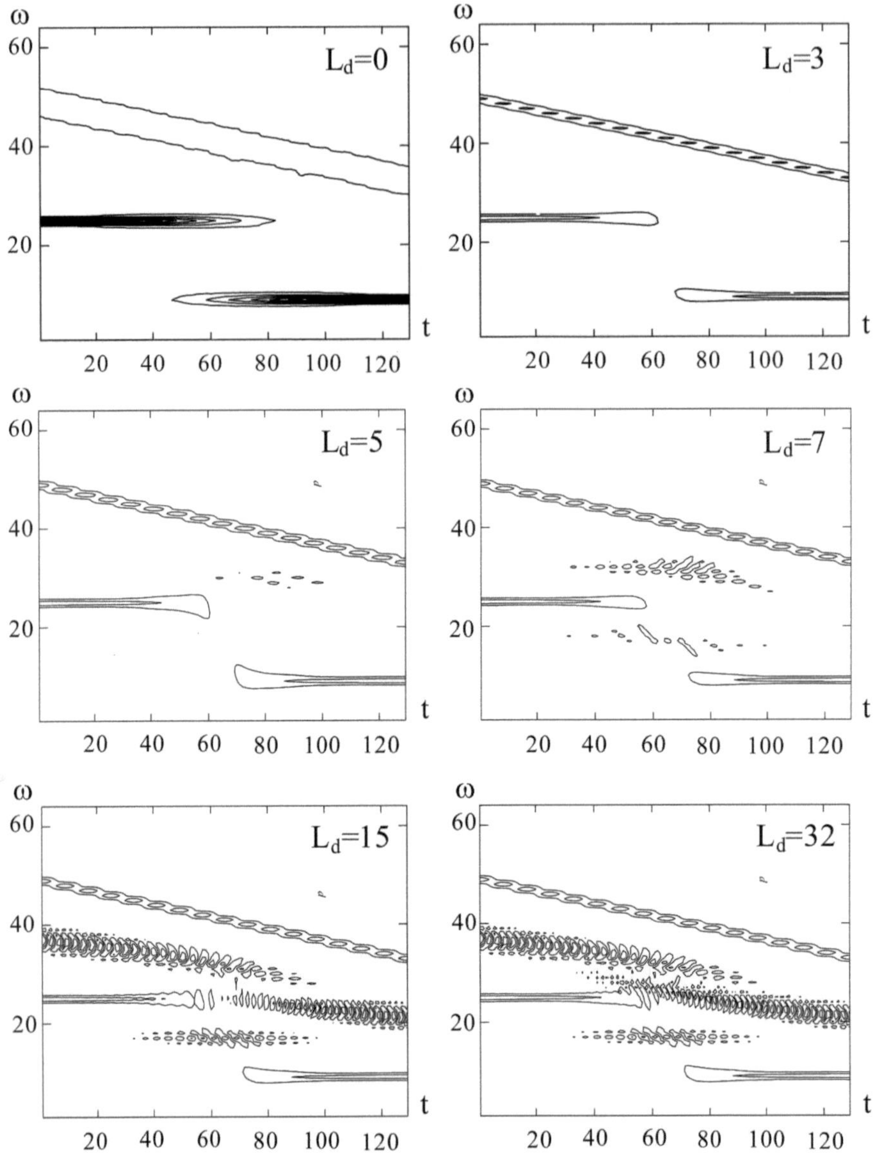


**Fig. 1.14** The time-frequency representations of a multicomponent signal obtained by using SM for different values of window width (defined as $2L_d + 1$)

In addition, we consider the time-frequency representation of a speech signal obtained using the spectrogram and the SM, as shown in Fig. 1.15.

The time-frequency representation obtained with the SM provides better temporal and frequency resolution, which allows us to obtain more accurate description and analysis of speech components.

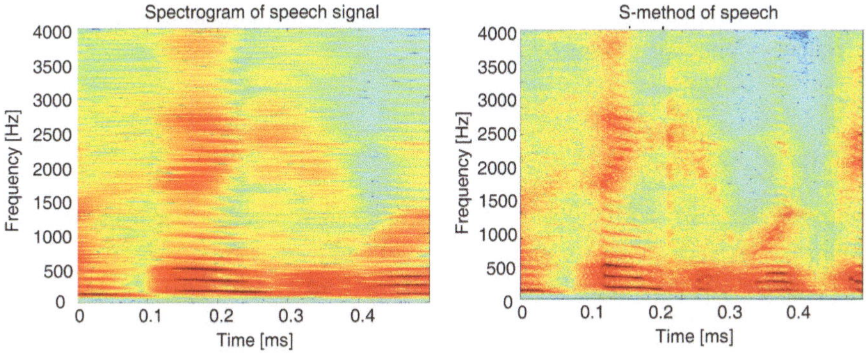


**Fig. 1.15** The spectrogram and the SM of speech signal

## 1.7 Time-Varying Filtering

For nonstationary signals, the time-varying filtering provides more efficient processing compared to the approaches performed in either the time or frequency domain separately. The time-varying filtering has been defined as:

$$H\{x(t)\} = \int_{-\infty}^{\infty} h(t, t-\tau)x(\tau)d\tau, \tag{1.36}$$

where $h(t,\tau)$ is the impulse response of the time-varying system $H$. In order to get undistorted frequency modulated signals, when time-varying filtering is applied, a slightly modified relation is used:

$$H\{x(t)\} = \int_{-\infty}^{\infty} h\left(t+\frac{\tau}{2}, t-\frac{\tau}{2}\right)x(t+\tau)d\tau. \tag{1.37}$$

The optimal system form can be obtained by minimizing the mean squared error:

$$H_{opt} = \arg\min_{H} E\{|f(t) - H\{x(t)\}|^2\}, \tag{1.38}$$

where $x(t) = f(t) + \nu(t)$, $f(t)$ is a signal and $\nu(t)$ is a noise. The time-varying transfer function in the Wigner distribution framework has been defined as the Weyl symbol mapping of the impulse response into the time-frequency plane:

$$L_H(t,\omega) = \int_{-\infty}^{\infty} h\left(t+\frac{\tau}{2}, t-\frac{\tau}{2}\right)e^{-j\omega\tau}d\tau. \tag{1.39}$$

Assuming that the signal and noise are uncorrelated, the optimal filter in the time-frequency domain is defined by:

$$L_H(t,\omega) = \frac{\overline{WD}_{ff}(t,\omega)}{\overline{WD}_{ff}(t,\omega) + \overline{WD}_{\nu\nu}(t,\omega)} = 1 - \frac{\overline{WD}_{\nu\nu}(t,\omega)}{\overline{WD}_{xx}(t,\omega)}, \tag{1.40}$$

where $\overline{WD}_{xx}(t,\omega) = \overline{WD}_{ff}(t,\omega) + \overline{WD}_{\nu\nu}(t,\omega)$. This form corresponds to the well-known Wiener filter in the stationary cases. Observe that: $\overline{WD}_{\nu\nu}(t,\omega) = 0 \Rightarrow L_H(t,\omega) = 1$, while $\overline{WD}_{ff}(t,\omega) = 0 \Rightarrow L_H(t,\omega) = 0$. Here, $\overline{WD}$ represents the mean value of the Wigner distributions for different realizations of the signal and noise. As a consequence of averaging, the cross-terms will be significantly reduced, as well as the noise. However, in practice the time-varying filtering should be often performed on a single noisy realization. It means that a cross-terms free distribution should be used (e.g., the S-method) for multicomponent signals instead of the Wigner distribution. The approximate filter, based on the cross-terms free distribution and a single realization, provides satisfying results in many real applications.

Let us now observe a case when we can assume that the Wigner spectrum of the observed signal $f(t)$ lies inside a time-frequency region $R$, while the noise is mainly outside this region (there might be a small amount of noise inside $R$ that is negligible compared to the noise outside $R$). Then the support function can be defined as:

$$L_H(t,\omega) = \begin{cases} 1, & \text{for } (t,\omega) \in R, \\ 0, & \text{for } (t,\omega) \notin R. \end{cases} \tag{1.41}$$

In numerical implementations, the lag window $w(\tau)$ is introduced in the filtering definition leading to the pseudo form of the Eq. (1.36):

$$H\{x(t)\} = \int_{-\infty}^{\infty} h\left(t + \frac{\tau}{2}, t - \frac{\tau}{2}\right) w(\tau) x(t+\tau) d\tau. \tag{1.42}$$

By using the Parseval's theorem, the previous form of time-varying filtering can be written in the form:

$$H\{x(t)\} = \frac{1}{2\pi} \int_{-\infty}^{\infty} L_H(t,\omega) STFT(t,\omega) d\omega. \tag{1.43}$$

## 1.8 Robust Statistics in the Time-Frequency Analysis

In the presence of impulse noise or heavy-tailed noises, the standard time-frequency representations and distributions do not provide satisfactory results. Namely, the standard time-frequency distributions, which are based on the standard Fourier transform of the signal or its autocorrelation function, are optimal for the case of signals corrupted by Gaussian noise. However, the standard form is not optimal for the case of impulse or heavy-tailed disturbances. Therefore, more optimal forms of time-frequency distributions for these types of signals can be derived by applying the concept of robust estimation theory developed by Huber.

Generally, the Fourier transform can be defined as a solution of the minimization problem:

$$X(k) = \arg\min_{\mu} \{I(k,\mu)\}, \tag{1.44}$$

where:

$$I(k,\mu) = \sum_{n=0}^{N-1} F\{\varepsilon(k,\mu)\} = \sum_{n=0}^{N-1} |\varepsilon(k,\mu)|^L = \sum_{n=0}^{N-1} \left|x(n)e^{-j2\pi nk/N} - \mu\right|^L. \tag{1.45}$$

The value of $X(k)$ is equal to the value of $\mu$ that minimizes $I(k,\mu)$. The function $F\{\varepsilon\}$ is called the loss function of error $\varepsilon$. The loss function is usually obtained using the maximum likelihood (ML) approach where the form of $F$ corresponds to the probability density function (pdf) of noise:

$$F\{\varepsilon\} = -\log p_{\nu}(\varepsilon). \tag{1.46}$$

For instance, let us observe the case of Gaussian noise with the pdf given in the form:

$$p_{\nu}(\varepsilon) \sim e^{-|\varepsilon|^2}. \tag{1.47}$$

According to Eqs. (1.46) and (1.47), the loss function will be of the form:

$$F\{\varepsilon\} = |\varepsilon|^2, \tag{1.48}$$

or in other words we have:

$$I(k,\mu) = \sum_{n=0}^{N-1} \left|x(n)e^{-j2\pi nk/N} - \mu\right|^2. \tag{1.49}$$

The minimization procedure is applied as follows:

$$\begin{aligned} \frac{\partial I(k,\mu)}{\partial \mu^*} = 0 \;&\Rightarrow 2\sum_{n=0}^{N-1}\left(x(n)e^{-j2\pi nk/N} - \mu\right) = 0 \\ &\Rightarrow \mu = \frac{1}{N}\sum_{n=0}^{N-1} x(n)e^{-j2\pi nk/N}. \end{aligned} \tag{1.50}$$

Note that this form corresponds to the standard Fourier transform definition:

$$X(k) = \frac{1}{N}\sum_{n=0}^{N-1} x(n)e^{-j2\pi nk/N}. \tag{1.51}$$

In practice, the ML estimation approach is very sensitive to the assumed pdf model. Therefore, the imprecision in noise pdf modelling may significantly affect the result. Therefore, we can conclude that the ML estimation approach provides good results if the noise model is known in advance, which is a rare case. Hence, the robust estimates from the Huber's theory are introduced instead of ML estimates. For instance, we can observe the worst case of noise and determine its robust estimate. For a wide class of impulsive/heavy-tailed pdfs, the following loss function is recommended (corresponding to the Laplacian pdf):

$$F\{\varepsilon\} = |\varepsilon| = \sqrt{\mathrm{Re}^2\{\varepsilon\} + \mathrm{Im}^2\{\varepsilon\}}, \tag{1.52}$$

which is usually considered in the simplified form as follows:

$$F\{\varepsilon\} = |\mathrm{Re}(\varepsilon)| + |\mathrm{Im}(\varepsilon)|. \tag{1.53}$$

This form of the loss function leads to the robust M-Fourier transform. Since this type of a loss function does not produce a closed-form solution, the iterative procedures are used in calculation of the robust M-Fourier transform.

Let us consider the case of STFT. In analogy with the Fourier transform, the standard STFT form is obtained as:

$$STFT(n,k) = \arg\min_{\mu} \{I(n,k,\mu)\}, \tag{1.54}$$

where:

$$I(n,k,\mu) = \sum_{m=-N/2}^{N/2-1} \left|x(n+m)e^{-j2\pi mk/N} - \mu\right|^2. \tag{1.55}$$

Note that, for the sake of simplicity, the rectangular window (with unit amplitude) is assumed and thus $w(m)$ is omitted. By applying the minimization procedure with respect to $\mu$ as in Eq. (1.50), the standard STFT is obtained:

$$STFT(n,k) = \sum_{m=-N/2}^{N/2-1} x(n+m)e^{-j2\pi mk/N}. \tag{1.56}$$

In the case of the loss function $F\{\varepsilon\} = |\varepsilon|$, the robust STFT can be obtained. As a solution of the optimization problem Eq. (1.54), the nonlinear set of equations is given by:

$$STFT_M(n,k) = \frac{\sum_{m=-N/2}^{N/2-1} \left(x(n+m)e^{-j2\pi mk/N}\right) \Big/ \left|x(n+m)e^{-j2\pi mk/N} - STFT_M(n,k)\right|}{\sum_{m=-N/2}^{N/2-1} 1/\left|x(n+m)e^{-j2\pi mk/N} - STFT_M(n,k)\right|}. \tag{1.57}$$

This problem can be solved using an iterative procedure (for each $n$ and each $k$), which is computationally very demanding and limits the application. There are alternative approaches for calculating the robust STFT in order to overcome the drawbacks of the iterative procedure.

In the case of signals corrupted by impulse noise, the commonly used alternative approach is based on median filter, obtained as a central element in the sorted sequence. The median filter is considered in more detail in Chap. 4. In the case of the STFT, we can observe the vector $x(n+m)e^{-j2\pi mk/N}$ for $m = [-N/2, \ldots, N/2 - 1]$, having independent real and imaginary part. Now, using the loss function $F\{\varepsilon\} = |\mathrm{Re}(\varepsilon)| + |\mathrm{Im}(\varepsilon)|$ and a marginal median filter form, we can define the marginal median STFT form as follows:

$$\begin{aligned} STFT_M(n,k) = \; & median\left(\mathrm{Re}\left\{x(n+m)e^{-j2\pi mk/N}, \text{ for } m \in \left[-\frac{N}{2}, \frac{N}{2}\right)\right\}\right) \\ & + j median\left(\mathrm{Im}\left\{x(n+m)e^{-j2\pi mk/N}, \text{ for } m \in \left[-\frac{N}{2}, \frac{N}{2}\right)\right\}\right). \end{aligned} \tag{1.58}$$

For cases when the noise is a mixture of Gaussian and impulse noise (or heavy-tailed noise), the resulting pdf function would have a complex form which is not suitable for practical applications. Hence, a more optimal form that can fit for different types of noise is defined and it is known as the L-estimate robust transform. The L-estimate STFT is defined as:

$$\begin{aligned} & STFT_L(n,k) = \sum_{i=0}^{N-1} a_i\left(g_i(n,k) + j \cdot h_i(n,k)\right), \\ & g_i(n,k) \in G(n,k), \quad h_i(n,k) \in H(n,k), \end{aligned} \tag{1.59}$$

where,

$$G(n,k) = \{\mathrm{Re}(x(n+m)e^{-j2\pi mk/N})\}: \quad m \in [\frac{-N}{2}, \frac{N}{2}),$$
$$H(n,k) = \{\mathrm{Im}(x(n+m)e^{-j2\pi mk/N})\}: \quad m \in [\frac{-N}{2}, \frac{N}{2}). \tag{1.60}$$

The elements: $g_i(n,k)$ and $h_i(n,k)$ are sorted in non-decreasing order as: $g_i(n,k) \leq g_{i+1}(n,k)$ and $h_i(n,k) \leq h_{i+1}(n,k)$, respectively. The coefficients $a_i$ can be written in analogy with the α-trimmed filter (from the nonlinear digital filter theory), as follows:

$$a_i = \begin{cases} \dfrac{1}{N(1-2\alpha)+4\alpha}, & \text{for } i \in [N\alpha, \ \ldots, N - N\alpha], \\ 0, & \text{otherwise.} \end{cases} \tag{1.61}$$

where $N$ is even, while the parameter α takes values within the range [0,1/2]. For $\alpha = 0$ the standard STFT is obtained, while for $\alpha = 1/2$ we obtain the marginal median STFT. Larger values of $\alpha$ provide better reduction of heavy-tailed noise, while smaller values of $\alpha$ better preserve the spectral characteristics. Accordingly, the choice of $\alpha$ should provide a good trade-off between these two requirements.

Similarly to the STFT, we can define robust forms of other time-frequency representations/distributions. For example, the WD can be defined using the following optimization problem:

$$WD(n,k) = \arg\min_{\mu} \{I(n,k,\mu)\} = \arg\min_{\mu} \sum_{m=-N/2}^{N/2-1} F(\varepsilon(n,k,\mu)), \tag{1.62}$$

with the error function defined as:

$$\varepsilon(n,k) = \left\{ \mathrm{Re}\left\{ x(n+m)x^*(n-m)e^{-j4\pi km/N} \right\} - \mu, \ m \in \left[ -\frac{N}{2}, \frac{N}{2} \right) \right\}. \tag{1.63}$$

The median form of the WD can be calculated by the formula:

$$WD_M(n,k) = median\left\{ \mathrm{Re}\{x(n+m)x^*(n-m)e^{-j4\pi km/N}\}, \ m \in \left[ -\frac{N}{2}, \frac{N}{2} \right) \right\}. \tag{1.64}$$

Similarly, the L-estimate WD can be calculated using:

$$\begin{aligned} WD_L(n,k) &= \sum_{i=0}^{N-1} a_i s_i(n,k),\ s_i(n,k) \in S(n,k), \\ S(n,k) &= \left\{ \mathrm{Re}\left\{ x(n+m)x^*(n-m)e^{-j4\pi km/N} \right\} : m \in \left[ -\frac{N}{2}, \frac{N}{2} \right) \right\} \end{aligned} \tag{1.65}$$

where the elements $s_i(n,k)$ are sorted in non-decreasing order.

# 1.9 Wavelet Transform

## *1.9.1 Continuous Wavelet Transform*

Wavelets are mathematical functions formed by scaling and translation of basis functions $\psi(t)$ in the time domain. $\psi(t)$ is also called the mother wavelet and satisfies the following conditions:

1. The total area under the curve $\psi(t)$ is equal to zero:

$$\int_{-\infty}^{\infty} \psi(t)dt = 0. \tag{1.66}$$

2. The function has finite energy, i.e., it is square-integrable:

$$\int_{-\infty}^{\infty} |\psi(t)|^2 dt < \infty. \tag{1.67}$$

The wavelet is defined by:

$$\psi_{a,b}(t) = \frac{1}{\sqrt{|a|}} \psi\left(\frac{t-b}{a}\right), \tag{1.68}$$

where $a$ and $b$ are two arbitrary real numbers used as scaling and translation parameters, respectively. The factor $\sqrt{|a|}$ also represents a normalization factor, which allows the energy of the wavelet function to remain independent of parameter $a$. For the values $0<a<1$, the basis function shrinks in time, while for $a>1$ it spreads in time.

The wavelet transform of the signal *f*(*t*) is mathematically described by the expression:

$$W(a,b) = \int_{-\infty}^{\infty} \psi_{a,b}(t) f(t) dt. \tag{1.69}$$

*W*(*a*,*b*) is called the continuous wavelet transform (CWT), where $a$ and $b$ are continuous variables and *f*(*t*) is a continuous function. The inverse wavelet transform is obtained as:

$$f(t) = \frac{1}{C} \int_{-\infty}^{\infty} \int_{-\infty}^{\infty} \psi_{a,b}(t) W(a,b) da db, \tag{1.70}$$

where:

$$C = \int_{-\infty}^{\infty} \frac{|\Psi(\omega)|^2}{\omega} d\omega, \tag{1.71}$$

and $\Psi(\omega)$ is the Fourier transform of $\psi(t)$. The inverse continuous wavelet transform exists if the parameter $C$ is positive and finite.

### 1.9.2 Wavelet Transform with Discrete Wavelet Functions

In practical applications, the parameters $a$ and $b$ are discretized (i.e., scaling and translation are performed in discrete steps). The discretization of parameter $a$ is done using powers of fixed dilation parameter $a_0>1$:

$$a = a_0^{-j}, \text{where} \quad j \in \mathbf{Z},$$

while for $b$ we use:

$$b = k b_0 a_0^{-j}, k \in \mathbf{Z}, b_0 > 0.$$

By using the discrete parameters $a$ and $b$, we obtain the discretized family of wavelets:

$$\psi_{j,k}(t) = {a_0}^{j/2} \psi\left(a_0^j t - k b_0\right). \tag{1.72}$$

The corresponding wavelet transform is then given by:

$$W_{j,k}^{d} = a_0{}^{j/2} \int_{-\infty}^{\infty} f(t)\psi\left(a_0^{j}t - kb_0\right)dt. \tag{1.73}$$

If $a_0 = 2$ and $b_0 = 1$ are used, we achieve the dyadic sampling. The corresponding signal decomposition is called the dyadic decomposition. In such a case, the discrete wavelet functions with the given parameters form a set of orthonormal basis functions:

$$\psi_{j,k}(t) = 2^{j/2}\psi\left(2^{j}t - k\right). \tag{1.74}$$

Therefore, the dyadic wavelet transform is calculated as:

$$W_{j,k}^{d} = 2^{j/2} \int_{-\infty}^{\infty} f(t)\psi\left(2^{j}t - k\right)dt. \tag{1.75}$$

### 1.9.3 Wavelet Families

Wavelets are widely applied in image processing, biomedical signal processing, audio signal processing, just to name a few. Let us mention some of the most commonly used wavelets:

- The Haar wavelet is the oldest and the simplest wavelet.
- Daubechies wavelets represent a set of orthonormal wavelets of limited duration. For example, Daubechies D4 wavelet has 4 coefficients; D8 has 8 coefficients, and so on. Note that the Haar wavelet is actually the Daubechies wavelet of the first order.
- Bi-orthogonal wavelets are widely used in image compression. For example, JPEG2000 compression algorithm is based on bi-orthogonal Le Gall (5,3) and Cohen–Daubechies–Feauveau (CDF) (9,7) wavelets.
- The Mexican hat wavelet is a wavelet function that equals to the second derivative of the Gaussian function.
- Symlets are symmetric wavelets, created as a modification of the Daubechies wavelet.
- The Morlet wavelet is based on a modulated Gaussian function.

### 1.9.4 Multiresolution Analysis

The wavelet transform is generally based on the possibility to represent the components of certain function at different resolution levels, which has been also known as the multiresolution analysis. Let us observe the multiresolution representation of a function $f(t)$. We can write:

$$f(t) = f_J(t) + \sum_{j=J}^{\infty} \Delta f_j(t) \quad \text{or} \quad f(t) = \sum_{j=-\infty}^{\infty} \Delta f_j(t), \tag{1.76}$$

where $J$ denotes the lowest resolution scale. Hence, the function $f(t)$ can be represented as a sum of approximation at the certain scale (resolution) $J$ and a sum of details that are complement to the approximation $f_J(t)$ at levels $j>J$. An illustration of approximation functions and details at a few scales $(J, J+1, J+2)$ for a sinusoidal function $f(t)$ is given in Fig. 1.16.

The multiresolution analysis can be extended to the decomposition of the Hilbert space $L^2(\mathbf{R})$. According to Eq. (1.76) we may write:

$$L^2(\mathbf{R}) = V_J(t) + \sum_{j=J}^{\infty} \Delta W_j(t), \tag{1.77}$$

where $V_J$ denotes the approximation space at the resolution scale $J$, while the terms in summation correspond to wavelet spaces for $j \geq J$. Therefore, we can say that the Hilbert space can be observed as a composition of the approximation space $V_J$ and the infinite set of wavelet spaces $W_j$, $(j = J, \ldots, \infty)$.

Consequently, we might observe that the Hilbert space can be actually represented as a decomposition of approximation spaces $V_j$, $j \in \mathbb{Z}$ and $\ldots \subset V_{-2} \subset V_{-1} \subset V_0 \subset V_1 \subset V_2 \subset \ldots$, where each approximation space represents the scaled version (binary scaling that causes shrinking or dilation) of the basic space $V_0$. The approximation $V_{j+1}$ contains more details compared to $V_j$, which are modelled by the wavelet space $W_j$: $V_{j+1} = V_j \oplus W_j$ (Fig. 1.17), where $\oplus$ denotes the orthogonal summation.

Now, we can expand the space $L^2(\mathbf{R})$ as follows:

$$L^2(\mathbf{R}) = V_0 \oplus W_0 \oplus W_1 \oplus \ldots \tag{1.78}$$

or equivalently for an arbitrary starting scale $J$:

$$L^2(\mathbf{R}) = V_J \oplus W_J \oplus W_{J+1} \oplus \ldots \tag{1.79}$$

or:

$$L^2(\mathbf{R}) = \ldots \oplus W_{-2} \oplus W_{-1} \oplus W_0 \oplus W_1 \oplus W_2 \oplus \ldots, \tag{1.80}$$

which represents the space as a composition of wavelet spaces alone.

The basic space $V_0$ is generated using the scaling function $\varphi(t)$. Hence, the basis of the space $V_0$ consists of the functions $\varphi(t)$ and its translations $\varphi(t\text{-}k)$.

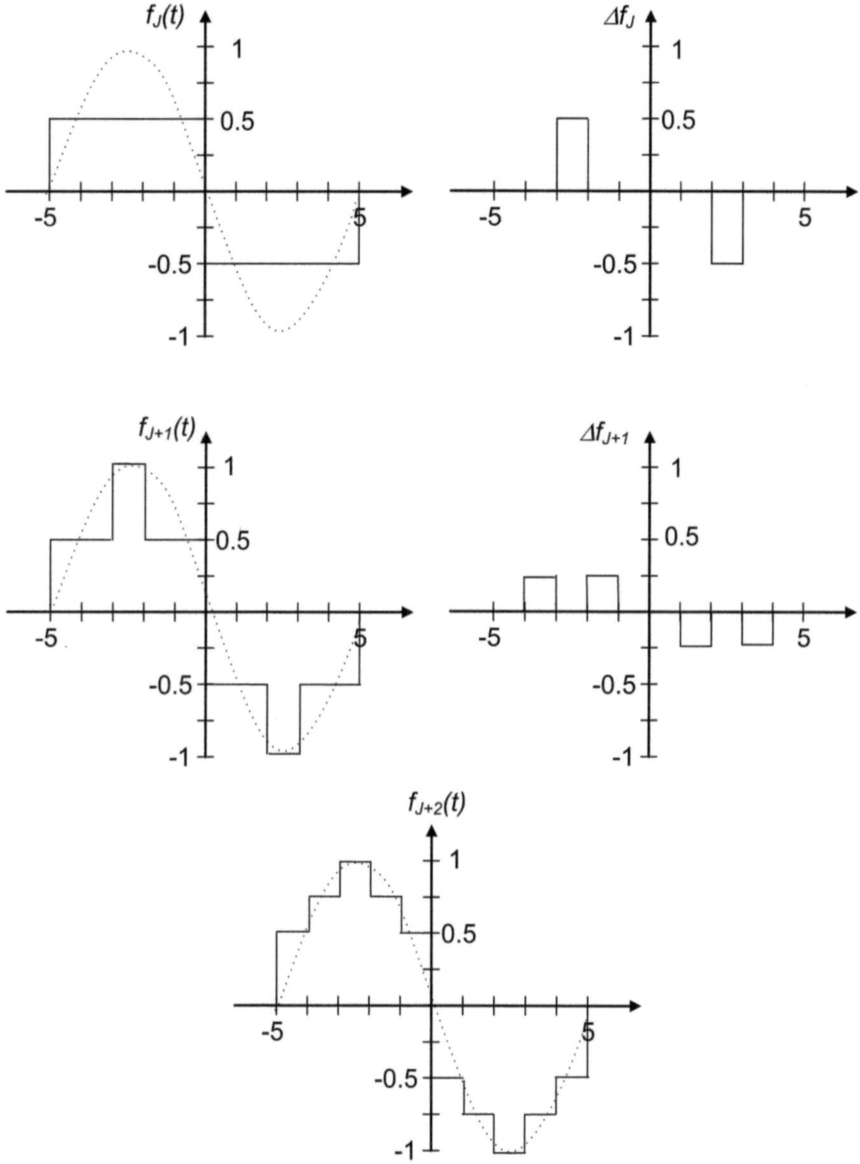


**Fig. 1.16** The approximation function $f_J(t)$ and its details at levels $J$, $J+1$, and $J+2$

Consequently, a set of basis functions of the space $V_j$ is composed of integer translations and binary scaling of the function $\varphi(t)$:

$$\varphi_{j,k}(t) = 2^{j/2}\varphi\left(2^j t - k\right), \; j,k \in \mathbb{Z} \tag{1.81}$$

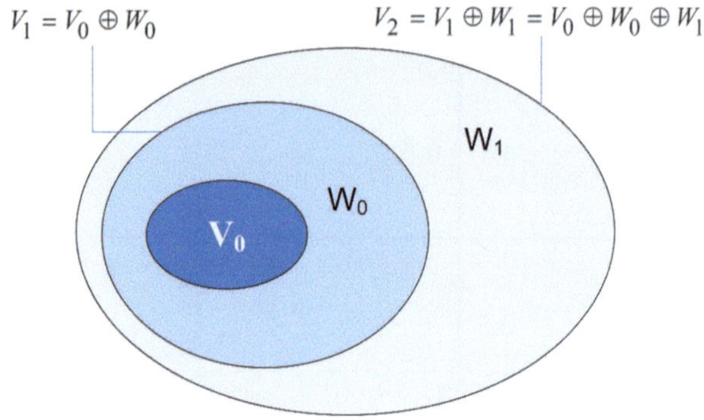


**Fig. 1.17** Approximation (scaling) and wavelet spaces

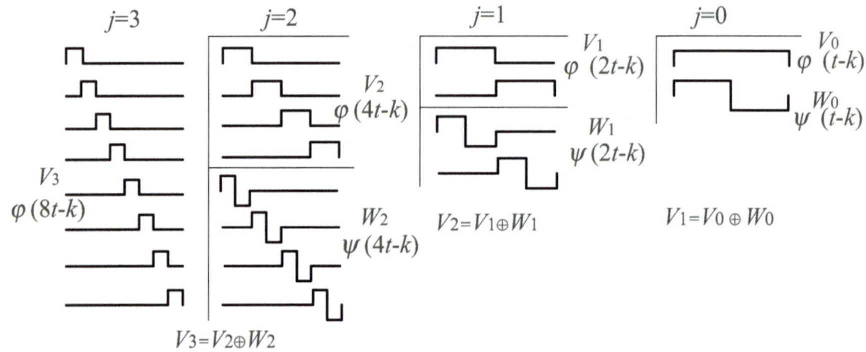


**Fig. 1.18** An illustration of scaling functions and wavelets decomposition of $V_3$, $V_2$, and $V_1$ (Haar wavelets are used)

Having in mind that the shape of $\varphi_{j,k}(t)$ changes with $j$, $\varphi(t)$ is usually called a scaling function. Similarly to approximation spaces, the wavelet space $W_j$ is generated by the scaled and translated wavelet functions:

$$\psi_{j,k}(t) = 2^{j/2}\psi\left(2^j t - k\right), \; j,k \in \mathbb{Z}. \tag{1.82}$$

A simple example of functions $\varphi(t)$ and $\psi(t)$ are the Haar scaling and the wavelet function, given by:

$$\varphi(t) = \begin{cases} 1, \; 0 \le t < 1 \\ 0, \; \text{otherwise} \end{cases} \qquad \psi(t) = \begin{cases} 1, \; 0 \le t < 0.5, \\ -1, \; 0.5 \le t < 1. \end{cases} \tag{1.83}$$

The Haar scaling functions and the wavelet functions of spaces $V_3$, $V_2$, and $V_1$ are illustrated in Fig. 1.18.

Since the spaces $V_0$ and $W_0$ are contained within the $V_1$, then the functions $\varphi(t) \in V_0$ and $\psi(t) \in W_0$ belong also to the space $V_1$. Consequently, $\varphi(t)$ and $\psi(t)$

can be represented using the basis/expansion function of space $V_1$: $\varphi_{1,k}(t) = \sqrt{2}\varphi(2t - k), for\ j = 1.$

In other words, we might say that the basis/expansion functions from a certain space can be derived using the expansion functions from a double-resolution space. The corresponding representation is called the dilation equation:

$$\varphi(t) = \sum_k s(k)\sqrt{2}\varphi(2t - k), \tag{1.84}$$

while the wavelet equation is:

$$\psi(t) = \sum_k d(k)\sqrt{2}\varphi(2t - k). \tag{1.85}$$

Starting from the coefficients $s(k)$ (which satisfy certain conditions), we solve the dilation equation to obtain the basis function $\varphi_{j,k}(t)$ of space $V_j$. Then the coefficients $d(k)$ should be determined (usually depending on the choice of $s(k)$), and thus, the basis $\psi_{j,k}(t)$ of wavelet space $W_j$ is determined.

#### 1.9.4.1 Function decomposition into Multiresolution Subspaces

Generally speaking, a function $f(t)$ can be expressed as a linear combination of expansion functions as follows:

$$f(t) = \sum_k s(k)\varphi_k(t), \tag{1.86}$$

where $s(k)$ are expansion coefficients and $\varphi_k(t)$ are expansion functions. We assume that the expansion functions are obtained using integer translation and binary scaling of the function $\varphi(t)$ given by $\varphi_{j,k}(t)$ as in Eq. (1.81). Then the approximation or projection of the function $f(t)$ in the space $V_j$ can be expressed as:

$$f_j(t) = \sum_k s_j(k)\varphi_{j,k}(t) = \sum_k s_{j,k}\varphi_{j,k}(t), \tag{1.87}$$

where the expansion coefficients are defined as:

$$s_{j,k} = \left\langle f_j, \varphi_{j,k} \right\rangle = \int_t f(t)\varphi_{j,k}(t)dt. \tag{1.88}$$

Furthermore, the details that remain after the approximation in $V_j$ are modelled by:

$$\Delta f_j(t) = \sum_k d_j(k)\psi_{j,k}(t) = \sum_k d_{j,k}\psi_{j,k}(t), \tag{1.89}$$

where the coefficients of the details are obtained as:

$$d_{j,k} = \left\langle f_j, \psi_{j,k} \right\rangle = \int_t f(t)\psi_{j,k}(t)dt. \tag{1.90}$$

Consequently, the approximation of the function $f(t)$ on the finer scale $j+1$, can be obtained as follows:

$$f_{j+1}(t) = f_j(t) + \Delta f_j(t). \tag{1.91}$$

Therefore, we may observe that $f_j(t) \to f(t)$ for $j \to \infty$. Note that if $f(t) \in V_j \Rightarrow f(t), f(2t), f(t-k), f(2t-k) \in V_{j+1}$. Using Eqs. (1.87) and (1.89), and for an arbitrary scale $J$, we can write the expression for the wavelet series expansion of function $f(t)$:

$$f(t) = \sum_k s_{J,k}\varphi_{J,k}(t) + \sum_{j=J}^{\infty}\sum_k d_{j,k}\psi_{j,k}(t), \tag{1.92}$$

where $s_{J,k}$ are called the approximation coefficients, while $d_{j,k}$ are called the detail or wavelet coefficients.

The wavelet decomposition of a signal can be described by using a set of coefficients, each providing the information about time and frequency localization of the signal. However, the uncertainty principle prevents us from a precise localization in both time and frequency. For example, the Haar wavelet is well localized in time, but supports a wide frequency band. The Mexican wavelet is well localized in the frequency domain, but not in the time domain.

We can conclude that the multiresolution analysis allows the discrete wavelet transform to decompose the signal into different subbands. The subbands at lower frequencies have a better frequency resolution and a poor time resolution, while the subbands at higher frequencies have a better time resolution and a poor frequency resolution, as illustrated in Fig. 1.19.

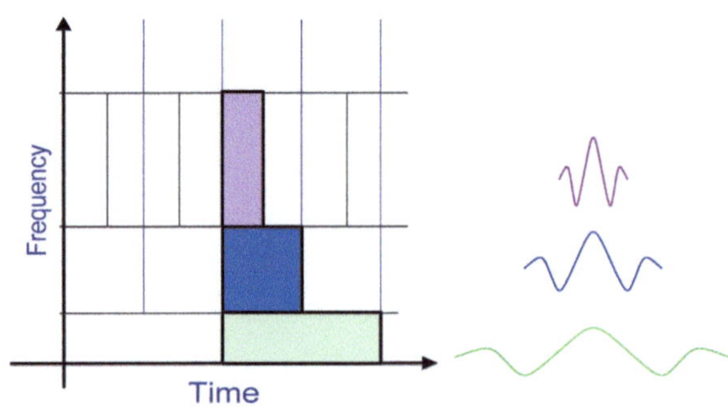


**Fig. 1.19** Time-frequency representation of wavelet

### *1.9.5 Haar Wavelet*

The Haar wavelet is the oldest and simplest type of wavelet function, given in the form:

$$\psi(t) = \begin{cases} 1, \ 0 \le t < 0.5, \\ -1, \ 0.5 \le t < 1. \end{cases} \tag{1.93}$$

The Haar wavelet is not continuous, and therefore not differentiable. Hence, the discontinuity of the wavelet function can cause small approximation smoothness in certain cases. However, this property can be an advantage for analysis of the transient signals (signals with steep and sudden transitions).

Let us consider its scaled and shifted version, in the form:

$$\begin{aligned} \psi_{j,k}(t) &= 2^{\frac{j}{2}}\psi\left(2^j t - k\right), \\ j &= 0, \pm 1, \pm 2, \ldots, \quad k = 0, \pm 1, \ldots, 2^j - 1, \end{aligned} \tag{1.94}$$

where the scaling parameter is $a = 2^{-j}$, while the translation parameter is $b = 2^{-j}k$. The parameter $j$ represents the scale. Greater $j$ values shrink the basis function in time. In addition, for each scale the basis function translates in time by $k$, as depicted in Fig. 1.20.

Note that the Haar wavelets are orthogonal functions:

$$< \psi_{j,k}, \psi_{j',k'} > = \int_{-\infty}^{\infty} \psi_{j,k}(t)\psi_{j',k'}(t)dt = \begin{cases} 1, \ for\, j = j', k = k' \\ 0, \ \text{otherwise} . \end{cases} \tag{1.95}$$

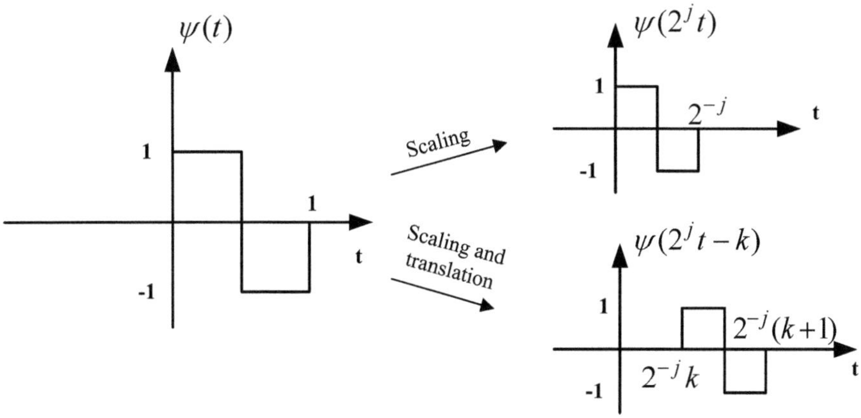


**Fig. 1.20** The wavelet functions

Recall that the Haar scaling function $\varphi(t)$ is given by:

$$\varphi(t)=\begin{cases}1, & 0\le t<1\\ 0, & \text{otherwise}\end{cases}. \tag{1.96}$$

According to Eq. (1.92), the Haar wavelets and scaling function can be used to represent the function $f(t)$ as follows:

$$f(t)=\sum_k s_{J,k}\varphi_{J,k}(t)+\sum_{j=J}^{\infty}\sum_k d_{j,k}\psi_{j,k}(t)=\sum_{j=-\infty}^{\infty}\sum_k d_{j,k}\psi_{j,k}(t), \tag{1.97}$$

where $d_{j,k}$ denotes the Haar wavelet coefficients. Assume that we have a certain continuous function $f(t)$, and we need to consider just a set of its samples that will be used for the Haar wavelet decomposition. In order to use the discrete values of $f$, the samples on the scale $j$ can be taken as the mean values on the interval $[2^{-j}k,\, 2^{-j}(k+1)]$ of length $2^{-j}$, and we have that:

$$\begin{aligned} m_{j,k}&=\frac{1}{2^{-j}}\int_{2^{-j}k}^{2^{-j}(k+1)}f(t)dt\,\Bigg|\begin{matrix}\textit{substitute}\\ t=\tau+2^{-j}k\end{matrix}=2^j\int_0^{2^{-j}}f\left(\tau+2^{-j}k\right)d\tau\\ &=2^j\int_0^{2^{-j}}f\left(t+2^{-j}k\right)dt \end{aligned} \tag{1.98}$$

The previous relation can be written as follows (for $t=2^{-j}k\Rightarrow 2^jt-k=0$; while for $t=2^{-j}(k+1)\Rightarrow 2^jt-k=1$):

$$\begin{aligned} m_{j,k}&=2^j\int_{-\infty}^{\infty}f(t)\varphi\left(2^jt-k\right)dt=2^{j/2}\int_{-\infty}^{\infty}2^{j/2}f(t)\varphi\left(2^jt-k\right)dt\ \Rightarrow\\ m_{j,k}&=2^{j/2}s_{j,k}, \end{aligned}$$

where according to Eq. (1.88): $s_{j,k}=\langle f,\varphi_{j,k}\rangle=\int_{-\infty}^{\infty}f(t)2^{j/2}\varphi\left(2^jt-k\right)dt.$

Therefore, $m_{j,k}=2^{j/2}s_{j,k}$ is the mean value of function $f$ within the *k-th* time interval $[2^{-j}k, 2^{-j}(k+1)]$, while $s_{j,k}$ are assumed to be the approximation coefficients at scale $j$. In other words if $f$ is scaled by $2^{j/2}$ then the mean values $m_{j,k}$ will correspond to the approximation coefficients $s_{j,k}$. In the sequel, we will show that the approximation coefficients at scale $j$ can be used to derive approximation and detail coefficients at scale $j$-1. Since, the number of translations is twice higher at scale $j$ (due to the narrower interval) than at scale $j$-1 (because $k=0,\pm 1,\ldots,2^j-1$), we may write $s_{j,2k}$ at scale $j$, while at scale $j$-1 we can use the notation $s_{j-1,k}$. Consider now the difference between two adjacent samples at scale $j$:

$$s_{j,2k} - s_{j,2k+1} = \int_{-\infty}^{\infty} f(t)2^{j/2}\left(\varphi\left(2^j t - 2k\right) - \varphi\left(2^j t - (2k+1)\right)\right)dt. \tag{1.99}$$

The difference between the scaling functions is:

$$\begin{aligned}
\varphi\left(2^j t - 2k\right) - \varphi\left(2^j t - (2k+1)\right) &= \begin{cases} 1 - 0 = 1, & \left(2^j t - 2k\right) \in [0,1) \\ 0 - 1 = -1, & \left(2^j t - (2k+1)\right) \in [0,1) \end{cases} \\
&= \begin{cases} 1, & t \in \left[\frac{2k}{2^j}, \frac{2k+1}{2^j}\right) \\ -1, & t \in \left[\frac{2k+1}{2^j}, \frac{2k+2}{2^j}\right) \end{cases} \\
&= \begin{cases} 1, \ t \in \left[2^{-(j-1)}k, \frac{2k+1}{2^j}\right) \\ -1, \ t \in \left[\frac{2k+1}{2^j}, 2^{-(j-1)}(k+1)\right). \end{cases}
\end{aligned}$$

Therefore, we have:

$$\varphi\left(2^j t - 2k\right) - \varphi\left(2^j t - (2k+1)\right) = \psi\left(2^{j-1}t - k\right), \tag{1.100}$$

and the difference between the mean values can be expressed as:

$$\begin{aligned}
s_{j,2k} - s_{j,2k+1} &= \int_{-\infty}^{\infty} f(t)2^{j/2}\psi\left(2^{j-1}t - k\right)dt = \sqrt{2}\int_{-\infty}^{\infty} f(t)\left[2^{(j-1)/2}\psi\left(2^{j-1}t - k\right)\right]dt \\
&= \sqrt{2}d_{j-1,k},
\end{aligned} \tag{1.101}$$

or,

$$\frac{1}{\sqrt{2}}\left(s_{j,2k} - s_{j,2k+1}\right) = d_{j-1,k}. \tag{1.102}$$

Furthermore, let observe the sum of the approximation coefficients:

$$s_{j,2k} + s_{j,2k+1} = \int_{-\infty}^{\infty} f(t)2^{j/2}\left(\varphi\left(2^j t - 2k\right) + \varphi\left(2^j t - (2k+1)\right)\right)dt. \tag{1.103}$$

The sum of the scaling functions is:

$$\varphi(2^j t - 2k) + \varphi(2^j t - (2k+1)) = \begin{cases} 1+0=1, & (2^j t - 2k) \in [0,1) \\ 0+1=1, & (2^j t - (2k+1)) \in [0,1) \end{cases}$$
$$= \begin{cases} 1, & t \in \left[\frac{2k}{2^j}, \frac{2k+1}{2^j}\right) \\ 1, & t \in \left[\frac{2k+1}{2^j}, \frac{2k+2}{2^j}\right) \end{cases}$$
$$= \left\{1,\ t \in \left[2^{-(j-1)}k,\ 2^{-(j-1)}(k+1)\right)\right.$$
$$= \varphi(2^{j-1}t - k)$$

Finally, we may write:

$$s_{j,2k} + s_{j,2k+1} = \int_{-\infty}^{\infty} f(t) 2^{j/2} \varphi(2^{j-1}t - k) dt$$
$$= \sqrt{2} \int_{-\infty}^{\infty} f(t) \left[2^{(j-1)/2} \varphi(2^{j-1}t - k)\right] dt = \sqrt{2} s_{j-1,k}, \tag{1.104}$$

or equivalently:

$$\frac{1}{\sqrt{2}} \left(s_{j,2k} + s_{j,2k+1}\right) = s_{j-1,k}. \tag{1.105}$$

Let us illustrate how this simplified algorithm can be used to decompose a row of image pixels:

$$f = \{10, 12, 14, 16, 18, 20, 22, 24\}.$$

There are $n = 8$ elements in the row of image pixels representing the samples of function $f$. Therefore, $\sqrt{n} = \sqrt{8} = \sqrt{2^3} = \sqrt{2^j}$, i.e., we consider $j = 3$. First we can scale the function $f$ by $2^{j/2} = \sqrt{2^j}$:

$$f_{\text{scaled}} = \frac{10}{\sqrt{8}} \ \frac{12}{\sqrt{8}} \ \frac{14}{\sqrt{8}} \ \frac{16}{\sqrt{8}} \ \frac{18}{\sqrt{8}} \ \frac{20}{\sqrt{8}} \ \frac{22}{\sqrt{8}} \ \frac{24}{\sqrt{8}}.$$

The decomposition procedure is based on applying Eqs. (1.102) and (1.105) sequentially within the three steps. In order to simplify the notation, let us use the symbols $A$ and $B$ for the approximation coefficients on different scales:

$$s_{j-1,k} = \frac{s_{j,2k} + s_{j,2k+1}}{\sqrt{2}} = \frac{1}{\sqrt{2}}(A+B); \quad d_{j-1,k} = \frac{s_{j,2k} - s_{j,2k+1}}{\sqrt{2}} = \frac{1}{\sqrt{2}}(A-B).$$

**Step I:**

$$\begin{array}{cccccccc} A & B & A & B & A & B & A & B \\ \frac{10}{\sqrt{8}} & \frac{12}{\sqrt{8}} & \frac{14}{\sqrt{8}} & \frac{16}{\sqrt{8}} & \frac{18}{\sqrt{8}} & \frac{20}{\sqrt{8}} & \frac{22}{\sqrt{8}} & \frac{24}{\sqrt{8}} \end{array}$$

$$(\mathrm{A}+\mathrm{B})/\sqrt{2}: \begin{array}{cccc} \frac{22}{\sqrt{8}\sqrt{2}} & \frac{30}{\sqrt{8}\sqrt{2}} & \frac{38}{\sqrt{8}\sqrt{2}} & \frac{46}{\sqrt{8}\sqrt{2}} = \\ \frac{22}{4} & \frac{30}{4} & \frac{38}{4} & \frac{46}{4} \end{array}$$

$$(\mathrm{A}-\mathrm{B})/\sqrt{2}: \begin{array}{cccc} \frac{-2}{\sqrt{8}\sqrt{2}} & \frac{-2}{\sqrt{8}\sqrt{2}} & \frac{-2}{\sqrt{8}\sqrt{2}} & \frac{-2}{\sqrt{8}\sqrt{2}} = \\ \frac{-1}{2} & \frac{-1}{2} & \frac{-1}{2} & \frac{-1}{2} \end{array}$$

Therefore, after the first decomposition level we have:

$$\frac{22}{4} \quad \frac{30}{4} \quad \frac{38}{4} \quad \frac{46}{4} \quad \frac{-1}{2} \quad \frac{-1}{2} \quad \frac{-1}{2} \quad \frac{-1}{2}$$

**Step II:**

$$(\mathrm{A}+\mathrm{B})/\sqrt{2}: \begin{array}{cccc} A & B & A & B \\ \frac{22}{4} & \frac{30}{4} & \frac{38}{4} & \frac{46}{4} \end{array}$$

$$\frac{52}{4\sqrt{2}} \quad \frac{84}{4\sqrt{2}} = \frac{13}{\sqrt{2}} \quad \frac{21}{\sqrt{2}}$$

$$(\mathrm{A}-\mathrm{B})/\sqrt{2}: \quad \frac{-8}{4\sqrt{2}} \quad \frac{-8}{4\sqrt{2}} = \frac{-2}{\sqrt{2}} \quad \frac{-2}{\sqrt{2}}$$

After the second decomposition level we have:

$$\frac{13}{\sqrt{2}} \quad \frac{21}{\sqrt{2}} \quad \frac{-2}{\sqrt{2}} \quad \frac{-2}{\sqrt{2}} \quad \frac{-1}{2} \quad \frac{-1}{2} \quad \frac{-1}{2} \quad \frac{-1}{2}$$

**Step III:**

$$(\mathrm{A}+\mathrm{B})/\sqrt{2}: \begin{array}{cc} A & B \\ \frac{13}{\sqrt{2}} & \frac{21}{\sqrt{2}} \end{array}$$

$$\frac{34}{\sqrt{2}\sqrt{2}} = 17$$

$$(\mathrm{A}-\mathrm{B})/\sqrt{2}\colon \frac{-8}{\sqrt{2}\sqrt{2}} = -4$$

After the third decomposition level, the resulting sequence is obtained:

$$17 \quad -4 \quad \frac{-2}{\sqrt{2}} \quad \frac{-2}{\sqrt{2}} \quad \frac{-1}{2} \quad \frac{-1}{2} \quad \frac{-1}{2} \quad \frac{-1}{2}$$

Alternatively, we can apply even simpler procedure. First, we find the mean value of pairs. Then we calculate pixel differences representing the coefficients of details.

| | 10 12 | 14 16 | 18 20 | 22 24 | |
|---|---|---|---|---|---|
| (A+B)/2 → | 11 | 15 | 19 | 23 | **Mean values** |
| (A-B)/2 → | -1 | -1 | -1 | -1 | **Details coefficients** |

Clearly, the newly created pixel vector {11, 15, 19, 23, −1, −1, −1, −1} can be used to completely reconstruct the image row. In the next level, we use 4 mean values and obtain 2 new mean and 2 new detail coefficients.

| | 11 15 | 19 23 | |
|---|---|---|---|
| (A+B)/2 → | 13 | 21 | **Mean values** |
| (A-B)/2 → | -2 | -2 | **Details coefficients** |

The new vector has the values {13, 21, −2, −2, −1, −1, −1, −1}.

Then carry out the decomposition of the remaining two mean values, from which we get a vector {17,−4, −2, −2, −1, −1, −1, −1}. The last step is the wavelet transform normalization by using the parameter $2^{-j/2}$:

$$\left\{\frac{17}{\sqrt{2^0}}, \frac{-4}{\sqrt{2^0}}, \frac{-2}{\sqrt{2^1}}, \frac{-2}{\sqrt{2^1}}, \frac{-1}{\sqrt{2^2}}, \frac{-1}{\sqrt{2^2}}, \frac{-1}{\sqrt{2^2}}, \frac{-1}{\sqrt{2^2}}\right\}.$$

The same procedure can be applied to every image row, and thus the whole image can be decomposed. The mean values produce a lower resolution image from which the details are removed.

### 1.9.6 Daubechies Orthogonal Filters

The Daubechies filter family is often used for the construction of orthogonal discrete wavelets. Suppose that the filter bank consists of the analysis filters *h* and

$g$ and the synthesis filters $h'$ and $g'$ of the length $N$ ($N$ is even). The impulse responses of filters are then given by:

$$\begin{aligned} h &= (h(0), h(1), \ldots, h(N-1)), \\ g &= (g(0), g(1), \ldots, g(N-1)), \\ h' &= (h'(0), h'(1), \ldots, h'(N-1)), \\ g' &= (g'(0), g'(1), \ldots, g'(N-1)). \end{aligned}$$

The Daubechies filters satisfy the following conditions:

1. The vector $\boldsymbol{h}$ is normalized;
2. For each integer that satisfies $1 \leq n < N/2$, the vector formed by the first $2n$ elements of $h$ should be orthogonal to the vector containing the last $2n$ elements of the same $h$;
3. The filter $\boldsymbol{h}'$ is the flipped version of $\boldsymbol{h}$;
4. Vector $\boldsymbol{g}$ is formed based on $\boldsymbol{h}'$ by multiplying the vector elements with $-1$ on even positions;
5. Vector $\boldsymbol{g}'$ is obtained from $\boldsymbol{h}$ by inverting the sign of the elements on odd positions;
6. The frequency response of the filter is equal to $\sqrt{2}$ for $\omega = 0$:

$$H(0) = \sqrt{2}.$$

7. The *k-th* derivative of the filter is equal to zero for $\omega = \pi$:

$$H^{(k)}(\pi) = 0 \quad for \quad k = 0, 1, 2, \ldots, \frac{N}{2} - 1.$$

Suppose that the length of the filters is $N = 4$. Then, by using the above conditions we get:

**Condition 1**: $h^2(0) + h^2(1) + h^2(2) + h^2(3) = 1$,
**Condition 2**: $h(0)h(2) + h(1)h(3) = 0$,
**Condition 3**: $h' = (h(3), h(2), h(1), h(0))$,
**Condition 4**: $g = (h(3), -h(2), h(1), -h(0))$,
**Condition 5**: $g' = (-h(0), h(1), -h(2), h(3))$.
**Condition 6**: The filter $h$ in the frequency domain can be written in the form of Fourier series:

$$H(\omega) = h(0) + h(1)e^{-j\omega} + h(2)e^{-2j\omega} + h(3)e^{-3j\omega}.$$

It is necessary to construct a filter that satisfies the condition $H(0) = \sqrt{2}$. If we set $\omega = 0$, we get:

**Table 1.1** Coefficients of the Daubechies D4 low-pass analysis filter *h*

| $h(0)$ | $h(1)$ | $h(2)$ | $h(3)$ |
|---|---|---|---|
| –0.12941 | 0.224144 | 0.836516 | 0.482963 |

$$\sqrt{2} = h(0) + h(1) + h(2) + h(3).$$

The last condition is related to the *k*-th derivatives of the filter.

**Condition 7**:

$$\left.\begin{array}{l} H^{(0)}(\omega) = h(0) + h(1)e^{-j\omega} + h(2)e^{-2j\omega} + h(3)e^{-3j\omega} \\ H^{(0)}(\pi) = 0 \end{array}\right\} \boxed{h(0) - h(1) + h(2) - h(3) = 0}$$

$$\left.\begin{array}{l} H^{(1)}(\omega) = -jh(1)e^{-j\omega} - 2jh(2)e^{-2j\omega} - 3jh(3)e^{-3j\omega} \\ H^{(1)}(\pi) = 0 \end{array}\right\} \boxed{-h(1) + 2h(2) - 3h(3) = 0}$$

Hence, a system of equations is obtained which can be used to calculate the coefficients:

$$\boxed{\begin{array}{l} h^2(0) + h^2(1) + h^2(2) + h^2(3) = 1 \\ h(0)h(2) + h(1)h(3) = 0 \\ h(0) + h(1) + h(2) + h(3) = \sqrt{2} \\ h(0) - h(1) + h(2) - h(3) = 0 \\ -h(1) + 2h(2) - 3h(3) = 0 \end{array}}$$

The system has two solutions. The first solution represents the coefficients of a low-pass analysis filter (Table 1.1), and the second solution represents the coefficients of a low-pass synthesis filter.

### 1.9.7 Filter Banks

Using the multiresolution analysis, a signal can be decomposed into two parts: one representing the approximation of the original signal and the other containing information about the details. Thus, in analogy with the formula Eq. (1.92) from the multiresolution analysis, the signal can be represented as:

$$f_m(t) = \sum_n \alpha_{m+1,n}\varphi_{m+1,n} + \sum_n \beta_{m+1,n}\psi_{m+1,n}, \tag{1.106}$$

where $\alpha_{m+1,n}$ are the approximation coefficients at resolution $2^{m+1}$, while $\beta_{m+1,n}$ are the coefficients of details. The functions $\varphi_{m+1,n}$ and $\psi_{m+1,n}$ represent the scaling and wavelet function, respectively. In multiresolution analysis, the decomposition of signals using the discrete wavelet transform can be expressed in terms of finite impulse response filters (FIR filters) for computation of the wavelet coefficients.

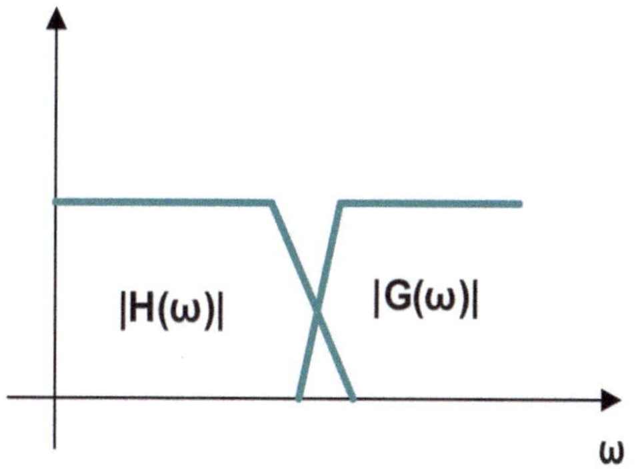


**Fig. 1.21** Analysis filters

The recursive realization of discrete wavelet transform in different levels can be written as:

$$\begin{aligned}\alpha_{m,n}(f) &= \sum_{k} h_{2n-k}\alpha_{m-1,k}(f),\\ \beta_{m,n}(f) &= \sum_{k} g_{2n-k}\alpha_{m-1,k}(f),\end{aligned} \tag{1.107}$$

where $h$ and $g$ are low-pass and high-pass filters (Fig. 1.21) often referred to as **analysis filters**: $h_i = 2^{1/2}\int \varphi(x-i)\varphi(2x)dx$, $g_i = (-1)^i h_{-i+1}$.

Since $h$ and $g$ are defined from the orthonormal basis functions, they provide exact reconstruction:

$$\alpha_{m-1,i}(f) = \sum_{n} h_{2n-i}\alpha_{m,n}(f) + \sum_{n} g_{2n-i}\beta_{m,n}(f). \tag{1.108}$$

Theoretically, for many orthonormal basis wavelet functions, there is a large number of filters that can be used for their implementation. In practice, finite impulse response filters (FIR filters) are used to implement the wavelets efficiently. The orthonormal wavelet functions may have infinite support causing the filters $h$ and $g$ to have infinitely many taps. For efficient implementation, it is preferred to have filters with small number of taps, achieved using bi-orthogonal basis functions.

**Synthesis filters** ($h'$ and $g'$) are used for the signal reconstruction. Namely the signal decomposition is done by using Eq. (1.107), while the reconstruction is given by the following expression:

$$\alpha_{m-1,i}(f) = \sum_{n} \alpha_{m,n}(f)h'_{2n-i} + \sum_{n} \beta_{m,n}(f)g'_{2n-i}. \tag{1.109}$$

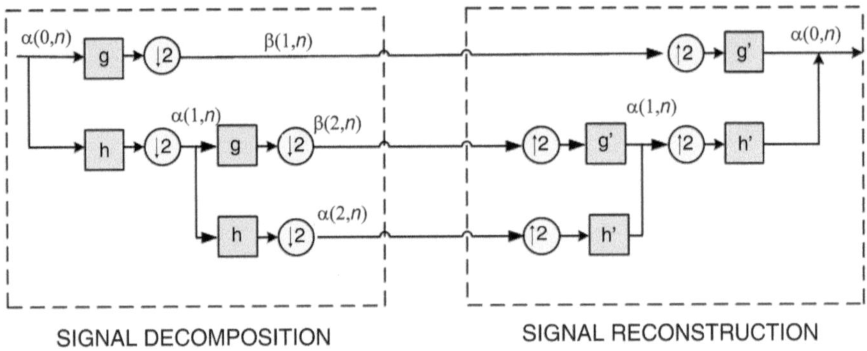


**Fig. 1.22** Wavelet decomposition and reconstruction of signals using a filter bank

If $(h',g')=(h,g)$ holds, the filters are orthogonal. Otherwise, they are bi-orthogonal. Figure 1.22 illustrates the concept of filter banks. The input signal $\alpha(0,n)$ is filtered in parallel by a low-pass filter $h$ and a high-pass filter $g$. The signal is down-sampled (by simply dropping the alternate output samples in each stream) after passing through the filters. Therefore, the output from the analysis filters at the first level of decomposition is given by:

$$(\downarrow \alpha(1,n))=(\ldots,\alpha(1,-6),\alpha(1,-4),\alpha(1,-2),\alpha(1,0),\alpha(1,2),\alpha(1,4),\alpha(1,6),\ldots).$$

Before passing through the synthesis filters, the signal has to be up-sampled:

$$(\uparrow \alpha(1,n))=(\ldots,\alpha(1,-6),0,\alpha(1,-4),0,\alpha(1,-2),0,\alpha(1,0),0,\alpha(1,2),0,a(1,4),0,\alpha(1,6),\ldots).$$

## 1.9.8 Two-Dimensional Signals

The two-dimensional discrete wavelet transform is generally used to decompose two-dimensional signals (e.g., images). Consider the two-dimensional separable scaling and wavelet functions. They can be represented as the product of one-dimensional functions $\varphi(x,y)=\varphi(x)\varphi(y)$ and $\psi(x,y)=\psi(x)\psi(y)$, enabling the application of the one-dimensional discrete wavelet transform separately to the rows and columns of a two-dimensional matrix. Several different approaches to analyze two-dimensional signals using the discrete wavelet transform are described below.

- **Standard wavelet decomposition**

The first step of decomposition involves creating a low-frequency subband $L_1$ and a high-frequency subband $H_1$. The same procedure is then carried out over low-frequency subband $L_1$ by forming subbands $L_2$ and $H_2$. We continue this procedure until we reach a desired number of subbands. The second step involves the same procedure for the columns. The end result is a low-pass coefficient in the upper left corner. The decomposition is illustrated in Fig. 1.23.

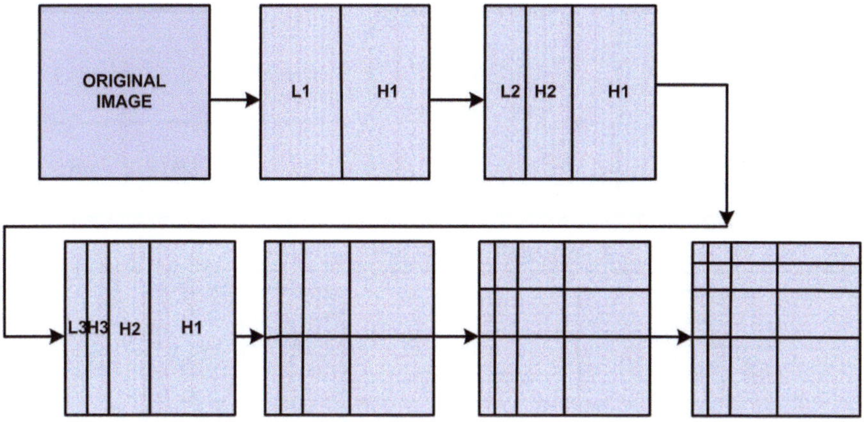


**Fig. 1.23** Standard wavelet decomposition

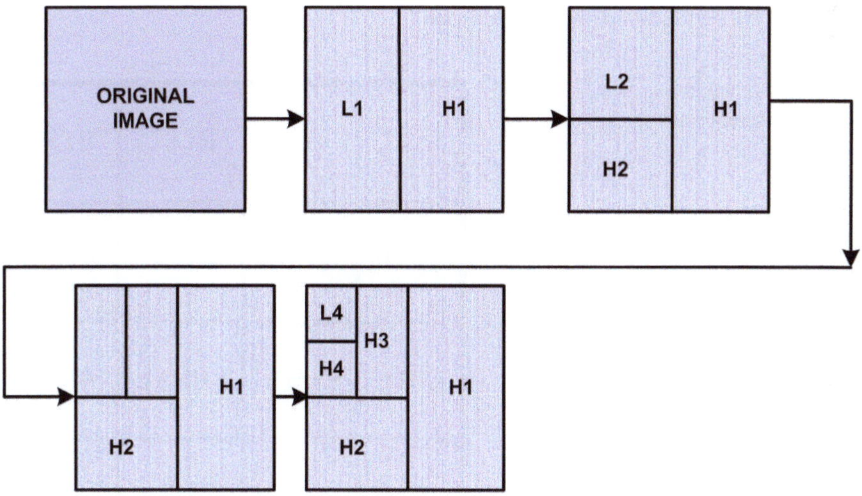


**Fig. 1.24** Quincunx decomposition

- **Quincunx decomposition**

This decomposition uses each low-frequency subband on the level $j$ ($L_j$) and divides it into subbands $L_{j+1}$ and $H_{j+1}$ (subbands on the level $j+1$). Figure 1.24 illustrates the Quincunx decomposition.

- **Pyramidal wavelet decomposition**

The most commonly used decomposition method in practical applications is the pyramidal decomposition as shown in Fig. 1.25. Suppose that the image dimensions are $M \times N$. Initially, the one-dimensional wavelet transform is performed for image rows and subbands $L_1$ and $H_1$ are obtained. Then the wavelet transform is

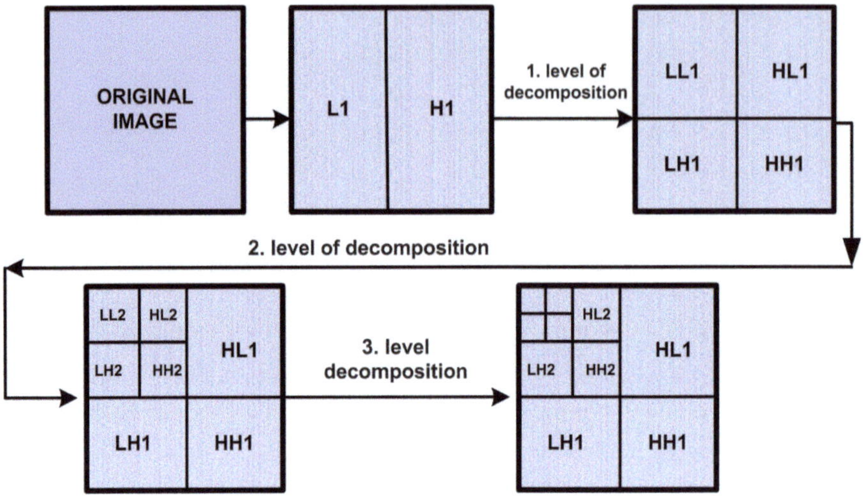


**Fig. 1.25** The pyramidal wavelet decomposition

**Fig. 1.26** The uniform wavelet decomposition

performed for each column resulting in four subbands $LL_1$, $LH_1$,, $HL_1$, $HH_1$ with dimensions equal to $M/2 \times N/2$. The $LL_1$ subband represents a version of the original image with lower resolution. The $LL_1$ subband is then further decomposed into subbands $LL_2$, $LH_2$, $HL_2$, and $HH_2$. If further decomposition is needed, it would be based on the low-pass subbands $LL_j$.

- **Uniform wavelet decomposition**

This decomposition is initially performed for rows and columns, producing the four subbands. Then, the same procedure is repeated for each subband and 16 new subbands are obtained. Figure 1.26 illustrates this process for two levels of decomposition.

## 1.10 Signal Decomposition Using Hermite Functions

Projecting signals using the Hermite functions is widely used in various image and signal processing applications (e.g., image filtering, texture analysis, speaker identification). The Hermite functions allow us to obtain good localization of the signals in both the signal and transform domains. These functions are defined as follows:

$$\psi_p(n) = \frac{(-1)^p e^{n^2/2}}{\sqrt{2^p p! \sqrt{\pi}}} \frac{d^p\left(e^{-n^2}\right)}{dn^p} = \frac{e^{-n^2} H_p(n)}{\sqrt{2^p p! \sqrt{\pi}}}, \tag{1.110}$$

where $H_p(n)$ is the $p$-th order Hermite polynomial. The definition of Hermite function can be done using the recursive formulas:

$$\begin{aligned} \Psi_0(n) &= \frac{1}{\sqrt[4]{\pi}} e^{-n^2/2}, \quad \Psi_1(n) = \frac{\sqrt{2}n}{\sqrt[4]{\pi}} e^{-n^2/2}, \\ \Psi_p(n) &= n\sqrt{\frac{2}{p}}\Psi_{p-1}(n) - \sqrt{\frac{p-1}{p}}\Psi_{p-2}(n), \ \forall p \geq 2. \end{aligned} \tag{1.111}$$

The Hermite functions are often calculated at the roots of the Hermite polynomials. The first few Hermite functions are illustrated in Fig. 1.27.

If $n_i$ and $n_j$ are roots of Hermite polynomial $H_{N+1}(n)$ such that $n_i \neq n_j$ then:

$$\sum_{p=0}^{N} \psi_p(n_i)\psi_p(n_j) = 0, \tag{1.112}$$

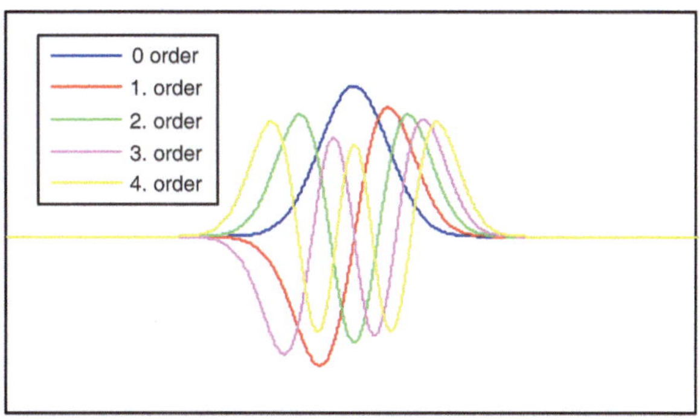


**Fig. 1.27** An illustration of the first few Hermite functions

and,

$$\sum_{p=0}^{N} \psi_p^2(n_i) = (N+1)\psi_N^2(n_i). \tag{1.113}$$

The normalized Hermite functions satisfy the bound:

$$|\psi_p(n)| \leq 0.816, \text{ for all } n \text{ and } p. \tag{1.114}$$

The Hermite functions form an orthogonal basis in the underlying signal space:

$$\sum_{n=0}^{N-1} \psi_p(n)\psi_q(n) = \begin{cases} 0, & p \neq q \\ 1, & p = q \end{cases}. \tag{1.115}$$

### 1.10.1 One-Dimensional Signals and Hermite Functions

Assume that $X$ is a discrete one-dimensional signal (of length $M$) that will be expanded by using the Hermite functions. The first step in the Hermite projection method is to remove the baseline, defined as:

$$x(i) = X(1) + \frac{X(M) - X(1)}{M} \cdot i, \tag{1.116}$$

where $i = 1, \ldots, M$. The baseline is then subtracted from the original signal:

$$f(i) = X(i) - x(i). \tag{1.117}$$

Then, the decomposition by using $N$ Hermite functions is defined as follows:

$$f(i) = \sum_{p=0}^{N-1} c_p \psi_p(i), \tag{1.118}$$

where the coefficients $c_p$ are obtained as:

$$c_p = \int_{-\infty}^{\infty} f(i)\psi_p(i)di. \tag{1.119}$$

The Gauss-Hermite quadrature technique can be used to calculate the Hermite expansion coefficients, as follows:

$$c_p \approx \frac{1}{M} \sum_{m=1}^{M} \mu_{M-1}^{p}(x_m) f(x_m), \tag{1.120}$$

**Table 1.2** Hermite polynomials of orders 1–10 and the corresponding zeros

| Hermite Polynomials | Zeros |
|---|---|
| $H_1(x) = 2x$ | 0 |
| $H_2(x) = 4x^2 - 2$ | $\pm 0.707$ |
| $H_3(x) = 8x^3 - 12x$ | $\pm 1.2247$, 0 |
| $H_4(x) = 16x^4 - 48x^2 + 12$ | $\pm 1.6507$, $\pm 0.5246$ |
| $H_5(x) = 32x^5 - 160x^3 + 120x$ | $\pm 2.0202$, $\pm 0.9586$, 0 |
| $H_6(x) = 64x^6 - 480x^4 + 720x^2 - 120$ | $\pm 2.3506$, $\pm 1.3358$, $\pm 0.4361$ |
| $H_7(x) = 128x^7 - 1344x^5 + 3360x^3 - 1680x$ | $\pm 2.6520$, $\pm 1.6736$, $\pm 0.8163$, 0 |
| $H_8(x) = 256x^8 - 3584x^6 + 13440x^4 - 13440x^2 + 1680$ | $\pm 2.9306$, $\pm 1.9817$, $\pm 1.1572$, $\pm 0.3812$ |
| $H_9(x) = 512x^9 - 9216x^7 + 48384x^5 - 80640x^3 + 30240x$ | $\pm 3.1910$, $\pm 2.2666$, $\pm 1.4686$, $\pm 0.7236$, 0 |
| $H_{10}(x) = 1024x^{10} - 23040x^8 + 161280x^6 - 403200x^4 + 302400x^2 - 30240$ | $\pm 3.4362$, $\pm 2.5327$, $\pm 1.7567$, $\pm 1.0366$, $\pm 0.3429$ |

where the points $x_m$ ($m = 1, \ldots, M$) are obtained as zeros of an $M$-th order Hermite polynomial: $H_M(x) = (-1)^M e^{x^2} \frac{d^M\left(e^{-x^2}\right)}{dx^M}$. Hence, the values of the function $f$ should be calculated at the zeros of Hermite polynomial. The Hermite polynomials of orders from 1 to 10, as well as the corresponding zeros, are given in the Table 1.2.

In the examples we will use the samples at the available time instants, not necessarily the zeros of the Hermite polynomial.

The constants $\mu_{M-1}^p(x_m)$ can be obtained using the Hermite functions:

$$\mu_{M-1}^p(x_m) = \frac{\psi_p(x_m)}{\left(\psi_{M-1}(x_m)\right)^2}. \tag{1.121}$$

Figure 1.28 depicts the original signal whose length is equal to 126 samples, and three reconstructed signals using 126, 90 and 70 Hermite functions, respectively. In the first case, the reconstructed signal is approximately equal to the original signal. However, the signal can be successfully reconstructed with a smaller number of Hermite functions (while losing some finer details). The fact that the signal can be represented with a fewer number of Hermite coefficients, makes them attractive for signal compression.

### *1.10.2 Hermite Transform and its Inverse Using Matrix Form Notation*

Based on the Gauss-quadrature rule, the direct Hermite transform can be also defined in the matrix form as follows:

$$\mathbf{C} = \mathbf{Hf}, \tag{1.122}$$

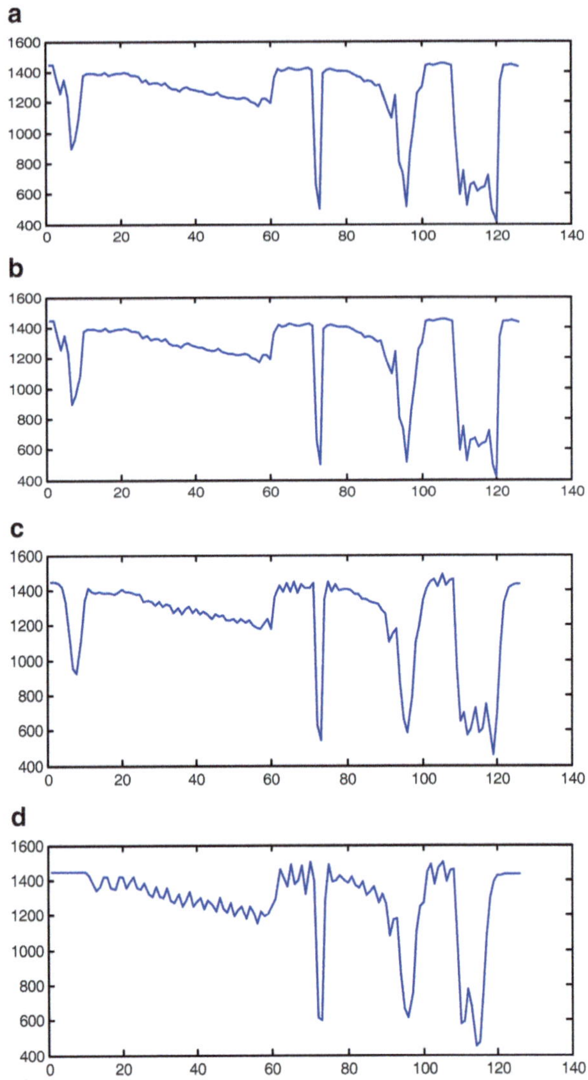

**Fig. 1.28** (**a**) The original signal of length 126 samples, (**b**) the reconstructed signal with 126 functions, (**c**) the reconstructed signal with 90 functions, (**d**) the reconstructed signal with 70 functions

where **C** is the vector of Hermite coefficients, **H** is the Hermite transform matrix. Without loss of generality, we assume that the number of samples is equal to the number of Hermite functions used for expansion: $M = N$. Thus, **f** is a signal of length $N$, which is represented using $N$ Hermite functions and corresponding coefficients. In the expanded form, Eq. (1.122) can be written as follows:

$$\underbrace{\begin{bmatrix} C(0) \\ C(1) \\ \dots \\ C(N-1) \end{bmatrix}}_{\mathbf{C}} = \frac{1}{N} \underbrace{\begin{bmatrix} \frac{\psi_0(0)}{(\psi_{N-1}(0))^2} & \frac{\psi_0(1)}{(\psi_{N-1}(1))^2} & \dots & \frac{\psi_0(N-1)}{(\psi_{N-1}(N-1))^2} \\ \frac{\psi_1(0)}{(\psi_{N-1}(0))^2} & \frac{\psi_1(1)}{(\psi_{N-1}(1))^2} & \dots & \frac{\psi_1(N-1)}{(\psi_{N-1}(N-1))^2} \\ \dots & \dots & \dots & \dots \\ \frac{\psi_{N-1}(0)}{(\psi_{N-1}(0))^2} & \frac{\psi_{N-1}(1)}{(\psi_{N-1}(1))^2} & \dots & \frac{\psi_{N-1}(N-1)}{(\psi_{N-1}(N-1))^2} \end{bmatrix}}_{\mathbf{H}} \underbrace{\begin{bmatrix} f(0) \\ f(1) \\ \dots \\ f(N-1) \end{bmatrix}}_{\mathbf{f}}$$

Note that in order to simplify the notations, the argument $x_n$ (zeros of the $N$-th order Hermite polynomials in $\mathbf{H}$ and $\mathbf{f}$) is replaced by the corresponding order number $n$. The inverse Hermite transform can be written as:

$$\mathbf{f} = \mathbf{\Psi C}, \tag{1.123}$$

or equivalently:

$$\underbrace{\begin{bmatrix} f(0) \\ f(1) \\ \dots \\ f(N-1) \end{bmatrix}}_{\mathbf{f}} = \underbrace{\begin{bmatrix} \psi_0(0) & \psi_1(0) & \dots & \psi_{N-1}(0) \\ \psi_0(1) & \psi_1(1) & \dots & \psi_{N-1}(1) \\ \dots & \dots & \dots & \dots \\ \psi_0(N-1) & \psi_1(N-1) & \dots & \psi_{N-1}(N-1) \end{bmatrix}}_{\mathbf{\Psi}} \underbrace{\begin{bmatrix} C(0) \\ C(1) \\ \dots \\ C(N-1) \end{bmatrix}}_{\mathbf{C}}$$

where $\mathbf{\Psi}$ is the inverse Hermite transform matrix.

Now, regarding the energy condition we might introduce a rough experimentally obtained approximation as follows:

$$\begin{aligned} \|C\|_2^2 = |C_0|^2 + \ldots + |C_{N-1}|^2 &= \frac{1}{N}\left( \left| \sum_{n=0}^{N-1} f(n) \frac{\psi_0(n)}{(\psi_{N-1}(n))^2} \right|^2 + \ldots + \left| \sum_{n=0}^{N-1} f(n) \frac{\psi_{N-1}(n)}{(\psi_{N-1}(n))^2} \right|^2 \right) \\ &\approx \frac{\sum_{n=0}^{N-1} |f(n)|^2}{\sum_{n=0}^{N-1} |\psi_{N-1}(n)|^2}. \end{aligned} \tag{1.124}$$

In other words, we can approximate the energy of coefficients in the Hermite domain as:

$$\|C\|_2^2 \approx \frac{\|\mathbf{f}\|_2^2}{\|\psi_{N-1}\|_2^2}. \tag{1.125}$$

The notation $\|\cdot\|_2^2$ represents the squared $\ell_2$ -norm: $\|\alpha\|_2^2 = \sum_i |\alpha_i|^2$.

### 1.10.3 Two-Dimensional Signals and Two-Dimensional Hermite Functions

For two-dimensional signals such as images, we use the two-dimensional Hermite functions defined as follows:

$$\Psi_{kl}(x,y) = \frac{(-1)^{k+l} e^{x^2/2+y^2/2}}{\sqrt{2^{k+l}k!l!\pi}} \frac{d^k\left(e^{-x^2}\right)}{dx^k} \frac{d^l\left(e^{-y^2}\right)}{dy^l}. \tag{1.126}$$

Some examples of two-dimensional Hermite functions are illustrated in Fig. 1.29.

Two-dimensional functions can be evaluated as a composition of one-dimensional Hermite functions:

$$\Psi_{kl}(x,y) = \Psi_k(x)\Psi_l(y) = \frac{(-1)^k e^{x^2/2}}{\sqrt{2^k k!\sqrt{\pi}}} \frac{d^k\left(e^{-x^2}\right)}{dx^k} \frac{(-1)^l e^{y^2/2}}{\sqrt{2^l l!\sqrt{\pi}}} \frac{d^l\left(e^{-y^2}\right)}{dy^l}.$$

Next we consider the Hermite projection method along one coordinate. For the two-dimensional signal of size $M_1 \times M_2$, the signal baseline is defined as:

$$b_y(x) = F(1,y) + \frac{F(M_1,y) - F(1,y)}{M_1} \cdot x, \tag{1.127}$$

where $F(x,y)$ is a two-dimensional signal, and $x = 1, \ldots, M_1$ and $y = 1, \ldots, M_2$. The baseline $b_y(x)$ is calculated for a fixed value of $y$. The corresponding matrix $b(x,y)$ contains all vectors $b_y(x)$ for $y = 1, \ldots, M_2$. Then, we subtract the baselines from the original signal:

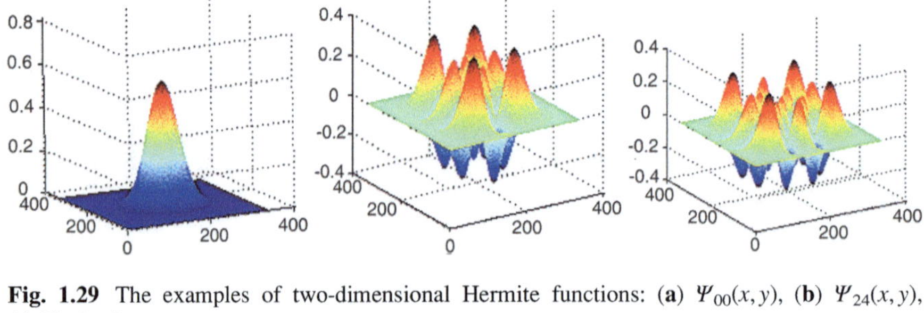


**Fig. 1.29** The examples of two-dimensional Hermite functions: (**a**) $\Psi_{00}(x,y)$, (**b**) $\Psi_{24}(x,y)$, (**c**) $\Psi_{44}(x,y)$

$$f(x, y) = F(x, y) - b(x, y). \tag{1.128}$$

The signal decomposition using $N$ Hermite functions is defined as follows:

$$f_y(x) = \sum_{p=0}^{N-1} c_p \psi_p(x), \tag{1.129}$$

where $f_y(x) = f(x, y)$ is valid for fixed $y$. The coefficients are equal to:

$$c_p = \int_{-\infty}^{\infty} f_y(x) \psi_p(x) dx. \tag{1.130}$$

Using the Gauss-Hermite quadrature rule, the coefficients can be calculated as follows:

$$c_p \approx \frac{1}{M_1} \sum_{m=1}^{M_1} \mu_{M_1-1}^{p}(x_m) f_y(x_m), \tag{1.131}$$

where constants $\mu_{M_1-1}^{p}(x_m)$ can be obtained by using Hermite functions as in the one-dimensional case.

To illustrate this concept, the original image and three versions of reconstructed images are shown in Fig. 1.30.

## 1.11 Generalization of the Time-Frequency Plane Division

Recall the definition of the discrete form of the STFT:

$$STFT(n, k) = \sum_{m=-N/2}^{N/2-1} x(n+m) e^{-j2\pi km/N}, \tag{1.132}$$

where $N$ is the length of a rectangular window, while $k = [-N/2, \ldots, N/2 - 1]$. The previous relation can be defined in the matrix form. First, let us define the $N \times N$ DFT matrix that corresponds to one windowed signal part:

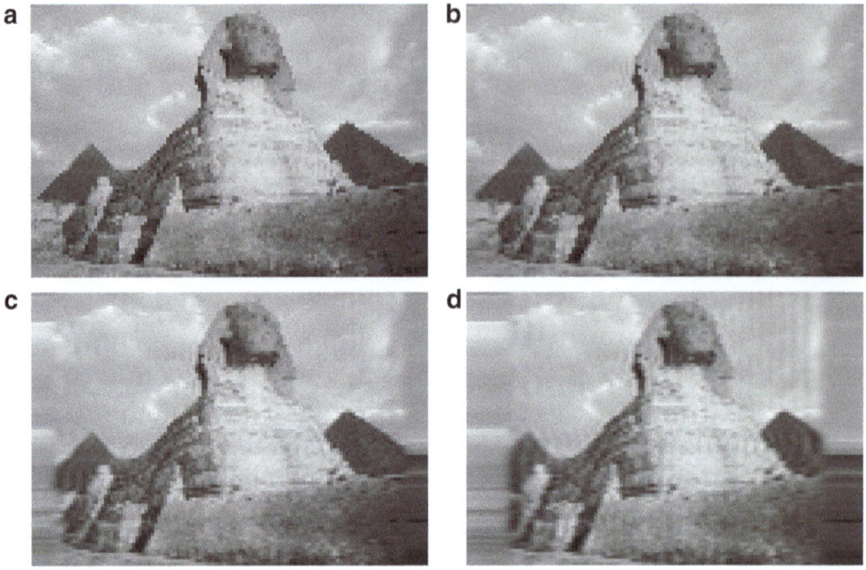

**Fig. 1.30** (**a**) The original image, reconstructed image by using Hermite projection method along the rows with: (**b**) 100 Hermite functions, (**c**) 80 Hermite functions, (**d**) 60 Hermite functions

$$\mathbf{W}_N = \begin{array}{c} {\scriptstyle m=-N/2} \\ \\ {\scriptstyle m=0} \\ \\ {\scriptstyle m=N/2-1} \end{array} \overset{\begin{array}{ccc} {\scriptstyle k=-N/2} & {\scriptstyle k=0} & {\scriptstyle k=N/2-1} \end{array}}{\begin{bmatrix} e^{-j\pi N/2} & \dots & 1 & \dots & e^{j\pi(N/2-1)} \\ \dots & & \dots & & \dots \\ 1 & \dots & 1 & \dots & 1 \\ \dots & & \dots & & \dots \\ e^{j\pi(N/2-1)} & \dots & 1 & \dots & e^{-j2\pi(N/2-1)^2/N} \end{bmatrix}}. \tag{1.133}$$

The nonoverlapping rectangular windows are assumed. Now, let us create the total transform matrix as follows:

$$\mathbf{W} = \mathbf{I}_{M/N} \otimes \mathbf{W}_N = \begin{bmatrix} \mathbf{W}_N & \mathbf{0} & \dots & \mathbf{0} \\ \mathbf{0} & \mathbf{W}_N & \dots & \mathbf{0} \\ \dots & \dots & \dots & \mathbf{0} \\ \mathbf{0} & \mathbf{0} & \dots & \mathbf{W}_N \end{bmatrix}, \tag{1.134}$$

where $\mathbf{I}$ is identity matrix of size $(M/N \times M/N)$, $\mathbf{0}$ is $N \times N$ zero matrix, $M$ is the total length of the signal, while $\otimes$ denotes the Kronecker product. The STFT can be defined using transform matrix $\mathbf{W}$ as:

$$\mathbf{STFT} = \mathbf{Wx} = \mathbf{WW}_M^{-1}\mathbf{X}, \tag{1.135}$$

where $\mathbf{x}$ is a time domain signal vector of size $M \times 1$, $\mathbf{X}$ is a Fourier transform vector of size $M \times 1$, while $\mathbf{STFT}$ is a column vector containing all STFT vectors $\mathrm{STFT}_{Ni}(n_i)$, $i=0,1,\ldots,K$.

If the windows are not rectangular, then the window function needs to be included in the STFT calculation as follows:

$$\mathbf{STFT} = \mathbf{WHx}, \tag{1.136}$$

where:

$$\mathbf{H} = \mathbf{I}_{M/N} \otimes \mathbf{H}_N = \begin{bmatrix} \mathbf{H}_N & \mathbf{0} & \ldots & \mathbf{0} \\ \mathbf{0} & \mathbf{H}_N & \ldots & \mathbf{0} \\ \ldots & \ldots & \ldots & \ldots \\ \mathbf{0} & \mathbf{0} & \ldots & \mathbf{H}_N \end{bmatrix}, \tag{1.137}$$

while the matrix $\mathbf{H}_N$ is a diagonal $N \times N$ matrix with the window values on the diagonal: $H_N(m,m) = w(m)$, $m = -N/2, \ldots, N/2-1$.

In the case of windows with variable length over time (time-varying windows), the smaller matrices within $\mathbf{W}$ or $\mathbf{H}$ will be of different sizes. Without loss of generality, assume a set of $K$ rectangular windows of sizes $N_1, N_2, \ldots, N_K$ instead of $\mathbf{W}_N$ we will have: $\mathbf{W}_{N_1}, \mathbf{W}_{N_2}, \ldots, \mathbf{W}_{N_K}$, where $K$ is the number of nonoverlapping windows used to represent the entire signal. Since the time-varying nonoverlapping STFT corresponds to a decimation-in-time DFT scheme, its calculation is more efficient than the DFT calculation of the entire signal. Note that there is a large number of combinations of time-varying nonoverlapping windows and consequently a large number of nonoverlapping STFTs. Then the STFT calculated using time-varying windows $\mathbf{W}_{N_i}$, $i=1,\ldots,K$ can be written as:

$$\mathbf{STFT} = \mathbf{Wx} = \begin{bmatrix} \mathbf{W}_{N_1} & \mathbf{0} & \ldots & \mathbf{0} \\ \mathbf{0} & \mathbf{W}_{N_2} & \ldots & \mathbf{0} \\ \ldots & \ldots & \ldots & \ldots \\ \mathbf{0} & \mathbf{0} & \ldots & \mathbf{W}_{N_K} \end{bmatrix} \begin{bmatrix} x(0) \\ x(1) \\ \ldots \\ x(M) \end{bmatrix}. \tag{1.138}$$

The zero matrices are denoted by $\mathbf{0}$. An illustration of the STFT with time-varying windows is shown in Fig. 1.31a). The signal length is $M=8$ samples, while:

$$\mathbf{W} = \begin{bmatrix} \mathbf{W}_2 & \mathbf{0} & \mathbf{0} & \mathbf{0} \\ \mathbf{0} & \mathbf{W}_4 & \mathbf{0} & \mathbf{0} \\ \mathbf{0} & \mathbf{0} & \mathbf{W}_1 & \mathbf{0} \\ \mathbf{0} & \mathbf{0} & \mathbf{0} & \mathbf{W}_1 \end{bmatrix}. \tag{1.139}$$

Note that for $\mathbf{W}_2$ (matrix of size $2 \times 2$, with elements calculated for $m=\{-1,0\}$ and $k=\{-1,0\}$), two signal samples ($x(0)$, $x(1)$) are captured by the window. The window should be centered at the middle point, but in this case $n=1$ is used as a central point. Thus, we obtained two STFT points: $\mathrm{STFT}_2(1,-1)$ and $\mathrm{STFT}_2(1,0)$.

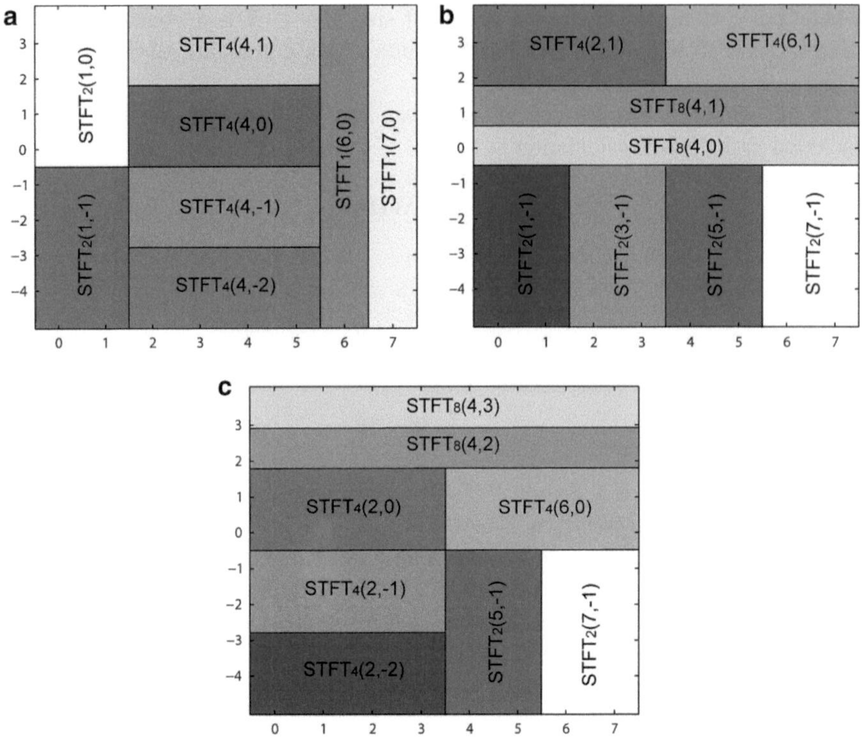


**Fig. 1.31** The STFTs calculated using: (**a**) time-varying window, (**b**) frequency-varying window, (**c**) time-frequency-varying window. The time index is on the horizontal axis, while the frequency index is on the vertical axis

The index denotes the window size, i.e., the number of samples captured by the window, the first coordinate is the central time sample, while the second coordinate is the frequency $k$.

Accordingly, for $\mathbf{W}_4$ (matrix of size $4 \times 4$, with elements calculated for $m = \{-2,-1,0,1\}$, $k = \{-2,-1,0,1\}$), which capture the next 4 samples on the positions $n = \{2,3,4,5\}$, we have: $\mathrm{STFT}_4(4,-2)$, $\mathrm{STFT}_4(4,-1)$, $\mathrm{STFT}_4(4,0)$, $\mathrm{STFT}_4(4,1)$. Note that the central point is $n = 4$. Similarly for $n = 6$ and $\mathbf{W}_1$ (only one value for $k$), we have $\mathrm{STFT}_1(6,0)$, and for $n = 7$ and $\mathbf{W}_1$ we obtain $\mathrm{STFT}_1(7,0)$.

The STFT may use frequency-varying windows as well, which is illustrated in Fig. 1.31b), but also the time-frequency-varying window (Fig. 1.31c).

## 1.12 Examples

1.1. (a) If the maximal signal frequency is $f_{\max} = 4$ KHz, what should be the maximal sampling interval $T_0$ according to the sampling theorem?

(b) For the signals whose sampling intervals are given as $T_1 = 2 \cdot T_0$ and $T_2 = T_0/2$, specify the maximal values of signal frequency having in mind that the sampling theorem has to be satisfied?

Solution:

(a) For the known maximal signal frequency, the sampling interval can be calculated according to the relation:

$$T_0 \leq \frac{1}{2 \cdot f_{\max}} = \frac{1}{2 \cdot 4 \cdot 10^3} = 125\mu s.$$

(b) For the sampling interval $T_1 = 2 \cdot T_0$, we have:

$$f_{\max} \leq \frac{1}{2 \cdot T_1} = \frac{1}{2 \cdot 2 \cdot T_0} = \frac{1}{2 \cdot 2 \cdot 125 \cdot 10^{-6}} = 2KHz.$$

In the case $T_2 = T_0/2$, the maximal frequency of the considered signal can be calculated as:

$$f_{\max} \leq \frac{1}{2 \cdot T_0/2} = \frac{1}{2 \cdot 125 \cdot 10^{-6}/2} = 8KHz.$$

1.2. Plot the Fourier transform of the signal $y = \sin(150\pi t)$, $t \in (-1,1)$, by using Matlab. The sampling interval should be set as: $T = 1/1000$.

Solution:
Firstly, we should check if the sampled signal satisfies the sampling theorem, i.e., we check if the condition $f \leq f_{max}$ is satisfied.

The maximal frequency of the signal is: $f_{max} \leq 1/(2T) = 500\text{Hz}$, while the sinusoidal signal frequency is obtained as: $2\pi f = 150\pi \Rightarrow f = 75Hz$, and satisfies the condition $f \leq f_{max}$.

For the calculation of the Fourier transform coefficients, we use *fft* Matlab function. In order to position the zero-frequency to the center of the array, the *fftshift* function is used as well. Matlab code is given in the sequel:

```
t=-1:1/1000:1;
y=sin(150*pi*t);
F=fft(y);
F=fftshift(F);
plot(abs(F))
```

1.3. Calculate the Fourier transform of the signal $y = \sin(150 \cdot \pi \cdot t)$, $t = -5{:}0.001{:}5$. Is the same sampling interval appropriate to satisfy the sampling theorem even for signals: $y = \sin(600 \cdot \pi \cdot t)$, $y = \sin(1800 \cdot \pi \cdot t)$? If the answer is yes, plot the spectra of the considered signals by using Matlab.

Solution:

$$f_{\max} = \frac{1}{2 \cdot \Delta t} = \frac{1}{2 \cdot 0.001} = 500Hz$$

$$\omega_1 = 2\pi f_1 = 150\pi \Rightarrow f_1 = 150/2 = 75Hz$$

$$\omega_2 = 2\pi f_2 = 600\pi \Rightarrow f_2 = 600/2 = 300Hz$$

In the third case the sampling theorem is not satisfied:

$$\omega_3 = 2\pi f_3 = 1800\pi \Rightarrow f_3 = 1800/2 = 900Hz$$

$f_3 > f_{max}$

**Matlab code:**

```
t=-5:0.001:5;
y1=sin(150*pi*t);
F1=fft(y1);
plot (abs(F1))
y2=sin(600*pi*t);
F2=fft(y2);
```

1.4. Consider the signal $y = \sin(150 \cdot \pi \cdot t)$ with an additive white Gaussian noise (zero mean value $\mu = 0$, and variance $\sigma^2 = 1$) and plot (in Matlab) the illustration of its spectrum via the Fourier transform calculation.

Solution:

```
y=sin(150*pi*t);
noise=randn(1,length(y));
yn=y+noise;
F=fftshift(fft(yn));
plot (abs(F))
```

1.5. Design a simple low-pass filter with a cutoff frequency $f_c = 1102{,}5$ Hz for the signal having 44000 samples, sampled at the frequency 22050 Hz.

Solution:
Let us denote the signal by $y$, while $f_s = 22050$ Hz represents the sampling frequency. The maximal signal frequency is: $f_{max} = f_s/2 = 11025$ Hz. The Fourier transform of $y$ contains 44000 coefficients: 22000 coefficients are located at the positive and 22000 coefficients at negative frequencies (Fig. 1.32).

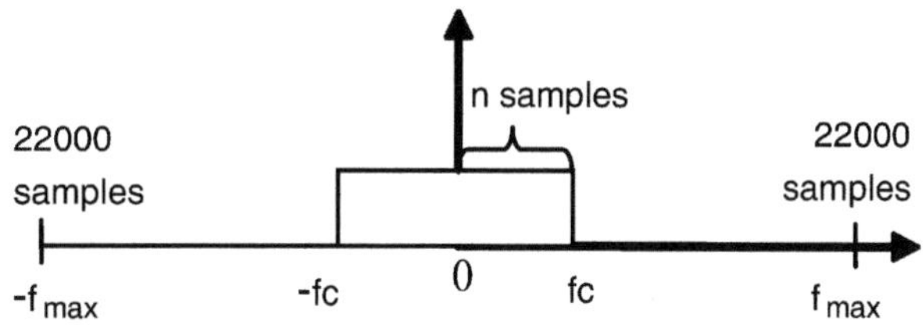


**Fig. 1.32** Illustration of the filter function

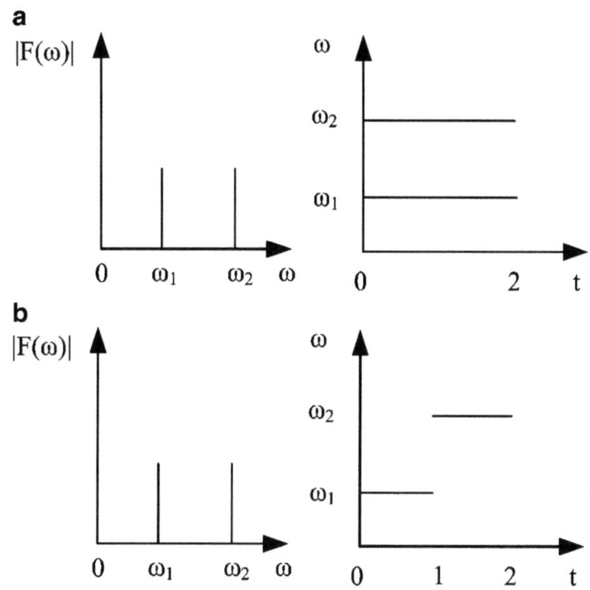


**Fig. 1.33** The Fourier transform and the ideal time-frequency distribution for signal: (**a**) $y_a$, (**b**) $y_b$

Hence, the following proportion holds:

$$f_{max}: f_c = 22000: n$$

and consequently we obtain that $n = 2200$.

The corresponding filter transfer function in the frequency domain can be defined as:

H = [zeros(1,22000–2200) ones(1,4400) zeros(1,22000–2200)];

In order to obtain the filtered signal, the Fourier transform of $y$ should be multiplied by the filter function H, and then the inverse DFT should be performed.

1.6. Make the illustrations of the Fourier transform (using only the absolute values) and the ideal time-frequency representation if the signal is given in the form:

(a) $y_a = e^{j\omega_1 t} + e^{j\omega_2 t}; t \in (0, 2)$,

(b) $y_b = y_1 + y_2,\ y_1 = e^{j\omega_1 t} \text{for} \, t \in (0, 1),\ \ y_2 = e^{j\omega_2 t} \text{for} \, t \in (1, 2)$.

We may assume that $\omega_1 < \omega_2$.

Solution is shown in Fig. 1.33.

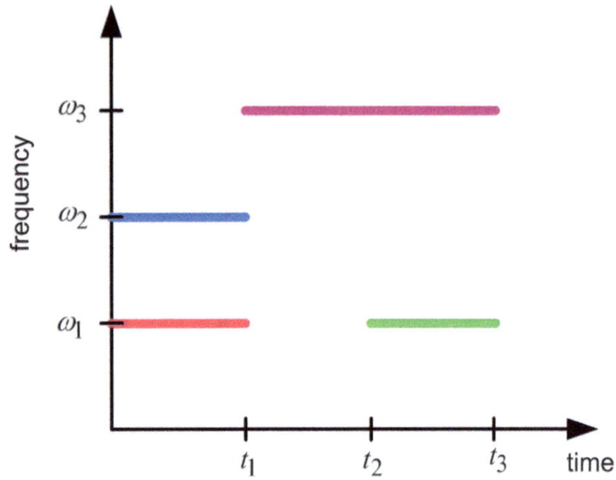


**Fig. 1.34** An ideal time-frequency representation of signal $f(t)$

1.7. Based on the ideal time-frequency representation of a certain signal $f(t)$, define each of the signal components and signal itself. Unit amplitudes and zero initial phases are assumed (Fig. 1.34).

Solution:
The analytic form of signal $f(t)$ can be defined as:

$$f(t) = \begin{cases} e^{j\omega_1 t} + e^{j\omega_2 t}, & t \in (0, t_1) \\ e^{j\omega_3 t}, & t \in (t_1, t_2) \\ e^{j\omega_1 t} + e^{j\omega_3 t}, & t \in (t_2, t_3) \end{cases}.$$

1.8. Consider a constant-frequency modulated signal and demonstrate in Matlab how the window width influences the resolution of the spectrogram. The signal is given in the form:

$$f(n) = \begin{cases} e^{j15nT}, & n = 1, \ldots, 127 \\ e^{j5nT}, & n = 128, \ldots, 256 \end{cases}, \text{ where T} = 0.25.$$

Solution:
The discrete version of the considered signal can be created in Matlab as follows:

```
T=0.25;
for n=1:256
   if n<128; f(n)=exp(j*15*n*T);
      else
      f(n)=exp(j*5*n*T);
   end
end
```

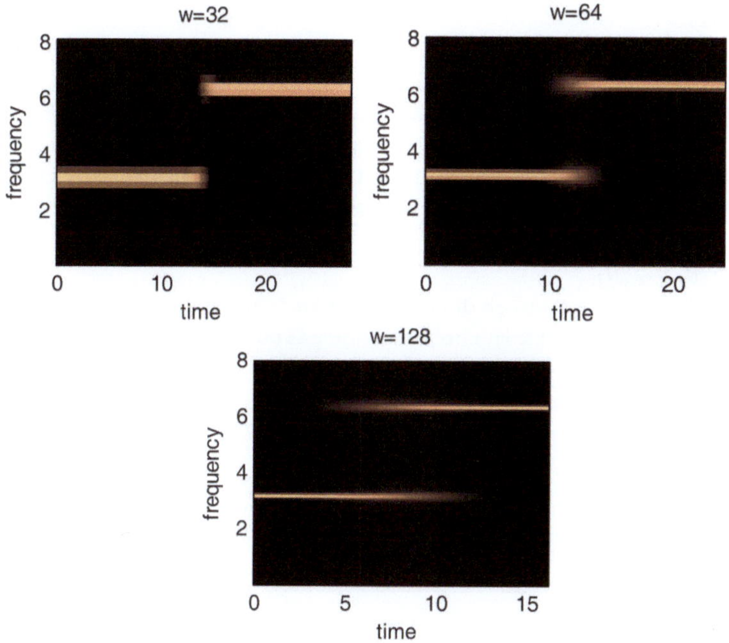


**Fig. 1.35** The spectrograms for different window widths

The spectrogram calculation in Matlab can be done by using the inbuilt function *spectrogram* as follows:

```
[s,F,T]=spectrogram(f,w,N_overlap);
imagesc(T,F,abs(s))
```

Hereby, to change the window width $w$ in a few realizations one should take, for instance, the following values: $w=32$, $w=64$, $w=128$ (Fig. 1.35). The parameter *N_overlap* specifies the number of overlapping samples between two windowed segments of $f$. It should be an integer smaller than window size $w$ (for integer value of $w$), e.g., $N\_overlap=w$-1 is used in this example.

1.9. Write the Matlab code which calculates the S-method. The signal should be loaded from the separate file *signal.m*, and it is defined as: $f(t)=e^{j(2\sin(\pi t)+11\pi t)}+e^{j2\pi\left(t^2+3t\right)}$. The signal length is $N=128$ samples, t = −2:2 with sampling interval $\Delta t=4/N$, the window width is $M=128$ samples, while $L_d=5$. The Gaussian window should be used.

Solution:
The Matlab file `signal.m` creates the signal:

```
function f=signal(t)
f=exp(-j*8*(cos(1*pi*t)+12*pi*t))+exp(j*2*pi*(6*t.^2-16*t));
```

The Matlab code for the S-method calculation is given in the sequel.

```
clear all
f=[];
M=128;  %window width
N=128;  %TF distribution is calculated for N=128 samples, while the
        signal length is equal to M/2+N+M/2=N+M (in addition to the
        observed N samples included are M/2 samples (half of the window
        width) from the left and M/2 samples to the right from the
        observed N samples)
t=-1:2/N:1-2/N;
Ld=5;   % parameter which determines the frequency domain
        % window width in the S-method calculation

% signal

for m=-M/2:1:M/2-1;
tau=2*m/N;
f=[f;signal(t+tau)];
end

% Calculating STFT and Spectrogram (SPEC)

for i=1:N
  w=gausswin(length(f(:,1)),3); % Gaussian window
  STFT(:,i)=fftshift(fft(f(:,i).*w));
  SPEC(:,i)=abs(STFT(:,i)).^2;
end

% Calculating the S-method (SM)

SFP=STFT;
SFN=STFT;
SM=SPEC;
for L=1:Ld
SFP=[zeros(1,N);SFP(1:N-1,:)];
SFN=[SFN(2:N,:);zeros(1,N)];
SM=SM+2*real([SFP.*conj(SFN)]);
end

% Plotting the spectrogram and the S-method

figure(1),imagesc(abs(SPEC))
figure(2),imagesc(abs(SM)) (Fig. 1.36)
```

1.10. Observe a general form of a constant amplitude signal with a phase function $\phi(t)$: $f(t) = Ae^{j\phi(t)}$. Prove that the second and higher phase derivatives cause the spreading of the concentration around the instantaneous frequency in the case of the STFT?

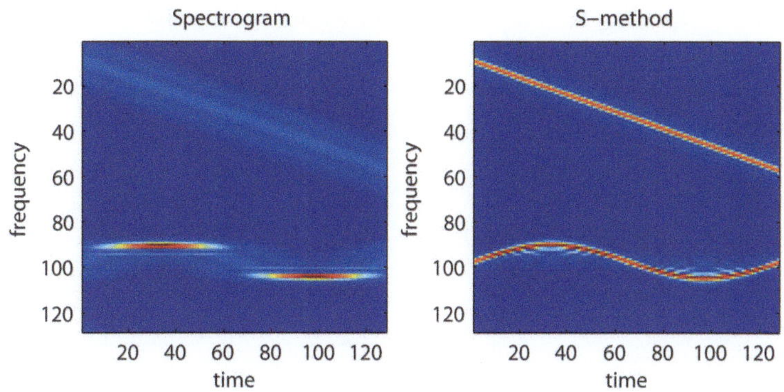


**Fig. 1.36** The spectrogram (*left*) and the S-method (*right*)

Solution:
The STFT of the signal $f(t)$ is defined as:

$$STFT(t,\omega)=\int_{-\infty}^{\infty} f(t+\tau)w(\tau)e^{-j\omega\tau}d\tau=\int_{-\infty}^{\infty} A\cdot e^{j\phi(t+\tau)}w(\tau)e^{-j\omega\tau}d\tau. \quad (1.140)$$

By applying the Taylor series expansion to the phase function, we obtain:

$$\phi(t+\tau)=\phi(t)+\phi'(t)\tau+\phi''(t)\tau^2/2!+\ldots \quad (1.141)$$

The short-time Fourier transform can be rewritten in the form:

$$STFT(t,\omega)=Ae^{j\phi(t)}\int_{-\infty}^{\infty} e^{j\phi'(t)\tau}w(\tau)e^{j\left(\phi''(t)\tau^2/2!+\ldots\right)}e^{-j\omega\tau}d\tau.$$

We can further develop the above expression as:

$$STFT(t,\omega)=Ae^{j\phi(t)}FT\left\{e^{j\phi'(t)\tau}\right\}*_{\omega}FT\{w(\tau)\}*_{\omega}FT\left\{e^{j\phi''(t)\tau^2/2!+\ldots}\right\}.$$

Finally, we obtain the STFT in the form:

$$STFT(t,\omega)=2\pi Ae^{j\phi(t)}\delta(\omega-\phi'(t))*_{\omega}W(\omega)*_{\omega}FT\left\{e^{j\phi''(t)\tau^2/2!+\ldots}\right\}. \quad (1.142)$$

Note that the last term in the expression for the STFT contains second and higher phase derivatives, and thus, we may conclude that this term will produce spreading of the concentration around the instantaneous frequency $\omega=\phi'(t)$.

1.11. For the signal from the previous example, analyze and derive the influence of the higher order phase derivatives in the case of the Wigner distribution.

Solution:
The Wigner distribution is defined as:

$$\begin{aligned} WD(t,\omega) &= \int_{-\infty}^{\infty} f(t+\tau/2)f^*(t-\tau/2)e^{-j\omega\tau}d\tau \\ &= \int_{-\infty}^{\infty} A^2 e^{j\phi(t+\tau/2)}e^{-j\phi(t-\tau/2)}e^{-j\omega\tau}d\tau. \end{aligned} \tag{1.143}$$

The Taylor series expansion of the moment phase function results in:

$$\begin{aligned} \phi(t+\tau/2)-\phi(t-\tau/2) = &\left(\phi(t)+\phi'(t)\tau/2+\sum_{k=2}^{\infty}\phi^{(k)}(t)(\tau/2)^k/k!\right)- \\ &-\left(\phi(t)-\phi'(t)\tau/2+\sum_{k=2}^{\infty}(-1)^k\phi^{(k)}(t)(\tau/2)^k/k!\right) \end{aligned}$$

By using the Taylor series expansion terms in the definition of the Wigner distribution, we obtain:

$$WD(t,\omega) = A^2\int_{-\infty}^{\infty} e^{\left(j\phi'(t)\tau+2\sum_{k=1}^{\infty}\frac{\phi^{(2k+1)}(t)}{(2k+1)!}\left(\frac{\tau}{2}\right)^{2k+1}-j\omega\tau\right)}d\tau, \tag{1.144}$$

or in other words:

$$WD(t,\omega) = 2\pi A^2\delta(\omega-\phi'(t))*_\omega FT\left\{e^{2j\sum_{k=1}^{\infty}\frac{\phi^{(2k+1)}}{(2k+1)!}\left(\frac{\tau}{2}\right)^{2k+1}}\right\}. \tag{1.145}$$

Hence, we may see that only the odd phase derivatives are included in the spread factor, causing inner-interferences and spreading of the concentration in the time-frequency domain.

1.12. Consider a signal in the form: $x(t) = \left(f(kt)e^{j\frac{At^2}{2}}\right)*\frac{1}{\sqrt{2\pi jB}}e^{j\frac{t^2}{2B}}$, where $*$ denotes the convolution. Prove that the Wigner distribution of $x(t)$ is equal to the Wigner distribution of $f(t)$ in the rotated coordinate system:

$$WD_x(t,\omega) = WD_f(t\cos\alpha - \omega\sin\alpha, \omega\cos\alpha + t\sin\alpha),$$

where $k = 1/\cos\alpha$, $B = \sin\alpha/\cos\alpha$, $A = -\sin\alpha\cos\alpha$.

Solution:
The signal $x(t)$ can be written as a convolution of two signals:

$$x(t) = f_1(t)^* f_2(t),$$

where $f_1(t) = f(kt)e^{jAt^2/2}$ and $f_2(t) = \frac{1}{\sqrt{2\pi jB}}e^{jt^2/2B}$.

It means that the Fourier transform of $x(t)$ can be written as:

$$X(\omega) = F_1(\omega)F_2(\omega) = F_1(\omega)e^{jB\omega^2/2}, \tag{1.146}$$

where,

$$F_2(\omega) = FT\left\{\frac{1}{\sqrt{2\pi jB}}e^{jt^2/2B}\right\} = e^{jB\omega^2/2}. \tag{1.147}$$

Furthermore, the Wigner distribution of $x(t)$ can be obtained by using $X(\omega)$ as follows:

$$\begin{aligned}
WD_x(t,\omega) &= \int_{-\infty}^{\infty} X(\omega+\theta/2)X^*(\omega-\theta/2)e^{j\theta t}d\theta \\
&= \int_{-\infty}^{\infty} F_1(\omega+\theta/2)F_2(\omega+\theta/2)F_1^*(\omega-\theta/2)F_2^*(\omega-\theta/2)e^{j\theta t}d\theta \\
&= \int_{-\infty}^{\infty} F_1(\omega+\theta/2)F_1^*(\omega-\theta/2)e^{-jB(\omega+\theta/2)^2/2+jB(\omega-\theta/2)^2/2}e^{j\theta t}d\theta \\
&= \int_{-\infty}^{\infty} F_1(\omega+\theta/2)F_1^*(\omega-\theta/2)e^{-jB\omega\theta}e^{j\theta t}d\theta
\end{aligned}$$

Hence, the following relation holds:

$$WD_x(t,\omega) = WD_{f_1}(t - B\omega, \omega). \tag{1.148}$$

Furthermore, we calculate the Wigner distribution $WD_{f_1}(t,\omega)$ of the signal $f_1(t)$ as follows:

$$\begin{aligned} WD_{f_1}(t,\omega) &= \int_{-\infty}^{\infty} f(k(t+\tau/2))f^*(k(t-\tau/2))e^{jA(t+\tau/2)^2/2}e^{-jA(t-\tau/2)^2/2}e^{-j\omega\tau}d\tau \\ &= \int_{-\infty}^{\infty} f(k(t+\tau/2))f^*(k(t-\tau/2))e^{jAt\tau}e^{-j\omega\tau}d\tau \\ &= \int_{-\infty}^{\infty} f(k(t+\tau/2))f^*(k(t-\tau/2))e^{-jk\tau((\omega-At)/k)}d\tau. \end{aligned}$$

The Wigner distribution $WD_{f_1}(t,\omega)$ can be thus expressed as:

$$WD_{f_1}(t,\omega) = WD_f(kt, \omega/k - At). \tag{1.149}$$

Consequently, from Eqs. (1.148) and (1.149) we have:

$$WD_x(t,\omega) = WD_f(kt - B(\omega/k - Akt), \omega/k - Akt), \tag{1.150}$$

or

$$WD_x(t,\omega) = WD_f[(1+BA)kt - B\omega/k, \omega/k - Akt].$$

By substituting the given parameters: $k = 1/\cos\alpha$, $B = \sin\alpha/\cos\alpha$, $A = -\sin\alpha\cos\alpha$, we obtain:

$$WD_x(t,\omega) = WD_f(t\cos\alpha - \omega\sin\alpha, \omega\cos\alpha + t\sin\alpha). \tag{1.151}$$

The rotation of the coordinate system is defined as:

$$\begin{bmatrix} t_r \\ \omega_r \end{bmatrix} = \begin{bmatrix} \cos\alpha & -\sin\alpha \\ \sin\alpha & \cos\alpha \end{bmatrix} \begin{bmatrix} t \\ \omega \end{bmatrix}.$$

1.13. By using the recursive procedure for the calculation of the Haar transform (un-normalized Haar transform can be used), perform the first level decomposition of a given $8 \times 8$ image. Use the one-dimensional decomposition of image rows in the first step, and then the decomposition of image columns.

| | | | | | | | |
|---|---|---|---|---|---|---|---|
| 10 | 10 | 10 | 10 | 26 | 10 | 10 | 10 |
| 10 | 10 | 10 | 10 | 26 | 10 | 10 | 10 |
| 10 | 10 | 10 | 10 | 26 | 10 | 10 | 10 |
| 10 | 10 | 10 | 10 | 26 | 10 | 10 | 10 |
| 10 | 10 | 10 | 10 | 26 | 10 | 10 | 10 |
| 10 | 10 | 10 | 10 | 26 | 10 | 10 | 10 |
| 18 | 18 | 18 | 18 | 26 | 18 | 18 | 18 |
| 10 | 10 | 10 | 10 | 26 | 10 | 10 | 10 |

Solution:
First we perform the first level decomposition along the image rows. Hence, for each row, it is necessary to calculate the mean values and differences (details). Then the resulted matrix should be used to perform the decomposition along columns. The low-frequency image content is obtained in the first quadrant, while the remaining parts contain image details.

| *Decomposition of rows* | *Decomposition of columns* |
|---|---|
| 10 10 18 10 0 0 8 0 | 10 10 18 10 0 0 8 0 |
| 10 10 18 10 0 0 8 0 | 10 10 18 10 0 0 8 0 |
| 10 10 18 10 0 0 8 0 | 10 10 18 10 0 0 8 0 |
| 10 10 18 10 0 0 8 0 | 14 14 20 14 0 0 6 0 |
| 10 10 18 10 0 0 8 0 | 0 0 0 0 0 0 0 0 |
| 10 10 18 10 0 0 8 0 | 0 0 0 0 0 0 0 0 |
| 18 18 22 18 0 0 4 0 | 0 0 0 0 0 0 0 0 |
| 10 10 18 10 0 0 8 0 | 4 4 2 4 0 0 −2 0 |

1.14. Consider the function $f(t)$ in the form:

$$f(t) = \begin{cases} t^2 + t, & 0 \le t < 1, \\ 0, & \text{otherwise.} \end{cases}$$

By using the Haar wavelets calculate the expansion coefficients:

$$s_{j_0}(k) = \int_t f(t)\varphi_{j_0,k}(t)dt, \; d_j(k) = \int_t f(t)\psi_{j,k}(t)dt, \quad \text{for } j_0 = 0.$$

Solution:

$$s_0(0) = \int_0^1 (t^2+t)\varphi_{0,0}(t)dt = \int_0^1 (t^2+t)dt = \frac{t^3}{3}\bigg|_0^1 + \frac{t^2}{2}\bigg|_0^1 = \frac{5}{6}$$

$$d_0(0) = \int_0^1 (t^2+t)\psi_{0,0}(t)dt = \int_0^{0.5} (t^2+t)dt - \int_{0.5}^1 (t^2+t)dt = -0.5$$

$$d_1(0) = \int_0^1 (t^2+t)\psi_{1,0}(t)dt = \int_0^{0.25} (t^2+t)\sqrt{2}dt - \int_{0.25}^{0.5} (t^2+t)\sqrt{2}dt = -\frac{3\sqrt{2}}{32}$$

$$d_1(1) = \int_0^1 (t^2+t)\psi_{1,1}(t)dt = \int_{0.5}^{0.75} (t^2+t)\sqrt{2}dt - \int_{0.75}^{1} (t^2+t)\sqrt{2}dt = -\frac{5\sqrt{2}}{32}$$

$$f(t) = \underbrace{\underbrace{\underbrace{\frac{5}{6}\varphi_{0,0}(t)}_{V_0} + \underbrace{\left(-\frac{1}{2}\psi_{0,0}(t)\right)}_{W_0}}_{V_1=V_0\oplus W_0} + \underbrace{\left(-\frac{3\sqrt{2}}{32}\psi_{1,0}(t) - \frac{5\sqrt{32}}{32}\psi_{1,1}(t)\right)}_{W_1}}_{V_2=V_1\oplus W_1=V_0\oplus W_0\oplus W_1} + \ldots$$

1.15. Consider the signal illustrated below and perform the Haar wavelet decomposition (e.g., unnormalized 3-level decomposition) (Fig. 1.37).

Solution (Fig. 1.38):

1.16. The signal in the form $y = \sin(\pi t/2)$, $t \in (-4, 4)$, should be sampled using the step $T = 1/2$ and then quantized. Perform the Haar decomposition on the resulting signal (up to the second level).

Solution (Fig. 1.39):

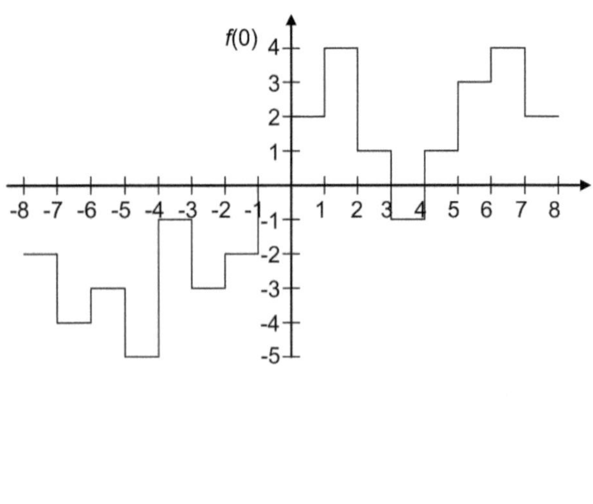


**Fig. 1.37** The signal *f*(0) before decomposition

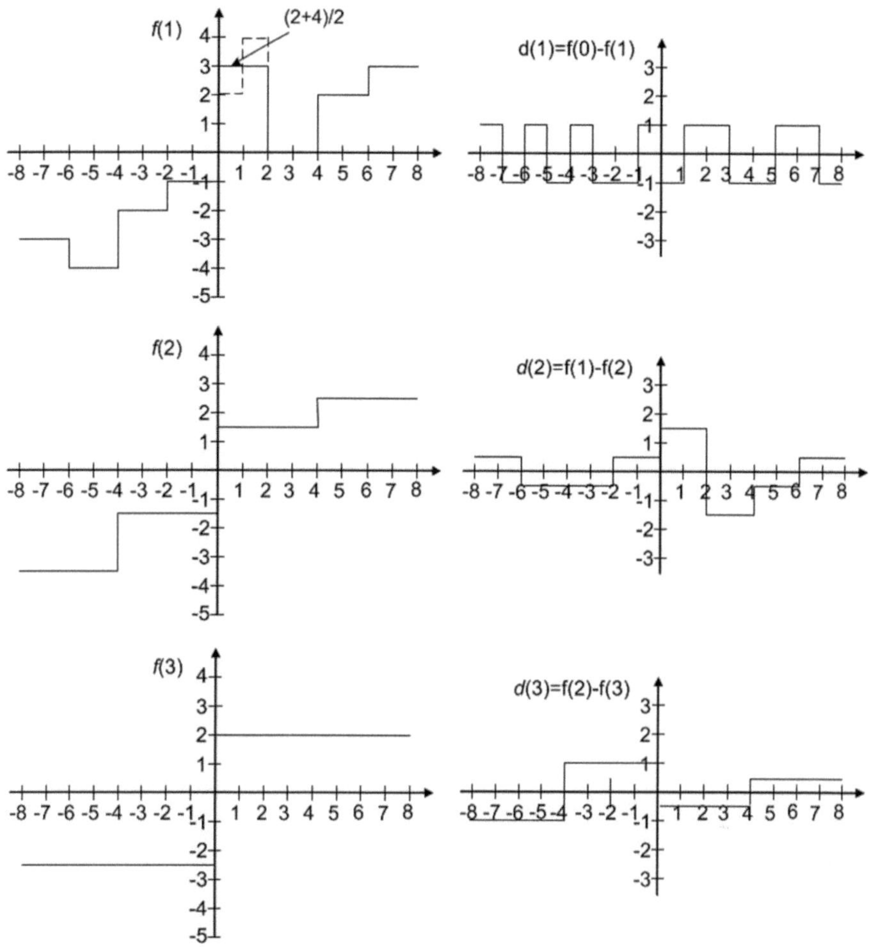


**Fig. 1.38** Signal decomposition

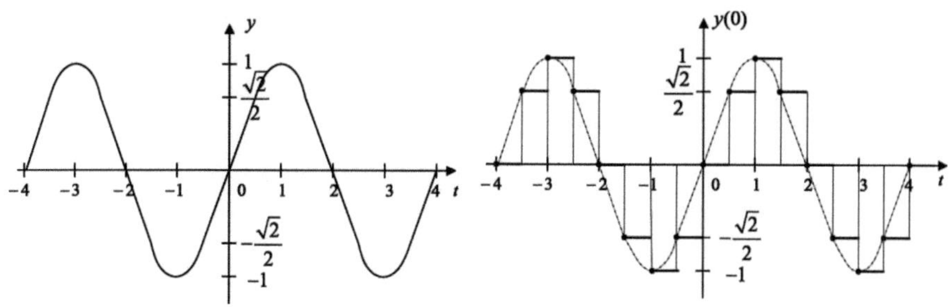


**Fig. 1.39** The original and quantized signal

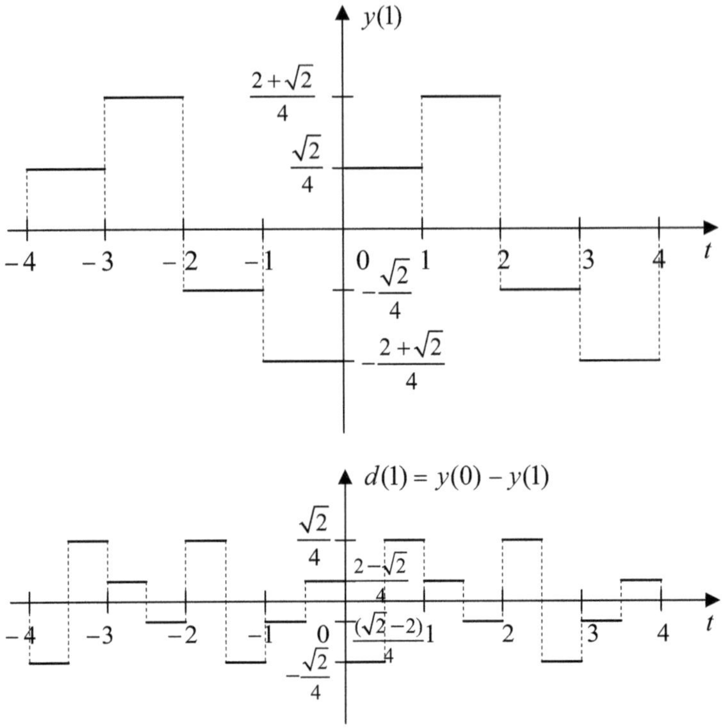


**Fig. 1.40** First level decomposition

The first level of decomposition is shown in Fig. 1.40.
The second level decomposition is shown in Fig. 1.41.

1.17. Perform the Haar decomposition for a signal in the form: $f_j(t) = \sum_{k=0}^{2^j-1} a_k \varphi_{j,k}(t)$,

where the scale is $j = 2$, while the approximation coefficients $\boldsymbol{a}$ are given by: $a = [2, -1, 3, 4]$. Recall that the details and approximation coefficients in the case of discrete Haar transform are calculated as:

$$d_k = \frac{1}{\sqrt{2}}(a_{2k} - a_{2k+1}) \text{ and } s_k = \frac{1}{\sqrt{2}}(a_{2k} + a_{2k+1}).$$

Solution:
For $j = 2$, the signal $f_j(t)$ can be written as:

$$f_2(t) = \sum_{k=0}^{3} a_k \varphi_{2,k}(t) = a_0\varphi_{2,0}(t) + a_1\varphi_{2,1}(t) + a_2\varphi_{2,2}(t) + a_3\varphi_{2,3}(t).$$

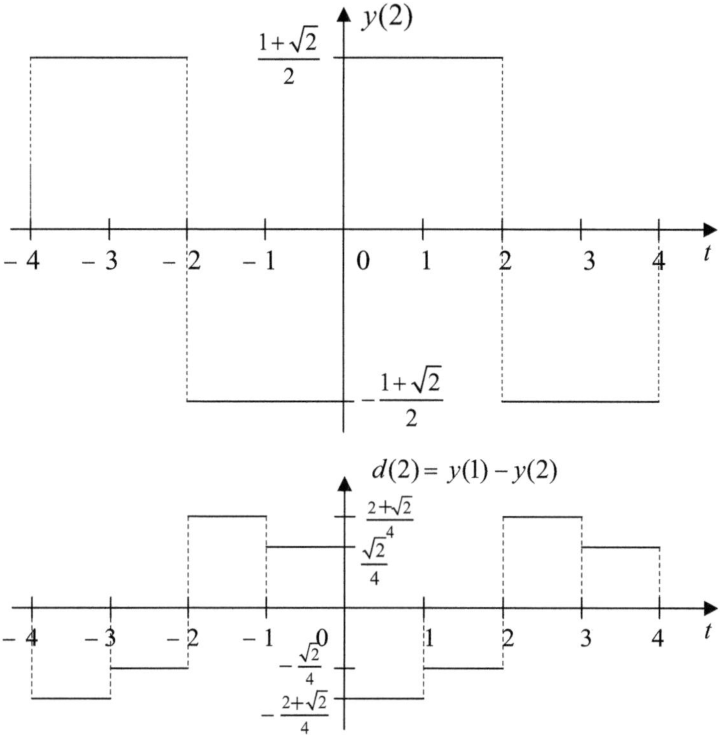


**Fig. 1.41** Second level decomposition

The basic scaling function is given by:
$\varphi_{j,k}(t) = 2^{j/2}\varphi(2^j t - k) = 2\varphi(4t - k)$, where $j = 2$ and $k = 0, \ldots, 2^{j-1}$.
Consequently, we have:

$$\varphi_{2,0}(t) = 2\varphi(4t) = \begin{cases} 2, & t \in (0, 1/4) \\ 0, & \text{otherwise} \end{cases} \qquad \varphi_{2,2}(t) = 2\varphi(4t-2) = \begin{cases} 2, & t \in (1/2, 3/4) \\ 0, & \text{otherwise} \end{cases}$$

$$\varphi_{2,1}(t) = 2\varphi(4t-1) = \begin{cases} 2, & t \in (1/4, 1/2) \\ 0, & \text{otherwise} \end{cases} \qquad \varphi_{2,3}(t) = 2\varphi(4t-3) = \begin{cases} 2, & t \in (3/4, 1) \\ 0, & \text{otherwise} \end{cases}$$

The function $f_2(t)$ is illustrated in Fig. 1.42.
The approximation on scale $j = 1$ is given by:

$$f_1(t) = \sum_{k=0}^{2^1-1} s_{1,k}\varphi_{1,k}(t) = s_{1,0}\varphi_{1,0}(t) + s_{1,1}\varphi_{1,1}(t)$$

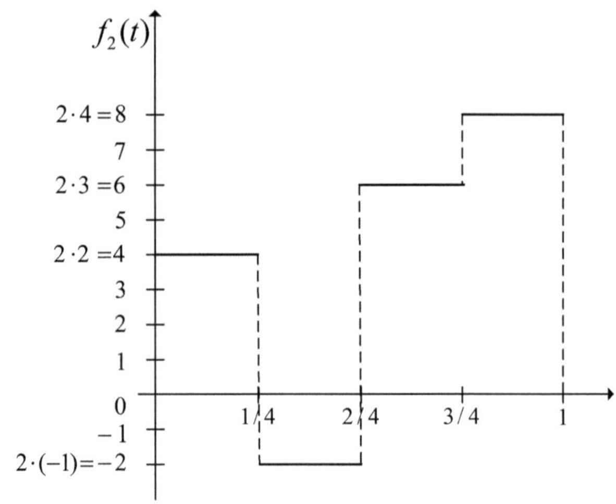


**Fig. 1.42** Illustration of the function $f_j(t)$ on the scale $j = 2$

The coefficients of the approximation are calculated as follows:

$$s_{1,0} = \frac{1}{\sqrt{2}}(a_0 + a_1) = \frac{1}{\sqrt{2}}(2 - 1) = \frac{\sqrt{2}}{2},$$

$$s_{1,1} = \frac{1}{\sqrt{2}}(a_2 + a_3) = \frac{1}{\sqrt{2}}(3 + 4) = \frac{7\sqrt{2}}{2}.$$

The scaling function on the scale $j = 1$ is of the form:

$$\varphi_{1,0}(t) = \sqrt{2}\varphi(2t) = \begin{cases} \sqrt{2}, & t \in (0, 1/2) \\ 0, & \text{otherwise} \end{cases}$$

$$\varphi_{1,1}(t) = \sqrt{2}\varphi(2t - 1) = \begin{cases} \sqrt{2}, & t \in (1/2, 1) \\ 0, & \text{otherwise} \end{cases}$$

From the above equations we can write the function $f_1(t)$:

$$f_1(t) = \frac{\sqrt{2}}{2}\varphi_{1,0}(t) + \frac{7\sqrt{2}}{2}\varphi_{1,1}(t).$$

Furthermore, the details on the scale $j = 1$ are calculated as follows:

$$g_1(t) = \sum_{k=0}^{2^1-1} d_{1,k}\psi_{1,k}(t) = d_{1,0}\psi_{1,0}(t) + d_{1,1}\psi_{1,1}(t),$$

where $d_{1,0}$ and $d_{1,1}$ are:

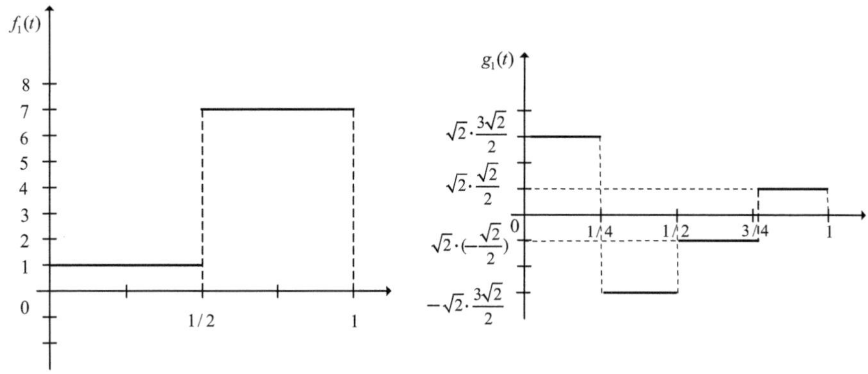


**Fig. 1.43** Illustration of the function $f_1(t)$ and details $g_1(t)$ on the scale $j = 1$

$$d_{1,0} = \frac{1}{\sqrt{2}}(a_0 - a_1) = \frac{1}{\sqrt{2}}(2+1) = \frac{3\sqrt{2}}{2},$$

$$d_{1,1} = \frac{1}{\sqrt{2}}(a_2 - a_3) = \frac{1}{\sqrt{2}}(3-4) = -\frac{\sqrt{2}}{2},$$

while $\psi_{1,0}(t)$ and $\psi_{1,1}(t)$ are defined using $\psi_{j,k}(t) = 2^{j/2}\psi(2^j t - k)$:

$$\psi_{1,0}(t) = 2^{1/2}\psi(2t) = \begin{cases} \sqrt{2}, & t \in (0, 1/4) \\ -\sqrt{2}, & t \in (1/4, 1/2) \end{cases}$$

$$\psi_{2,1}(t) = 2^{1/2}\psi(2t-1) = \begin{cases} \sqrt{2}, & t \in (1/2, 3/4) \\ -\sqrt{2}, & t \in (3/4, 1) \end{cases}$$

The illustrations of the function $f_1(t)$ and details $g_1(t)$ are given in Fig. 1.43. The approximation and details on the scale $j = 0$ are given by:

$$f_0(t) = \sum_{k=0}^{2^0-1} s_{0,k}\varphi_{0,k}(t) = s_{0,0}\varphi_{0,0}(t),$$

$$s_{0,0} = \frac{\sqrt{2}}{2}(s_{1,0} + s_{1,1}) = \frac{\sqrt{2}}{2}\left(\frac{\sqrt{2}}{2} + \frac{7\sqrt{2}}{2}\right) = 4.$$

The basic scaling function is in the form:

$$\varphi_{0,0}(t) = \begin{cases} 1, & t \in (0,1) \\ 0, & \text{otherwise} \end{cases}.$$

The details are defined by the function:

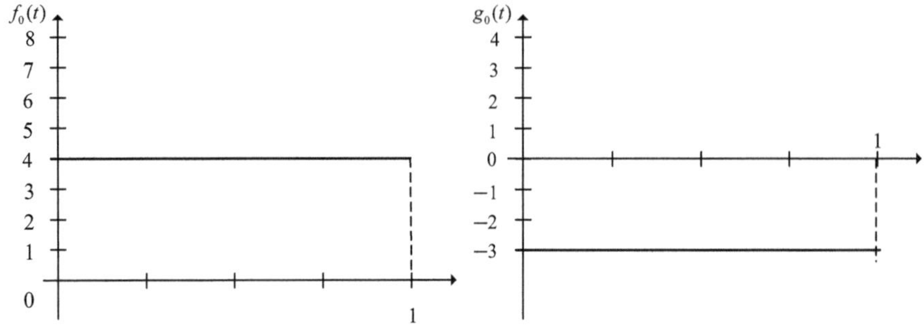


**Fig. 1.44** Illustration of the function $f_0(t)$ and details $g_0(t)$ on the scale $j=0$

$$g_0(t) = \sum_{k=0}^{2^0-1} d_{0,k}\psi_{0,k}(t) = d_{0,0}\psi_{0,0}(t),$$

$$d_{0,0} = \frac{\sqrt{2}}{2}(s_{1,0} - s_{1,1}) = \frac{\sqrt{2}}{2}\left(\frac{\sqrt{2}}{2} - \frac{7\sqrt{2}}{2}\right) = -3,$$

or in other words: $g_0(t) = -3\psi_{0,0}(t)$. The illustrations of $f_0(t)$ and $g_0(t)$ are given in Fig. 1.44.

1.18. Starting from the dilation equation:

$$\varphi(t) = \sum_{k=0}^{N-1} s(k)\sqrt{2}\varphi(2t-k), \tag{1.152}$$

and using the filter coefficients $h(k)$, where $s(k) = \sqrt{2}h(k)$ and $\sum_{k=0}^{N-1} h(k) = 1$, show that the Fourier transform $\Phi(\omega)$ of scaling function $\varphi(t)$ is equal to the product of filter frequency responses:

$$\Phi(\omega) = \prod_{i=1}^{\infty} H\left(\frac{\omega}{2^i}\right).$$

Solution:
The Fourier transform of the scaling function can be calculated as:

$$\begin{aligned}\Phi(\omega) &= \int_{-\infty}^{\infty} \varphi(t)e^{-j\omega t}dt = 2\sum_{k=0}^{N-1} h(k)\int_{-\infty}^{\infty} \varphi(2t-k)e^{-j\omega t}dt \\ &= \sum_{k=0}^{N-1} h(k)\int_{-\infty}^{\infty} \varphi(x)e^{-j\omega(x+k)/2}dx = \sum_{k=0}^{N-1} h(k)e^{-j\omega k/2}\int_{-\infty}^{\infty} \varphi(x)e^{-j\omega x/2}dx.\end{aligned} \tag{1.153}$$

From Eq. (1.153) we may observe that:

$$\Phi(\omega) = \Phi\left(\frac{\omega}{2}\right)\sum_{k=0}^{N-1} h(k)e^{-j\omega k/2} = \Phi\left(\frac{\omega}{2}\right)H\left(\frac{\omega}{2}\right) \quad (1.154)$$

Hence, by applying the recursion we obtain:

$$\Phi(\omega) = H\left(\frac{\omega}{2}\right)H\left(\frac{\omega}{4}\right)\ldots H\left(\frac{\omega}{2^n}\right)\Phi\left(\frac{\omega}{2^n}\right). \quad (1.155)$$

Having in mind that: $\lim_{n\to\infty}\Phi\left(\frac{\omega}{2^n}\right) = \Phi(0) = \int_{-\infty}^{\infty}\varphi(t)dt = 1$, the Fourier transform of the scaling function is obtained in the form:

$$\Phi(\omega) = \prod_{i=1}^{\infty} H\left(\frac{\omega}{2^i}\right). \quad (1.156)$$

1.19. Consider a case of Daubechies orthogonal filter of size $N=6$ (6-tap filter). Write the system of equations for the calculation of filter coefficients.

Solution:

**Condition 1**: $h^2(0) + h^2(1) + h^2(2) + h^2(3) + h^2(4) + h^2(5) = 1$ (1)
**Condition 2**: $h(0)h(2) + h(1)h(3) + h(2)h(4) + h(3)h(5) = 0$ (2)
**Condition 3**: $h = (h(5), h(4), h(3), h(2), h(1), h(0))$

$$h(0)h(4) + h(1)h(5) + h(2)h(6) + h(3)h(7) = 0$$
$$h(0)h(4) + h(1)h(5) = 0 \quad (3)$$

**Condition 4**: $g = (h(5), -h(4), h(3), -h(2), h(1), -h(0))$
**Condition 5**: $g' = (-h(0), h(1), -h(2), h(3), -h(4), h(5))$
**Condition 6**: $H(0) = \sqrt{2}$

$$H(\omega) = \sum_{k=0}^{5} h(k)e^{jk\omega} = h(0)e^0 + h(1)e^{j\omega} + h(2)e^{j2\omega} + h(3)e^{j3\omega} + h(4)e^{j4\omega} + h(5)e^{j5\omega}$$
$$H(0) = \sqrt{2} \quad \Rightarrow \quad h(0) + h(1) + h(2) + h(3) + h(4) + h(5) = \sqrt{2} \quad (4)$$

**Condition 7**: $H^{(k)}(\pi) = 0,\ for\ k = 0, 1, \ldots, N/2 - 1$

$$H^{(0)}(\pi) = 0 \quad \Rightarrow \quad h(0) - h(1) + h(2) - h(3) + h(4) - h(5) = 0 \quad (5)$$

$$H^{(1)}(\omega) = jh(1)e^{j\omega} + 2jh(2)e^{2j\omega} + 3jh(3)e^{3j\omega} + 4jh(4)e^{4j\omega} + 5jh(5)e^{5j\omega}$$
$$H^{(1)}(\pi) = 0 \quad \Rightarrow \quad h(1) - 2h(2) + 3h(3) - 4h(4) + 5h(5) = 0 \tag{6}$$

$$H^{(2)}(\omega) = -h(1)e^{j\omega} - 4h(2)e^{2j\omega} - 9h(3)e^{3j\omega} - 16h(4)e^{4j\omega} - 25h(5)e^{5j\omega}$$
$$H^{(2)}(\pi) = 0 \quad \Rightarrow \quad h(1) - 4h(2) + 9h(3) - 16h(4) + 25h(5) = 0 \tag{7}$$

The system of equations for the calculation of filter coefficients is given by:

$$h^2(0) + h^2(1) + h^2(2) + h^2(3) + h^2(4) + h^2(5) = 1 \tag{1}$$
$$h(0)h(2) + h(1)h(3) + h(2)h(4) + h(3)h(5) = 0 \tag{2}$$
$$h(0)h(4) + h(1)h(5) = 0 \tag{3}$$
$$h(0) + h(1) + h(2) + h(3) + h(4) + h(5) = \sqrt{2} \tag{4}$$
$$h(0) - h(1) + h(2) - h(3) + h(4) - h(5) = 0 \tag{5}$$
$$h(1) - 2h(2) + 3h(3) - 4h(4) + 5h(5) = 0 \tag{6}$$
$$h(1) - 4h(2) + 9h(3) - 16h(4) + 25h(5) = 0 \tag{7}$$

A solution of the system is:

$h(0) \approx 0.332671$, $h(1) \approx 0.806892$, $h(2) \approx 459878$, $h(3) \approx -0.135011$, $h(4) \approx -0.0854413$, $h(5) \approx 0.0352263$.

1.20. Determine the Hermite expansion coefficients, for a short discrete one-dimensional signal $X$ given below.

$$X = [1332.4 \quad 1313.4 \quad 1148.4 \quad 1243.2 \quad 735.7 \quad 861.9 \quad 1261.1 \quad 1438.1 \quad 1443.9 \quad 1454.1];$$

Solution:
The signal consists of $M = 10$ samples. After subtracting the baseline from the original signal $X$, we obtained the following vector which is used for Hermite expansion:

$$f = [-12.175 \quad -43.35 \quad -220.525 \quad -137.825 \quad -657.5 \quad -543.55 \quad -156.475 \quad 8.35 \quad 1.925 \quad 0];$$

The zeros of the Hermite polynomial of order ten are (Table 1.2):

$$x_m = [-3.4362 \quad -2.5327 \quad -1.7567 \quad -1.0366 \quad -0.3429 \quad 0.3429 \quad 1.0366 \quad 1.7567 \quad 2.5327 \quad 3.4362];$$

The first ten Hermite functions, calculated for $x_m$, are given below:

| | | | | | | | | | | |
|---|---|---|---|---|---|---|---|---|---|---|
| $\psi_0$ | 0.0021 | 0.0304 | 0.1605 | 0.4389 | 0.7082 | 0.7082 | 0.4389 | 0.1605 | 0.0304 | 0.0021 |
| $\psi_1$ | −0.0100 | −0.1089 | −0.3989 | −0.6434 | −0.3435 | 0.3435 | 0.6434 | 0.3989 | 0.1089 | 0.0100 |
| $\psi_2$ | 0.0328 | 0.2542 | 0.5871 | 0.3566 | −0.3830 | −0.3830 | 0.3566 | 0.5871 | 0.2542 | 0.0328 |
| $\psi_3$ | −0.0838 | −0.4368 | −0.5165 | 0.2235 | 0.3877 | −0.3877 | −0.2235 | 0.5165 | 0.4368 | 0.0838 |
| $\psi_4$ | 0.1753 | 0.5622 | 0.1331 | −0.4727 | 0.2377 | 0.2377 | −0.4727 | 0.1331 | 0.5622 | 0.1753 |
| $\psi_5$ | −0.3060 | −0.5098 | 0.3141 | 0.1100 | −0.3983 | 0.3983 | −0.1100 | −0.3141 | 0.5098 | 0.3060 |
| $\psi_6$ | 0.4471 | 0.2323 | −0.4401 | 0.3657 | −0.1382 | −0.1382 | 0.3657 | −0.4401 | 0.2323 | 0.4471 |
| $\psi_7$ | −0.5378 | 0.1575 | 0.1224 | −0.3044 | 0.3941 | −0.3941 | 0.3044 | −0.1224 | −0.1575 | 0.5378 |
| $\psi_8$ | 0.5058 | −0.4168 | 0.3041 | −0.1843 | 0.0617 | 0.0617 | −0.1843 | 0.3041 | −0.4168 | 0.5058 |
| $\psi_9$ | −0.3123 | 0.3491 | −0.3672 | 0.3771 | −0.3815 | 0.3815 | −0.3771 | 0.3672 | −0.3491 | 0.3123 |

Furthermore, the constants $\mu^p_{M-1}(x_m)$ are calculated by using the Hermite functions:

$$\mu^p_{M-1}(x_m) = \frac{\psi_p(x_m)}{\left(\psi_{M-1}(x_m)\right)^2}, \quad p = 0, \ldots, 9.$$

The obtained matrix is:

| | | | | | | | | | | |
|---|---|---|---|---|---|---|---|---|---|---|
| $\mu_9^0$ | 0.0210 | 0.2494 | 1.1904 | 3.0868 | 4.8662 | 4.8662 | 3.0868 | 1.1904 | 0.2494 | 0.0210 |
| $\mu_9^1$ | −0.1022 | −0.8934 | −2.9573 | −4.5253 | −2.3598 | 2.3598 | 4.5253 | 2.9573 | 0.8934 | 0.1022 |
| $\mu_9^2$ | 0.3362 | 2.0864 | 4.3533 | 2.5082 | −2.6317 | −2.6317 | 2.5082 | 4.3533 | 2.0864 | 0.3362 |
| $\mu_9^3$ | 0.8598 | −3.5851 | −3.8294 | 1.5719 | 2.6636 | −2.6636 | −1.5719 | 3.8294 | 3.5851 | 0.8598 |
| $\mu_9^4$ | 1.7980 | 4.6137 | 0.9867 | −3.3244 | 1.6333 | 1.6333 | −3.3244 | 0.9867 | 4.6137 | 1.7980 |
| $\mu_9^5$ | −3.1384 | −4.1838 | 2.3289 | 0.7735 | −2.7366 | 2.7366 | −0.7735 | −2.3289 | 4.1838 | 3.1384 |
| $\mu_9^6$ | 4.5848 | 1.9062 | −3.2628 | 2.5718 | −0.9492 | −0.9492 | 2.5718 | −3.2628 | 1.9062 | 4.5848 |
| $\mu_9^7$ | −5.5153 | 1.2929 | 0.9076 | −2.1412 | 2.7076 | −2.7076 | 2.1412 | −0.9076 | −1.2929 | 5.5153 |
| $\mu_9^8$ | 5.1871 | −3.4203 | 2.2549 | −1.2959 | 0.4237 | 0.4237 | −1.2959 | 2.2549 | −3.4203 | 5.1871 |
| $\mu_9^9$ | −3.2023 | 2.8647 | −2.7229 | 2.6520 | −2.6212 | 2.6212 | −2.6520 | 2.7229 | −2.8647 | 3.2023 |

The resulting vector of the Hermite expansion coefficients $c$ is (for the sake of simplicity the constants are written with two-decimal places):

$c = [-701.61 \;\; 90.30 \;\; 140.84 \;\; 77.5 \;\; -140.56 \;\; 2.08 \;\; 94.06 \;\; -54.75 \;\; -52.74 \;\; 88.06]$;

In order to verify the results, we can now reconstruct the signal using the Hermite expansion coefficients $c$. The samples of reconstructed signal are given below:

$X_r = [1332.4 \;\; 1313.4 \;\; 1148.4 \;\; 1243.2 \;\; 735.7 \;\; 861.9 \;\; 1261.1 \;\; 1438.1 \;\; 1443.9 \;\; 1454.1]$;

Note that the reconstructed signal is equal to the original one.

1.21. In this example, we provide the Matlab code for the Hermite projection method, which is used to obtain the illustrations in Fig. 1.28. For the sake of simplicity, instead of the signal values at zeros of Hermite polynomials $f(x_m)$, we can use original signal values $f(x)$.

Solution:

```
N=126;    % signal length
n=70;     % the number of Hermite functions

% the function that calculates the zeros of the Hermite polynomial
xm=hermite_roots(N);
% function that calculates Hermite functions
y=psi(n,xm);
% Loading a one-dimensional signal
load sig1.mat
x=signal1;

% Removing the baseline

i=1:N;
baseline=x(1)+(x(N)-x(1))/N.*i;
f=x-baseline;

% Calculating Hermite coefficients

for i=1:n
mi(i,:)=y(i,:)./(y(N,:)).^2;
Mi(i)=1/N*sum(mi(i,:).*f);
end
c=Mi;
ff=zeros(1,length(xm));
for ii=1:length(xm)
for i=1:n
  ff(ii)=ff(ii)+c(i)*y(i,ii);
end
end

% Signal reconstruction

ss=ff+baseline;
figure,plot((ss))
```

Matlab function *psi.m* that is used for the recursive calculation of the Hermite functions is given in the sequel:

```
function y=psi(n,x);
Psi=zeros(n,length(x));
psi0=1./(pi^(1/4)).*exp(-x.^2/2);
psi1=sqrt(2).*x./(pi^(1/4)).*exp(-x.^2/2);
```

```
Psi(1,:)=psi0; Psi(2,:)=psi1;

for i=2:180
Psi(i+1,:)=x.*sqrt(2/i).*Psi(i,:)-sqrt((i-1)/(i)).*Psi(i-1,:);
end
y=Psi;
```

1.22. Consider a signal with $M = 16$ samples, given by:

$$x(n) = e^{-j2\pi 4n/M},\ n = 0, 1, \ldots, 15.$$

Calculate the nonoverlapping STFTs using the following sets:

$$\mathbf{W}=\{\mathbf{W}_4,\mathbf{W}_4,\mathbf{W}_8\} \text{ and } \mathbf{W}=\{\mathbf{W}_2,\mathbf{W}_4,\mathbf{W}_4,\mathbf{W}_2,\mathbf{W}_1,\mathbf{W}_1\}.$$

Compare these two different representations using the concentration measure (smaller $\mu$ means better concentration):

$$\mu(S(n,k)) = \|\mathbf{S}\|_1 = \sum_n \sum_k |S(n,k)|.$$

Solution:
The two considered transform matrices are given by:

$$\mathbf{W} = \begin{bmatrix} \mathbf{W}_4 & \mathbf{0} & \mathbf{0} \\ \mathbf{0} & \mathbf{W}_4 & \mathbf{0} \\ \mathbf{0} & \mathbf{0} & \mathbf{W}_8 \end{bmatrix}, \ \mathbf{W} = \begin{bmatrix} \mathbf{W}_2 & \mathbf{0} & \ldots & \mathbf{0} & \ldots & \mathbf{0} & \mathbf{0} \\ \mathbf{0} & \mathbf{W}_4 & & & & & \mathbf{0} \\ \ldots & & \mathbf{W}_4 & & & & \ldots \\ & & & \mathbf{W}_2 & & & \\ \mathbf{0} & & & & \mathbf{W}_2 & & \mathbf{0} \\ \ldots & & & & & \mathbf{W}_1 & \ldots \\ \mathbf{0} & \mathbf{0} & \ldots & \mathbf{0} & \ldots & & \mathbf{W}_1 \end{bmatrix}$$

The corresponding nonoverlapping STFTs are shown in Fig. 1.45, where the white color corresponds to zero value, and black color represents a maximal (absolute) value of the component in one STFT.

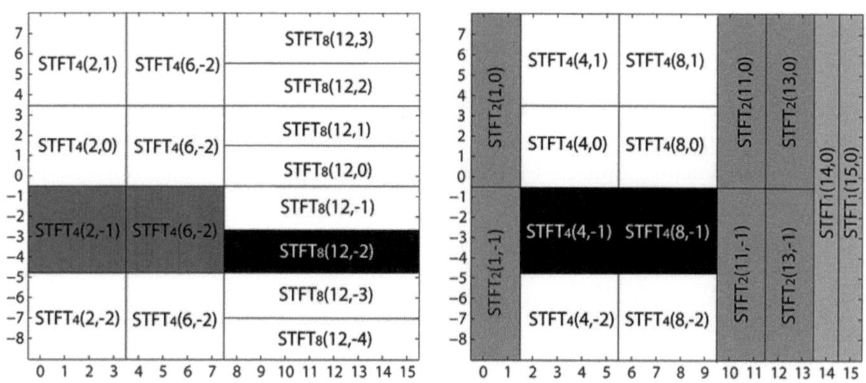


**Fig. 1.45** The nonoverlapping STFTs

The measure of concentration in the first case is 256, while in the second case $\mu = 295.76$.

1.23. For a signal defined as follows:

$$x(m) = \begin{cases} e^{-j2\pi 3n/8}, \; n = \{0,\ldots,7\}, m = \{0,\ldots,7\} \\ 0.25\dfrac{e^{-n^2}H_6(n)}{\sqrt{2^6 6!\sqrt{\pi}}}, n = \{0,\ldots,7\}, m = \{8,\ldots,15\} \end{cases}$$

where $H_6(n)$ is the 6th order Hermite polynomial, calculate a suitable representation by combining the STFT and the Hermite transform (HT).

Solution:

It can be observed that the first part of the signal representation is a complex sinusoid which can be efficiently represented by the Fourier transform. The second half of the signal corresponds to the Hermite function of order 6: $\psi_6(n) = \dfrac{e^{-n^2}H_6(n)}{\sqrt{2^6 6!\sqrt{\pi}}}$ and thus the HT can be used to represent this part of signal.

The combined STFT and Hermite transform is then given by:

$$\mathbf{X} = \mathbf{Z}\mathbf{x},$$

where the transform matrix $\mathbf{Z}$ is based on the combined STFT and HT and it is given by:

$$\mathbf{Z} = \begin{bmatrix} \mathbf{W}_8 & \mathbf{0} \\ \mathbf{0} & \mathbf{H}_8 \end{bmatrix}.$$

Note that $\mathbf{W}_8$ is $8 \times 8$ Fourier transform matrix while $\mathbf{H}_8$ is $8 \times 8$ Hermite transform matrix. For the comparison we may also calculate the representation based on nonoverlapping STFTs using the matrix:

$$\mathbf{W} = \begin{bmatrix} \mathbf{W}_8 & \mathbf{0} \\ \mathbf{0} & \mathbf{W}_8 \end{bmatrix},$$

instead of combined matrix $\mathbf{Z}$. The representation obtained using the STFT with matrix $\mathbf{W} = \{\mathbf{W}_8, \mathbf{W}_8\}$ is given in Fig. 1.46a, while the representation obtained using combined Fourier transform and Hermite transform basis is shown in Fig. 1.46b.

As in the previous example, zero values within the representation are illustrated using white color, while the maximal value of the representation is illustrated using black color. Now, we may observe that the representation

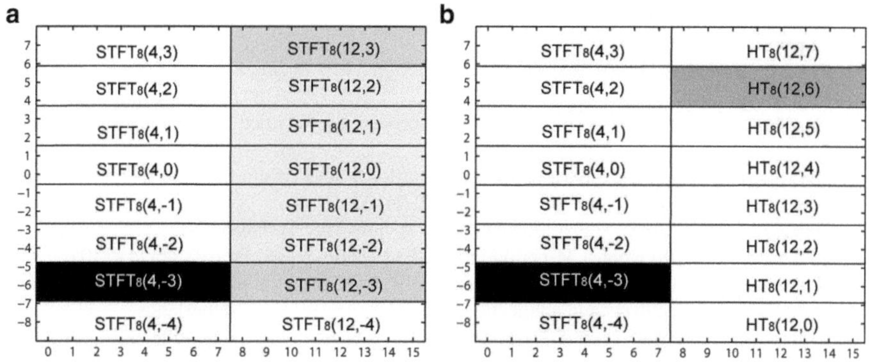


**Fig. 1.46** (**a**) Nonoverlapping STFT, (**b**) combined representation

obtained using the STFT with matrix $\mathbf{W} = \{\mathbf{W_8}, \mathbf{W_8}\}$ (Fig. 1.46a) is well concentrated for the first part of the signal resulting in only one non-zero value $\mathrm{STFT}_8(4,-3)$, which is not the case with the representation obtained for the second part of signal defined by the time instants $\{8, 9, \ldots, 15\}$. In the case of combined representation shown in Fig. 1.46b, the concentration of representation corresponding to the second part of the signal is improved using $\mathbf{H}_8$ instead of $\mathbf{W}_8$ (note that only one non-zero component is obtained given by $\mathbf{H}_8(12,6)$, where according to the signal definition, the order of Hermite function is 6, while the central instant is $m = 12$). The first part of the signal is represented using $\mathbf{W_8}$ as in previous case. Thus, we kept the same notations $\mathbf{STFT_8}$ $(\mathbf{x,y})$ in Fig. 1.46b, although in this case it would be more appropriate to use $\mathbf{FT}_8$ (Fourier transform) instead of $\mathbf{STFT_8}$.

# References

1. Acharya T, Ray AK (2005) Image processing: principles and applications. John Wiley & Sons, Hoboken, NJ
2. Acharya T, Tsai PS (2005) JPEG2000 standard for image compression concepts, algorithms and VLSI architectures. John Wiley & Sons, Hoboken, NJ
3. Amin MG, Williams WJ (1998) High spectral resolution time-frequency distribution kernels. IEEE Trans Signal Process 46(10):2796–2804
4. Bastiaans MJ, Alieva T, Stankovic LJ (2002) On rotated time-frequency kernels. IEEE Signal Process Lett 9(11):378–381
5. Boashash B, Ristic B (1998) Polynomial time-frequency distributions and time-varying higher order spectra: Application to the analysis of multicomponent FM signals and to the treatment of multiplicative noise. Signal Process 67(1):1–23
6. Boashash B (2003) Time-frequency analysis and processing. Elsevier, Amsterdam
7. Cohen L (1989) Time-frequency distributions-a review. Proc IEEE 77(7):941–981
8. Daubechies I (1992) Ten Lectures on Wavelets. Society for industrial and applied mathematics
9. Djurović I, Stanković L, Böhme JF (2003) Robust L-estimation based forms of signal transforms and time-frequency representations. IEEE Trans Signal Process 51(7):1753–1761

10. Dudgeon D, Mersereau R (1984) Multidimensional digital signal processing. Prentice Hall, Upper Saddle River, NJ
11. Fugal L (2009) Conceptual Wavelets in Digital Signal Processing. Space and Signal Technical Publishing
12. González RC, Woods R (2008) Digital image processing. Prentice Hall, Upper Saddle River, NJ
13. Hlawatsch F, Boudreaux-Bartels GF (1992) Linear and quadratic time-frequency signal representations. IEEE Signal Process Mag 9(2):21–67
14. Huber PJ (1981) Robust statistics. John Wiley & Sons Inc., Hoboken, NJ
15. Katkovnik V (1998) Robust M-periodogram. IEEE Trans Signal Process 46(11):3104–3109
16. Katkovnik V, Djurovic I, Stankovic L (2003) Robust time- frequency distributions. Time-frequency signal analysis and applications. Elsevier, Amsterdam, Netherlands
17. Kortchagine D, Krylov A (2000) Projection filtering in image processing. Proc. of Tenth International Conference on Computer Graphics and Applications (GraphiCon'2000): 42-45
18. Kortchagine D, Krylov A (2005) Image database retrieval by fast Hermite projection method. Proc. of Fifteenth International Conference on Computer Graphics and Applications (GraphiCon'2005): 308-311
19. Krylov A, Kortchagine D (2006) Fast Hermite projection method. In Proc. of the Third International Conference on Image Analysis and Recognition (ICIAR 2006), 1: 329-338
20. Leibon G, Rockmore DN, Park W, Taintor R, Chirikjian GS (2008) A fast Hermite transform. Theor Comput Sci 409(2):211–228
21. Mallat S (1999) A wavelet tour of signal processing, 2nd edn. Academic press, San Diego, CA
22. Oppenheim A (1978) Applications of digital signal processing. Prentice Hall, Upper Saddle River, NJ
23. Orović I, Orlandić M, Stanković S, Uskoković Z (2011) A virtual instrument for time-frequency analysis of signals with highly non-stationary instantaneous frequency. IEEE Trans Instr Measur 60(3):791–803
24. Orović I, Stanković S, Stanković LJ, Thayaparan T (2010) Multiwindow S-method for instantaneous frequency estimation and its application in radar signal analysis. IET Signal Process 4(4):363–370
25. Percival DB, Walden AT (2006) Wavelet methods for time series analysis. Cambridge University Press, Cambridge
26. Radunović D (2009) Wavelets from math to practice. Springer-Verlag, Berlin, Academic mind, Belgrade
27. Ruch D, van Fleet PJ (2009) Wavelet theory: an elementary approach with applications. John Wiley & Sons, Hoboken, NJ
28. Sandryhaila A, Saba S, Puschel M, Kovacevic J (2012) Efficient compression of QRS complexes using Hermite expansion. IEEE Trans Signal Process 60(2):947–955
29. Stanković L (1994) A method for time-frequency signal analysis. IEEE Trans Signal Process 42(1):225–229
30. Stanković L (1994) Multitime definition of the wigner higher order distribution: L-Wigner distribution. IEEE Signal Process Lett 1(7):106–109
31. Stankovic L, Dakovic M, Thayaparan T (2012) Time-frequency signal analysis with applications. Artech House, Boston
32. Stanković L, Stanković S, Daković M (2014) From the STFT to the Wigner distribution. IEEE Signal Process Mag 31(3):163–174
33. Stanković S (2010) Time-Frequency Analysis and its Application in Digital Watermarking (Review paper). EURASIP Journal on Advances in Signal Processing, Special Issue on Time-Frequency Analysis and its Application to Multimedia signals, Vol. 2010, Article ID 579295, 20 pages
34. Stanković S (2015) Time-Frequency Filtering of Speech Signals in Hands-Free Telephone Systems. 2nd Edition of the Monograph: Time-Frequency Signal Analysis and Processing, ed. B. Boashash, Elsevier

35. Stanković S, Orović I, Krylov A (2010) Video Frames Reconstruction based on Time-Frequency Analysis and Hermite projection method. EURASIP Journal on Advances in Signal Processing, Special Issue on Time-Frequency Analysis and its Application to Multimedia signals, Article ID 970105, p. 11
36. Stanković S, Orović I, Ioana C (2009) Effects of Cauchy integral formula on the precision of the IF estimation. IEEE Signal Process Lett 16(4):327–330
37. Stanković S, Stanković L (1997) An architecture for the realization of a system for time-frequency signal analysis. IEEE Trans Circuits Syst Part II 44(7):600–604
38. Stanković S, Tilp J (2000) Time-varying filtering of speech signals using linear prediction. Electron Lett 36(8):763–764
39. Stollnitz EJ, DeRose TD, Salesin DH (1995) Wavelets for computer graphics: a primer, part 1. IEEE Comput Graph Appl 15(3):76–84
40. Stollnitz EJ, DeRose TD, Salesin DH (1995) Wavelets for computer graphics: a primer, part 2. IEEE Comput Graph Appl 15(4):75–85
41. Strutz T (2009) Lifting Parameterization of the 9/7 Wavelet Filter Bank and its Application in Lossless Image Compression. ISPRA'09, Cambridge, UK: 161–166
42. Strutz T (2009) Wavelet filter design based on the lifting scheme and its application in lossless image compression. WSEAS Trans Signal Process 5(2):53–62
43. Sydney Burus C, Gopinath RA, Guo H (1998) Introduction to wavelets and wavelet transforms: a primer. Prentice-Hall Inc, Upper Saddle River, NJ
44. Veterli M, Kovačević J (1995) Wavelets and subband coding. Prentice Hall, Upper Saddle River, NJ
45. Viswanath G, Sreenivas TV (2002) IF estimation using higher order TFRs. Signal Process 82 (2):127–132
46. Zaric N, Orovic I, Stankovic S (2010) Robust Time-Frequency Distributions with Complex-lag Argument. EURASIP Journal on Advances in Signal Processing, 2010(ID 879874), 10 pages
47. Zaric N, Stankovic S, Uskokovic Z (2013) Hardware realization of the robust time-frequency distributions. Annales des Telecommunications

# Chapter 2
# Digital Audio

## 2.1 The Nature of Sound

The sound is created as a result of wave fluctuations around the vibrating material. The propagation speed, frequency and sound pressure level are important sound features. For example, the sound propagation speed through the air under normal atmospheric conditions is 344 m/s. Since in this chapter we focus our attention to specific types of audio signals such as speech and music, let us consider their frequency characteristics. Music is defined as the sound that has a distinct periodicity. Its frequency ranges from 20 Hz to 20 KHz, while in the case of speech the frequency ranges from 50 Hz to 10 KHz. It is important to note that the human auditory system is most sensitive to frequencies from 700 Hz to 6600 Hz.

Let us observe what affects the perception of sound in the human auditory system. If we consider a closed room as shown in Fig. 2.1, the auditory system receives direct and reflected waves. Reflected waves are delayed in comparison to the direct waves. The number of reflected waves and their respective delays depend on the geometry of the room.

The position of the sound source is perceived based on the delays between the direct and reflected waves detected by left and right ear. The time delay between two ears is about 0.7 ms. Here, it is interesting to mention some effects that appear as a result of the stereo nature of the human auditory system. For example, if one signal channel is delayed for 15 ms with respect to the other, it will be perceived as a signal with lower amplitude, although both signals are actually of the same amplitude. Hence, this effect can be reduced by increasing the amplitude of delayed signal. However, the auditory system registers two different sounds if the delay exceeds 50 ms.

The sound pressure level (SPL) is another key characteristic of audio signals. The SPL is the ratio of the measured sound pressure to the reference pressure ($P_0 = 20$ μPa). The reference pressure denotes the lowest sound pressure level that



S. Stanković et al., *Multimedia Signals and Systems*,
DOI 10.1007/978-3-319-23950-7_2

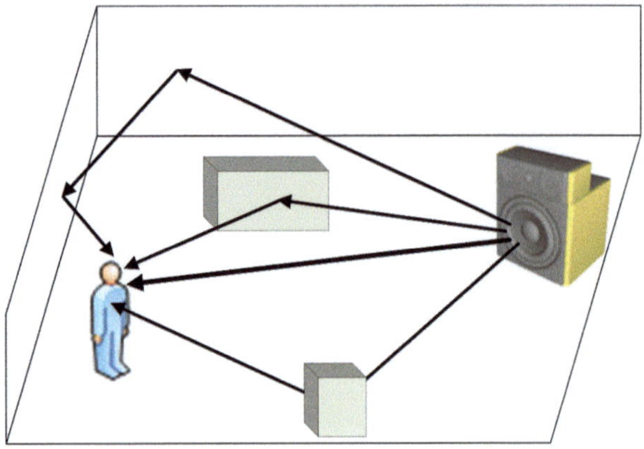

Fig. 2.1 An illustration of sound propagation within a closed room

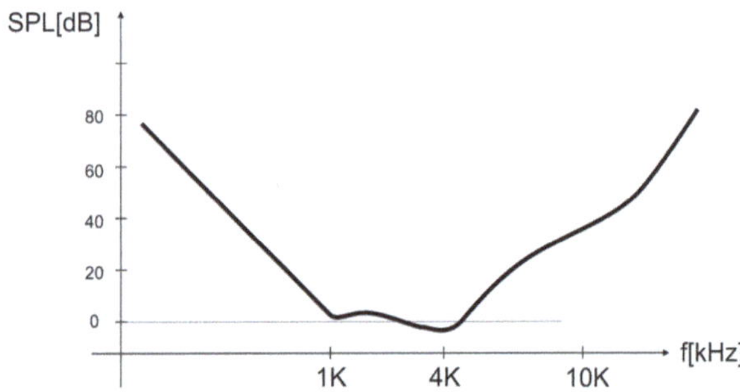


Fig. 2.2 The Fletcher curve

can be registered by the auditory system in a noise-free environment. The sound pressure is calculated as follows:

$$SPL = 20\log_{10}\frac{P}{P_0}\ [dB]. \tag{2.1}$$

In addition to these characteristics of acoustic signals, the Fletcher curve shown in Fig. 2.2 is a measure of SPL over the frequency spectrum for which a listener perceives a constant loudness when presented with pure steady tones. From Fig. 2.2, it can be observed that the human auditory system has a nonlinear sensitivity to the frequency.

## 2.2 Development of Systems for Storing and Playback of Digital Audio

The first system for audio recording and playback dates back to 1877 (the Edison phonograph). The first gramophone dates back to 1893. Electrical playback systems began replacing mechanical systems in 1925. The broadcast of AM (amplitude modulated) audio signals began in 1930. The LP (Long Play) system with a playback time of about 25 min was developed in 1948. This brief review of some of the inventions testifies that the audio industry has developed significantly over the last 100 years. For example, the first gramophones could play recordings about 2 min long, and the system used 78 revolutions per minute. The frequency range of the system was 200 Hz–3 KHz and its dynamic range was 18 dB. The later systems had the extended frequency range (30 Hz–15 KHz), with the dynamic range being 65 dB.

Efforts to improve the performance of audio devices have led to the use of tape recorders during the 1960s and 1970s. The development of Compact disc (CD) began during 1970s, when Mitsubishi, Sony, and Hitachi demonstrated the Digital Audio Disc (DAD). DAD was 30 cm in diameter. Philips and Sony continued to work together on this system. As a result, they produced a disc with a diameter of 12 cm in the early 1980s. The capacity of the disc was 74 min. A further development of the CD technology led to the development of mini discs, Digital Versatile Discs (DVD), Super Audio CDs (SACD).

Along with the development of digital audio devices, there was a growing need to develop systems for Digital Audio Broadcasting (DAB). The used bandwidth is 1.54 MHz. The frequency blocks are arranged as: 12 frequency blocks in the range 87–108 MHz, 39 blocks in the VHF band (174–240 MHz) and 23 frequency blocks in the L band (1452–1492 MHz). An example of DAB system is given in Fig. 2.3, showing the general principle of combining different signals and their transmission in digital form.

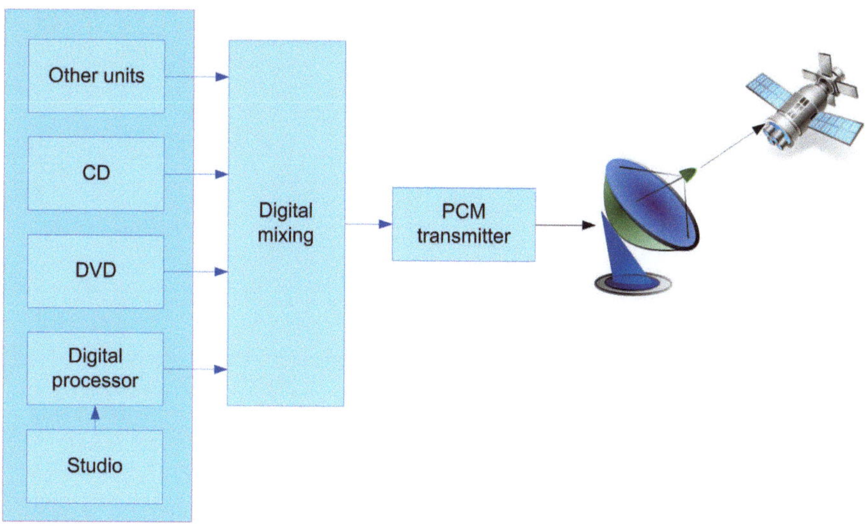


**Fig. 2.3** Block diagram of a DAB system

## 2.3 Effects of Sampling and Quantization on the Quality of Audio Signal

Sampling is the first step in digitalization of analog signals. Recall that sampling causes periodic extensions in the frequency domain. If the discretization is performed according to the sampling theorem, then the basic part of the signal spectrum will not overlap with periodic extensions. However, if the sampling rate is not sufficiently high, then there is a spectrum overlap (or aliasing) (Fig. 2.4).

The signal spectrum is extracted by using antialiasing filters with steep transition from pass to stop regions (a filter example is shown in Fig. 2.5). Note that filters with steep transitions are usually the higher order ones.

In many real applications, it is necessary to use more economic versions of antialiasing filters of lower orders. Therefore, the sampling rate is increased beyond what is required by the sampling theorem in order to allow for less steeper transitions. For example, the sampling frequency used for a CD is equal to 44.1 KHz, although the maximum frequency we want to reproduce is 20 KHz.

A sample and hold circuit that can be used for sampling of analog signals is shown in Fig. 2.6. A switching element is controlled by the signal *fs*, which defines the sampling frequency. The operational amplifier provides high resistance, and thus a large time constant for the capacitor C to discharge. Thus, the voltage on the capacitor is changed slightly between the two control pulses *fs*.

**Fig. 2.4** Aliasing effects

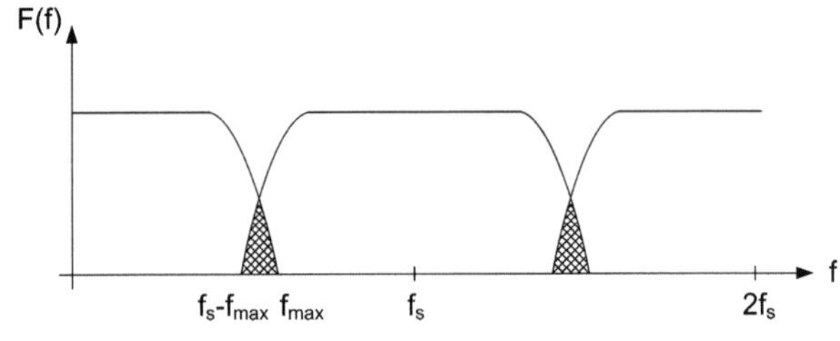


**Fig. 2.5** An example of antialiasing filter with a steep transition

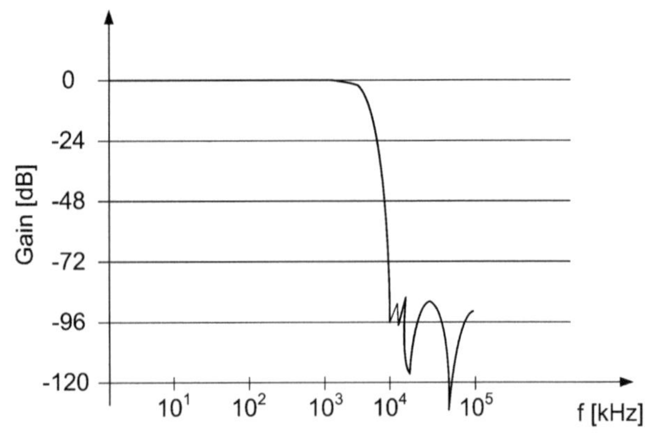

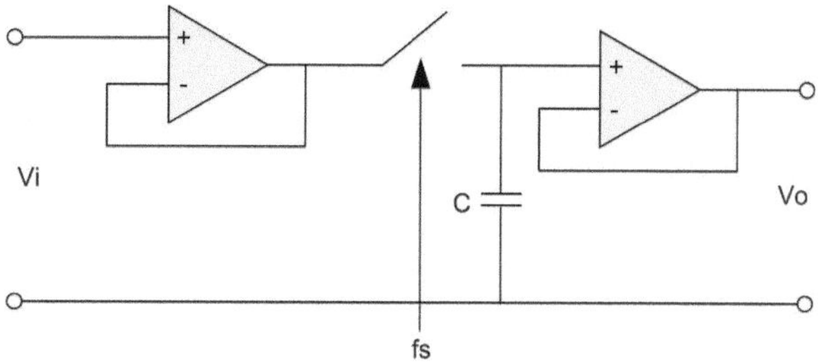


**Fig. 2.6** A circuit for signal sampling

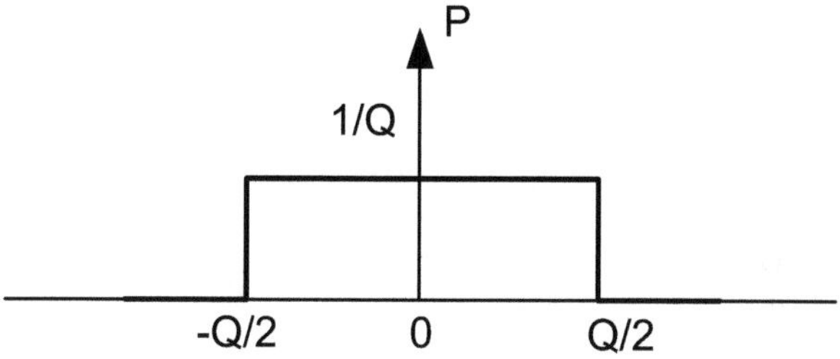


**Fig. 2.7** The probability density function of the quantization error

The next step after the sampling process is the quantization. Analog signals can have infinitely many different values, but a number of quantization levels is limited. As a result, the signal after quantization can meet only a certain degree of accuracy, as defined by the number of quantization levels. In other words, the quantization introduces the quantization noise. A relationship between the signal-to-noise ratio (S/N or SNR) and the number of bits (which is determined by the number of quantization levels) can be easily determined. Suppose that the probability density function of quantization error is uniform, as shown in Fig. 2.7.

The number of quantization levels in an $n$-bit system is denoted as $M = 2^n$. Now, consider a sinusoidal signal with the amplitude $V/2$. Then, the quantization interval is $Q = V/(M-1)$. Since the quantization noise is uniformly distributed in the range $[-Q/2, Q/2]$, the quantization noise power is equal to:

$$N = \frac{2}{Q}\int_{0}^{Q/2} x^2 dx = \frac{2}{Q}\frac{(Q/2)^3}{3} = \frac{Q^2}{12}. \tag{2.2}$$

On the other hand, the power of a sinusoidal signal is equal to:

$$P = \frac{1}{2\pi}\int_{0}^{2\pi}\left(\frac{V}{2}\right)^2 \sin^2 x\, dx = \frac{1}{2\pi}\frac{V^2}{4}\int_{0}^{2\pi}\frac{1-\cos 2x}{2}dx = \frac{V^2}{8}. \tag{2.3}$$

Therefore, S/N in the $n$-bit system is given by:

$$S/N = \frac{P}{N} = \frac{V^2/8}{V^2/(2^{2n}/12)} = \frac{3}{2}2^{2n}, \tag{2.4}$$

or equivalently:

$$S/N\,[dB] = 10\log\frac{S}{N} = 10\log\frac{3}{2} + 10\log 2^{2n} = 1.76 + 6n. \tag{2.5}$$

For example, if we use 16 bits to quantize the signal, then S/N ≈98 dB.

### 2.3.1 Nonlinear Quantization

The previous section discussed a uniform quantization approach (where each quantization interval $Q$ is identical). However, we can assign the quantization levels in a nonlinear manner. For instance, the quantization levels can be adjusted according to the input signal amplitude, such that a small amplitude signal will have smaller quantization intervals, and vice versa.

A process of nonlinear quantization of a variable $x$ can be described as follows. First $x$ is transformed (compressed) by using the nonlinear function $f(x)$, which is then linearly quantized. The quantized values are then processed (expanded) by the inverse nonlinear function $f^{-1}$. Lastly, for a nonlinear quantizer we have:

$$Q(x) = f^{-1}(Q_u(f(x))), \tag{2.6}$$

where $Q_u(x)$ denotes a linear quantizer. A typical function for nonlinear quantization is the $A$-law which is defined as follows:

$$F(x) = \begin{cases} Ax/(1 + \ln A) \text{ for } 0 < x \le V/A, \\ V(1 + \ln(Ax/V))/(1 + \ln A) \quad \text{for } V/A \le x \le V, \end{cases} \tag{2.7}$$

where $A$ is a constant that controls the compression ratio, while the peak magnitude of the input signal is labelled as $V$. In practice, $A = 87.6$ is often used.

Figure 2.8 depicts the process of nonlinear quantization. The $x$-axis represents the normalized amplitude of the input signal, while the $y$-axis represents the values of quantization intervals. For example, when the signal amplitude drops four times (−12 dB), the quantization interval is 3/4$Q$.

The concept of nonlinear quantization is applied in other schemes such as the floating-point conversion, which is used in professional audio systems. The principle of floating-point conversion is shown in Fig. 2.9.

This system is based on the principle of a logarithmic scale. Namely, the signal is sent through several parallel circuits with different gains ensuring that the input to

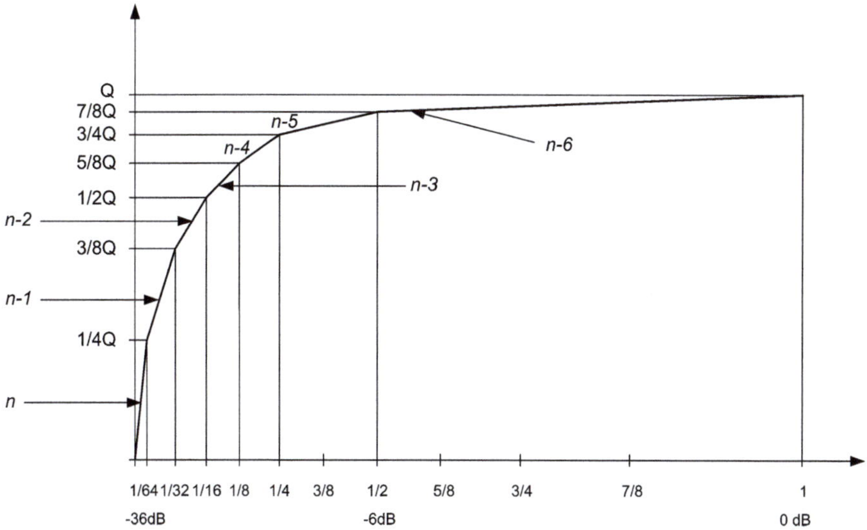


**Fig. 2.8** Nonlinear quantization

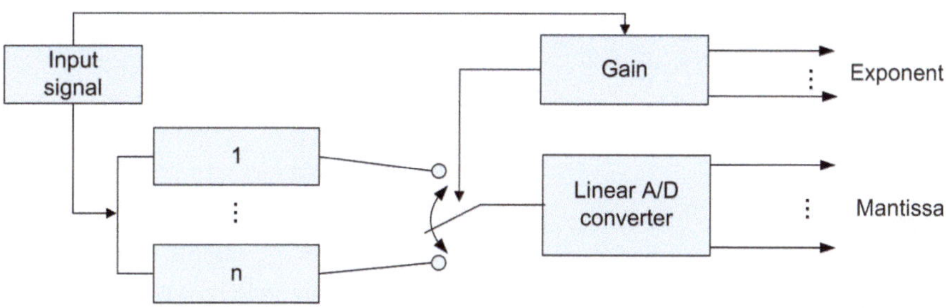


**Fig. 2.9** Floating-point conversion

linear A/D converter is always a signal whose level is suitable for linear conversion. The converted part of the signal is called the mantissa.

Information on the signal amplitude is provided through the second part of the system, whose output is a binary value called the exponent. Note that with three bits of the exponent we can achieve a conversion of signals with the following gains: 0, 6, 12, 18, 24, 30, 36 and 42 dB. Hence, we can effectively digitize signals with very different amplitude levels, which is often a practical demand for audio signals. A typical S/N curve for a signal based on an 8-bit mantissa and a 3-bit exponent is illustrated in Fig. 2.10.

It should be noted that although it is a 11-bit system, the S/N is between 42 and 48 dB, and its maximum value is defined by the mantissa.

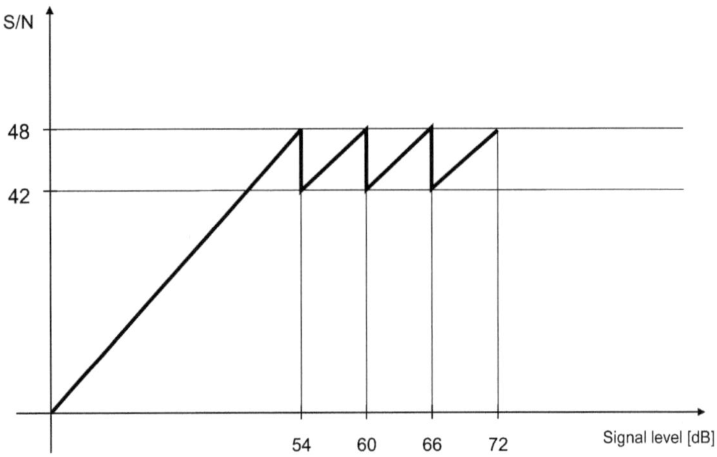


**Fig. 2.10** S/N ratio for a considered floating-point converter (8-bit mantissa and 3-bit exponent)

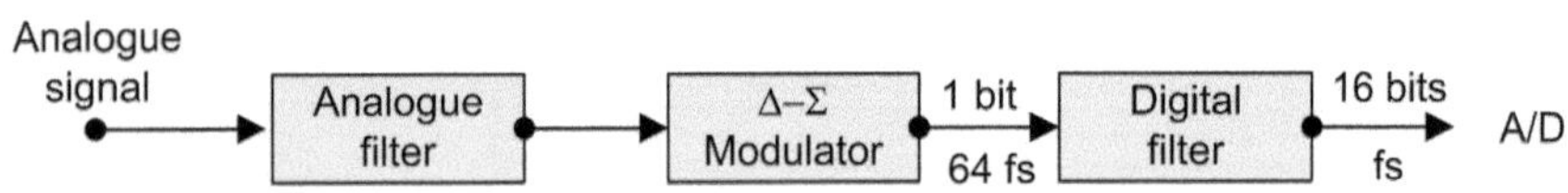


**Fig. 2.11** Single-bit A/D converter

### *2.3.2 Block Floating-Point Conversion*

This is a special case of floating-point conversion, used when a low bandwidth is required. Namely, the exponent is not associated with every sample, but it is done for a block of successive samples. In this way, a considerable bit rate reduction is enabled. This technique is also known as near-instantaneous companding.

### *2.3.3 Differential Pulse Code Modulation (DPCM)*

Using the previous conversion techniques, we analyze each sample separately in order to prepare it for transmission. In the case of differential pulse code modulation, the differences between neighboring samples are transmitted.

This modulation is a form of predictive coding in which the prediction for the current sample is carried out on the basis of the previous sample. It is particularly efficient when a small sampling period is used, since the differences between adjacent samples are very small and practically related to a single bit (the least significant bit). Sigma delta converters are used for this type of conversion. Note that the serial bit stream is impractical, and therefore digital filters (decimation filters) are usually applied to convert the serial stream into a multibit format (e.g., 16 bits for the CD system). A block scheme of a single-bit A/D converter is shown in Fig. 2.11.

### 2.3.4 Super Bit Mapping

In the CD technology, audio signals are usually encoded with 16 bits. In some cases (e.g., professional audio studios), 20 bits are used for encoding of audio signal. Then, the super bit mapping is used to convert 20-bit signals to 16-bit signals. The additional four bits are used to increase the accuracy of the least significant bits of the 16-bit signal. Super bit mapping takes the advantage of the nonlinear frequency response of human auditory system. The noise shaping technique is applied to distribute digital quantization noise in the areas of frequency response where the ear is much less sensitive (higher and lower frequencies). Using this technique, the perceptual quality equivalent to 20-bit sound is available on a standard compact disc.

## 2.4 Speech Signals

The system for generating speech signals is illustrated in Fig. 2.12. We can see that the lungs initialize the air flow through the trachea and larynx to the mouth. The lips form a longitudinal wave that will spread further through the air.

Note that the air flow is modulated by passing through the larynx and the vocal folds. Therefore, the vocal folds generate waves that pass through the mouth and the nasal cavity. The observed system for the voice production can be viewed through two subsystems called glottal and vocal tract. The glottal tract (up to the beginning of the pharynx) generates waves under the influence of the vocal folds, while the vocal tract works as a set of resonators and filters, which modulate and shape the wave in order to make specific sounds.

**Fig. 2.12** An illustration of the speech generation system

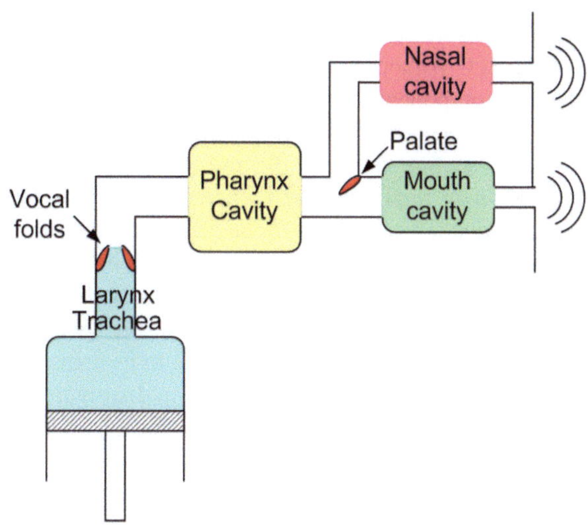

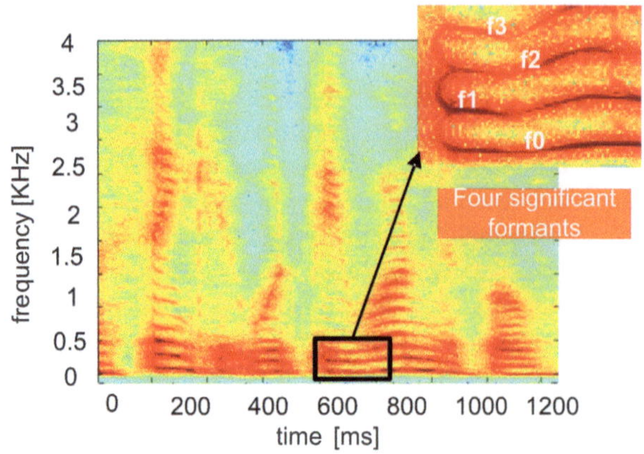


**Fig. 2.13** Time-frequency representation of speech formants

As the speech sounds can be divided into vowels and consonants, it is necessary to describe how they are formed within the speech production system. When generating vowels, the vocal folds resonate and produce quasi-periodic oscillating impulses that continue to be shaped in the vocal tract where the oral cavity acts as a resonator. During this process, some of the frequencies are attenuated, while others are amplified. By examining the spectrum of vowels, we can notice some harmonics that dominate over other components. These harmonics are called the formants and they actually represent the resonant frequencies of vocal tract. When analyzing the speech signal, we can often observe the first four formants. The structure of formants in the time-frequency domain is shown in Fig. 2.13.

The strongest formants for the vowel *A* range from 700 Hz to 1000 Hz. For the vowel *I* these formants are in the range of 200 Hz–400 Hz and 2200 Hz–3200 Hz, while for the vowel *O* they are restricted to frequencies from 400 Hz to 800 Hz.

Consonants can be divided into voiced and voiceless consonants. In the case of voiced consonants, the vocal folds produce noise, which is then modulated in the vocal tract. Although the noise spectrum is mainly spread and continuous, the specific components representing a certain form of formants appear as well. Voiceless consonants occur only in the oral cavity, when the vocal folds are not active.

Let us define some of the most important features of the formant, since it represents an important voice characteristic. The formant frequency is the maximum frequency within the frequency band defined by the formant. The formant bandwidth is defined as the frequency region in which the amplification differs less than 3 dB from the amplification at the peak (central) frequency of the formant.

Having in mind the characteristics of the speech production system, the speech signal can have a variety of values due to its continuous nature. However, from the perceptual point of view, we are able to distinguish just a finite number of sounds, since there is a limited set of meaningful information contained in speech. In this way, we consider only the functional units called phonemes. Note that the same phoneme can occur in different forms, which have no impact on its meaning. In other words, the strength and timbre of the voice will not affect the understanding of phonemes and will not change its functional value.

### 2.4.1 Linear Model of Speech Production System

Based on the previous analysis, we can model the speech production system as shown in Fig. 2.14.

The transfer functions of the glottal tract, the vocal tract, and the lips are denoted by $G(z)$, $V(z)$, and $L(z)$, respectively. $e(n)$ is the input excitation signal, which can be modelled as a train of Dirac impulses for voiced sounds or Gaussian noise for unvoiced sounds. Based on the system in Fig. 2.14, we can write:

$$S(z) = E(z)G(z)V(z)L(z). \tag{2.8}$$

By introducing the inverse filter:

$$A(z) = \frac{1}{G(z)V(z)L(z)}, \tag{2.9}$$

where $A(z)$ has a form of all-zero filter $A(z) = 1 + \sum_{i=1}^{p} a_i z^{-i}$, we can write the following relation:

$$E(z) = A(z)S(z). \tag{2.10}$$

In other words, if $z^{-1}$ is interpreted as the unit delay operator: $z^{-1}s(n) = s(n-1)$, then the previous relation can be written as the autoregressive model of order $p$:

$$s(n) + \sum_{i=1}^{p} a_i s(n-i) = e(n). \tag{2.11}$$

We can model every 700 Hz with one pair of poles.

Let us consider now the impact of the glottal tract and mouth. The speech production system can be observed from the glottal wave $g(n)$. Moreover, the characteristics of the glottal wave are known and given by:

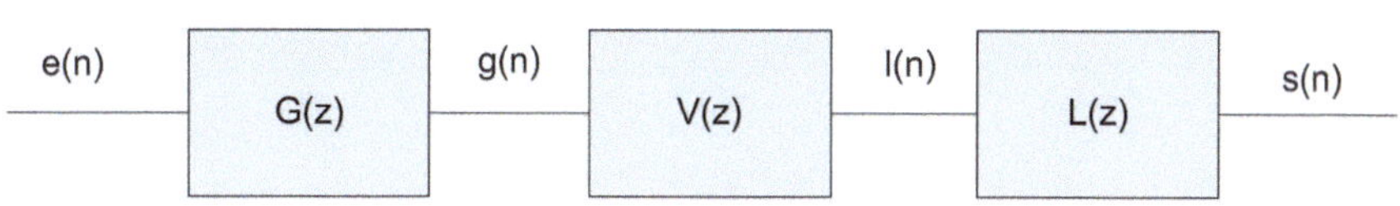


**Fig. 2.14** A model of the speech production system

$$g(t) = \begin{cases} \sin^2 \dfrac{\pi t}{2T_p}, & \text{for } 0 \le t \le T_p, \\ \cos \dfrac{\pi(t - T_p)}{2T_n}, & \text{for } T_p < t \le T_c,\ T_c = T_p + T_n, \\ 0, & \text{for } T_c < t \le T, \end{cases} \tag{2.12}$$

where $T_p = 3.25$ ms, $T_n = 1.25$ ms and the pitch period (time interval between two consecutive periodic excitation cycles) is $T = 8$ ms. The glottal tract can be modelled by the following transfer function:

$$Hg(z) = \frac{1}{(1 - qz^{-1})^2}, \tag{2.13}$$

which attenuates –12 dB/oct. (for $q \approx 1$). The influence of radiation from the lips can be approximated by:

$$L(z) = 1 - z^{-1}. \tag{2.14}$$

Since a linear model of the speech production system is assumed, the transfer functions $L(z)$ and $V(z)$ in Fig. 2.14 can replace the positions. Thus, as the input of $V(z)$ we have:

$$\left(1 - z^{-1}\right)g(n) = g(n) - g(n-1) = g'(n), \tag{2.15}$$

where $g'(n)$ is a differentiated glottal wave. When considering the remaining part of the system, we get:

$$s(n) = V(z)g'(n). \tag{2.16}$$

Next, an additional differentiation can be performed, which will result in:

$$s'(n) = V(z)g''(n). \tag{2.17}$$

Assuming that:

$$V(z) = \frac{1}{A_p(z)} = \frac{1}{1 + \sum_{i=1}^{p} a_i z^{-i}}, \tag{2.18}$$

we can obtain the final model of the speech production system:

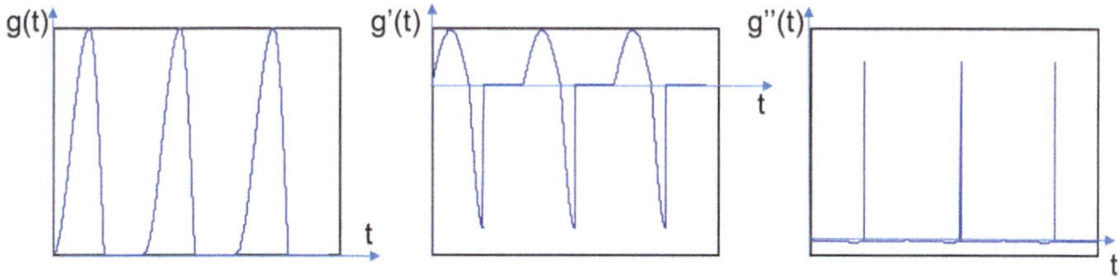


**Fig. 2.15** Excitation signals $g(t)$, $g'(t)$, $g''(t)$

$$s'(n)A_p(z) = s'(n) + \sum_{i=1}^{p} a_i s'(n-i) = g''(n). \tag{2.19}$$

It is important to emphasize that $g''(n)$ is the excitation signal that can be approximated in the form of the Dirac pulse train (Fig. 2.15). Then the signal $s'(n)$ is the pre-emphasized signal $s(n)$, with no influence of the glottal wave and radiation. This model also represents the auto-regressive model of the order $p$, as the one defined by Eq. (2.11).

## 2.5 Voice Activity Analysis and Detectors

Recall that different speech sounds are formed by forcing the air through the vocal system. They could be classified as voiced and unvoiced speech sounds, as shown in Fig. 2.16. Voiced speech parts are generated by vocal folds vibrations that cause the periodical air oscillations. As a result, a sequence of air pulses is created which excites the vocal tract and produces the acoustically filtered output. On the other hand, the unvoiced sounds are usually generated by forcing the air through certain constrictions in the vocal tract.

The voiced sounds are characterized by a significant periodicity in the time domain, with the fundamental frequency referred to as pitch frequency. The unvoiced sounds have a more noisy-like nature. Also, the voiced parts are characterized by significantly higher energy compared to the unvoiced sounds. As mentioned before, the voiced sounds contain formants in the frequency domain. Formants are very important in the speech analysis and applications (e.g., speech coding). Frequency components of unvoiced sounds are generally low energy components located mostly at the high frequencies. Due to the significant differences between voiced and unvoiced speech parts, some applications employ the sounds classification as a pre-processing step. The classification of voiced and unvoiced sounds can be done by using voice activity detectors. These detectors are based on voice activity indicators (energy, zero-crossing rate, prediction gain, etc.) combined with thresholding to decide between voiced and unvoiced option. Some of the existing voice activity indicators are described in the sequel.

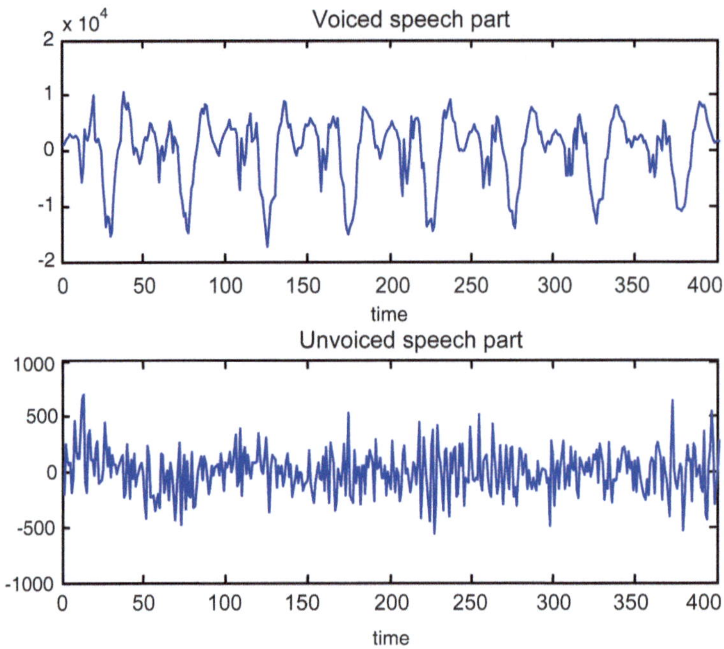


**Fig. 2.16** An illustration of different speech parts

*Energy*

Before processing, the speech signals are usually divided into frames with a certain number of samples. The length of the frame is determined such that the statistical signal characteristics are almost constant within the frame. The simplest way to make differentiation between the voiced and unvoiced parts is the frame energy which is defined as:

$$E(n) = \sum_{k=n-N+1}^{n} s^2(k), \tag{2.20}$$

where $s$ denotes the speech signal, $N$ is the length of frame, while $n$ is the end point of the frame. The voiced parts have the energy that is several times higher than the unvoiced parts energy.

Instead of energy, one can use the magnitudes of the frame samples:

$$MA(n) = \sum_{k=n-N+1}^{n} |s(k)|. \tag{2.21}$$

*Zero-Crossing Rate*

Due to the presence of low-frequency pitch component, the voiced sounds are characterized by a low zero-crossing rate compared to the unvoiced sounds. For a certain frame, the zero-crossing rate can be calculated as follows:

$$ZC(n) = \frac{1}{2} \sum_{k=n-N+1}^{n} |\operatorname{sgn}(s(k)) - \operatorname{sgn}(s(k-1))|. \tag{2.22}$$

*Prediction Gain*

As previously mentioned, the linear prediction algorithm is commonly used in the analysis and synthesis of speech signals. This method provides the extraction of certain sound characteristics that can be used for the voiced/unvoiced speech classification. The prediction of discrete signal $s(n)$ based on the $M$ samples can be defined as:

$$\hat{s}(k) = -\sum_{i=1}^{M} a_i s(k-i), \quad k = n-N+1, \ldots, n, \tag{2.23}$$

where $a_i$, $i=1, \ldots, M$ are estimated linear prediction coefficients of the autoregressive model, while $M$ is the order of the prediction system. For a nonstationary signal such as speech, the linear prediction is performed separately for each frame.

The estimation of linear prediction parameters is based on the criterion of mean square prediction error:

$$J = E\{e^2(k)\} = E\left\{\left(s(k) + \sum_{i=1}^{M} a_i s(k-i)\right)^2\right\}. \tag{2.24}$$

The optimal linear prediction coefficients are obtained by solving the system of equations based on the partial derivatives of the error function $J$ with respect to parameters $a_m$, for $m = 1, 2, \ldots, M$:

$$\frac{\partial J}{\partial a_m} = 2E\left\{\left(s(k) + \sum_{i=1}^{M} a_i s(k-i)\right)s(k-m)\right\} = 0. \tag{2.25}$$

The prediction gain is defined as the ratio between the signal energy and the prediction error:

$$PG(n) = 10\log_{10}\left(\frac{\sum_{k=n-N+1}^{n} s^2(k)}{\sum_{k=n-N+1}^{n} e^2(k)}\right). \tag{2.26}$$

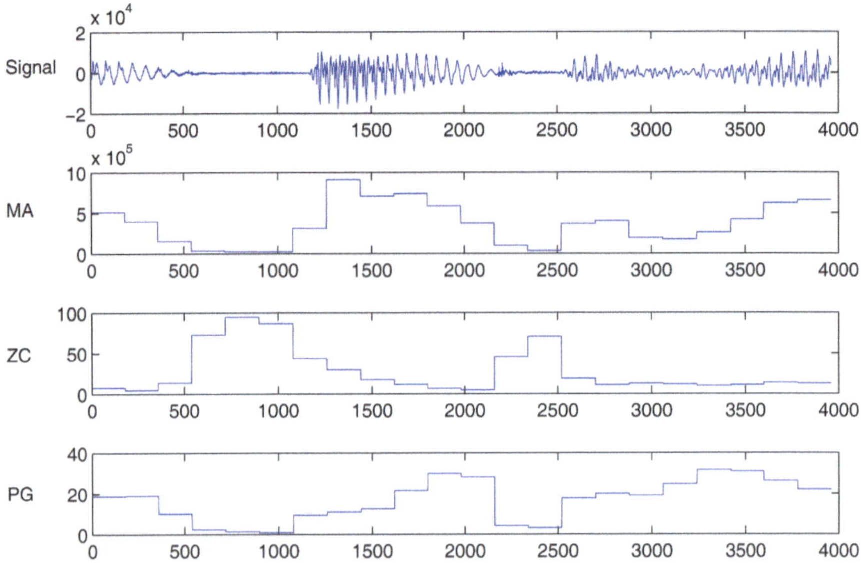


**Fig. 2.17** The outputs of the voice activity indicators based on the magnitudes (MA), zero-crossing rate (ZC), and prediction gain (PG)

This parameter can be used as an indicator of differences between the voiced and unvoiced speech parts. It is known that voiced sounds achieve higher prediction gain compared to the unvoiced ones for at least 3 dB. The periodicity of voiced frames causes a stronger correlation between the frame samples. On the other hand, random nature of unvoiced parts makes the prediction less efficient.

The outputs of the considered voiced/unvoiced sounds indicators for the frames with 180 samples (22.5 ms when the sampling rate is 8 KHz) are illustrated in Fig. 2.17.

The simple versions of the voice activity detectors assume one of these indicators as the input signal. As in the standard classification problems, here is also necessary to define suitable thresholds to separate the voiced and unvoiced speech parts. The thresholds setting is based on the analysis of large signal sets, with the aim to minimize the classification errors. In the practical applications, the considered detectors could be combined to improve the performance of the detection system.

### *2.5.1 Word Endpoints Detector*

The start and end points of words can be detected by using a word endpoints detector. One realization of this detector is based on the energy-entropy signal feature. The signal is firstly divided into time frames that are 8 ms long (e.g., 64 samples long for a speech signal sampled at 8 KHz). The energy of frame $E_i$ is calculated according to Eq. (2.20). On the other hand, the probability density function for the speech spectrum $S(\omega)$, is obtained by normalizing the frequency content within the frame. Hence, for the $i$-th frame we have:

$$p_i = \frac{S(\omega_i)}{\sum_{k=1}^{N} S(\omega_k)}, \tag{2.27}$$

where $N$ is the number of components within the frame. The energy-entropy feature can be calculated as follows:

$$EEF_i = (1 + |E_i \cdot H_i|)^{1/2}, \tag{2.28}$$

where $H_i$ represents the entropy of the $i$-th frame defined as:

$$H_i = \sum_{k=1}^{K} p_k \log p_k. \tag{2.29}$$

Energy-entropy features for the consecutive frames of speech signal are illustrated in Fig. 2.18.

By using the energy-entropy feature, the start and the end point of a spoken word can be determined as follows:

$$\begin{aligned} t_s &= \arg\min_i \{EEF(i) > T_1\}, 1 \le i \le N, \\ t_e &= \arg\max_i \{EEF(i) > T_2\}, 1 \le i \le N, \end{aligned} \tag{2.30}$$

where $N$ is the total number of considered speech frames, while $T_1$ and $T_2$ are thresholds for the start and end point, respectively. The thresholds can be set

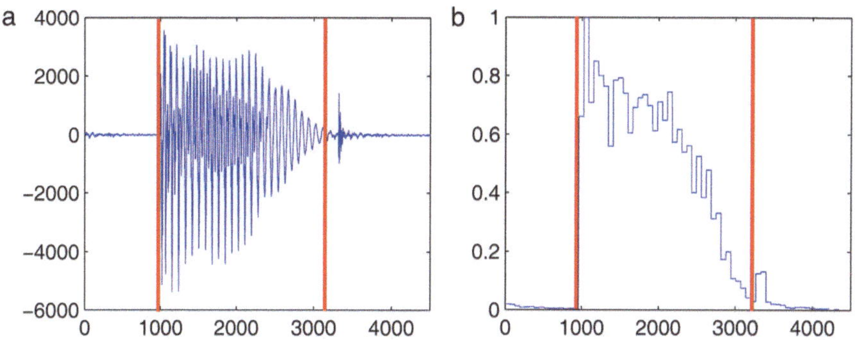


**Fig. 2.18** (**a**) Speech signal, (**b**) Energy-entropy feature for speech frames

empirically, based on various experiments with different speech signals and speakers. The typical values for the thresholds are $T_1 = 0.16$ and $T_2 = 0.17$. The resulting word endpoints are illustrated in Fig. 2.18.

## 2.6 Speech and Music Decomposition Algorithm

The singular value decomposition (SVD) has been used in numerous practical applications for characterization of signals and their components. The SVD has been applied on the time-frequency distributions to extract features used for signal characterization. Most of the procedures are based on the use of singular values. However, significant information about the patterns embedded in the matrix can be obtained by using the left and right singular vectors, especially those corresponding to the largest singular values. Namely, the left and right singular vectors contain the information about time and frequency domain information of the signal, respectively. Here, the SVD is used to extract speech and musical components from the auto-correlation function. The auto-correlation function is obtained by using the inversion of suitable time-frequency distribution, as described in the sequel.

### *2.6.1 Principal Components Analysis Based on SVD*

The SVD transforms the original correlated variables into the uncorrelated set of variables. It allows identifying the direction along which the data variations are dominant. For a certain matrix **S**, SVD is defined as follows:

$$\mathbf{S} = \mathbf{U}\boldsymbol{\Sigma}\mathbf{V}^T, \tag{2.31}$$

where $\boldsymbol{\Sigma}$ is a diagonal matrix of singular values. Matrix $\boldsymbol{\Sigma}$ is of the same size as **S**, and the values are sorted in decreasing order along the main diagonal. The **U** and **V** are orthonormal matrices whose columns represent left and right singular vectors, respectively. If **S** is $M \times N$ matrix ($M > N$), then the size of **U** is $M \times M$, $\boldsymbol{\Sigma}$ is an $M \times N$ matrix, while **V** is an $N \times N$ matrix. A memory-efficient method known as economy-sized SVD is computed as follows:

- Only $N$ columns of **U** are computed.
- Only $N$ rows of $\boldsymbol{\Sigma}$ are computed.

### 2.6.2 Components Extraction by Using the SVD and the S-Method

The audio signals, such as the speech and musical signals are multicomponent signals: $f(n) = \sum_c f_c(n)$. Let us consider the inverse Wigner distribution for a separately observed $c$-th signal component:

$$f_c(n+m)f_c^*(n-m) = \frac{1}{N+1}\sum_{k=-N/2}^{N/2} WD_c(n,k)e^{j\frac{2\pi}{N+1}k2m}. \tag{2.32}$$

By replacing $n + m = p$ and $n - m = q$, we obtain:

$$f_c(p)f_c^*(q) = \frac{1}{N+1}\sum_{k=-N/2}^{N/2} WD_c\left(\frac{p+q}{2},k\right)e^{j\frac{2\pi}{N+1}(p-q)k}. \tag{2.33}$$

The left hand side corresponds to the auto-correlation matrix:

$$R_c(p,q) = f_c(p)f_c^*(q),$$

where $f_c(p)$ is a column vector, whose elements represent the signal terms, and $f_c^*(q)$ is a row vector, with complex conjugate elements. For a sum of $M$ signal components, the total auto-correlation matrix becomes:

$$R(p,q) = \sum_{c=1}^{M} R_c(p,q) = \frac{1}{N+1}\sum_{k=-N/2}^{N/2}\sum_{c=1}^{M} WD_c\left(\frac{p+q}{2},k\right)e^{j\frac{2\pi}{N+1}(p-q)k}. \tag{2.34}$$

By using the S-method the previous relation can be written as:

$$R(p,q) = \frac{1}{N+1}\sum_{k=-N/2}^{N/2} SM\left(\frac{p+q}{2},k\right)e^{j\frac{2\pi}{N+1}(p-q)k}. \tag{2.35}$$

Furthermore, we observe the case when the time-frequency distribution is represented by a square matrix, i.e., time and frequency dimensions are the same. Consequently, the auto-correlation function $R(p,q)$ is given by the symmetric square matrix $\mathbf{R}$ with the symmetry axis along the main diagonal. Therefore, we have: $\mathbf{U} = \mathbf{V}$ are the matrices containing eigenvectors, while $\boldsymbol{\Sigma} = \boldsymbol{\Lambda}$ is the eigenvalue matrix:

$$\mathbf{U\Sigma V}^T = \mathbf{U\Lambda U}^T. \tag{2.36}$$

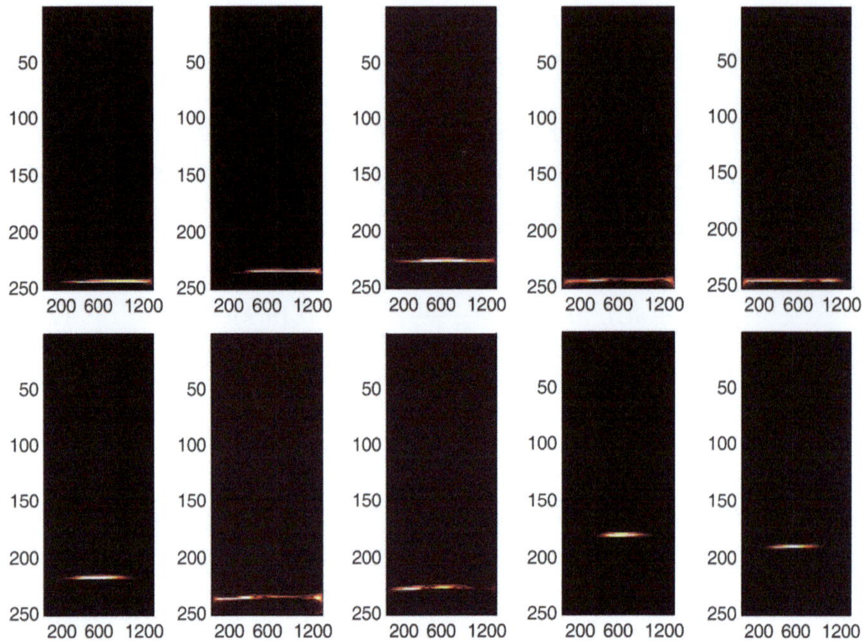


**Fig. 2.19** The formants isolated by using the eigenvalues decomposition method

Hence, the auto-correlation matrix **R** can be decomposed as follows:

$$\mathbf{R} = \sum_{j=1}^{M} \lambda_j \mathbf{u}_j(n)\mathbf{u}^*{}_j(n), \tag{2.37}$$

where $\lambda_j$ are the eigenvalues and $\mathbf{u}_j(n)$ are the eigenvectors of the autocorrelation matrix **R**. Note that the eigenvectors correspond to the signal components, while the eigenvalues are related to the components energy.

The speech formants, separated by using the eigenvalue decomposition, are shown in Fig. 2.19 (the formants at positive frequencies are shown). Now, it is possible to arbitrarily combine the components that belong to the low, middle or high-frequency regions. Consequently, an arbitrary time-frequency mask (Fig. 2.20) can be made and used in speech processing applications.

Let us consider a violin signal with a number of closely spaced components, as it can be seen from Fig. 2.21. The eigenvalue decomposition method is applied in the same way as in the case of speech signal. The extracted components are shown in Fig. 2.22. It is important to note that, due to the specific nature of audio signals, the perfect signal reconstruction from its separated components is not fully attainable.

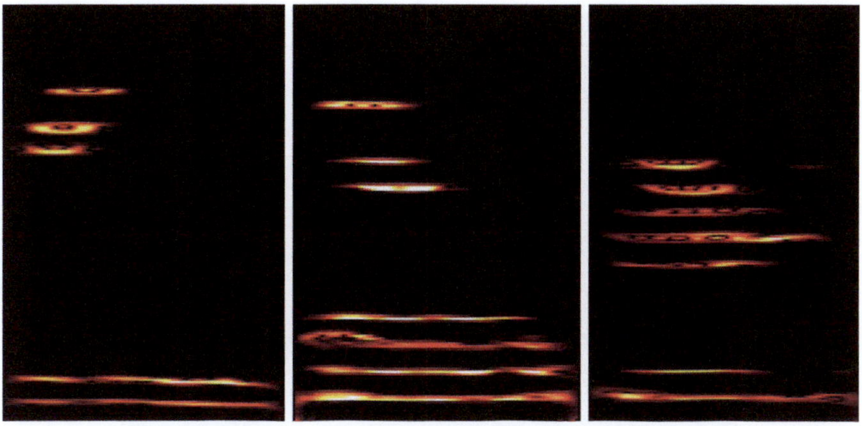

**Fig. 2.20** Illustrations of different components combinations selected by a few time-frequency masks

**Fig. 2.21** The time-frequency representation of the violin signal obtained by using the S-method

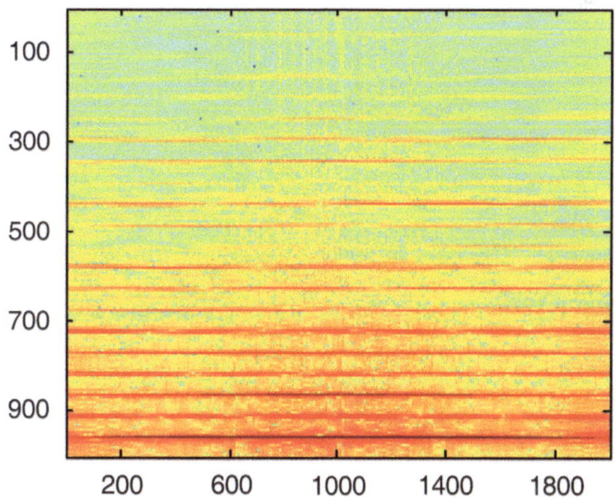


## 2.7 Psychoacoustic Effects

It was mentioned earlier that the ear is not equally sensitive to different frequencies. The sensitivity function (shown in Fig. 2.2) is obtained experimentally and is given by the following expression:

$$T(f) = 3.64\left(\frac{f}{1000}\right)^{-0.8} - 6.5e^{-0.6(f/1000-3.3)^2} + 10^{-3}\left(\frac{f}{1000}\right)^4 \; dB. \qquad (2.38)$$

Let us perform now a detailed analysis of the auditory system. It is composed of the outer (lobe) ear, the middle ear and the inner ear, as illustrated in Fig. 2.23. The

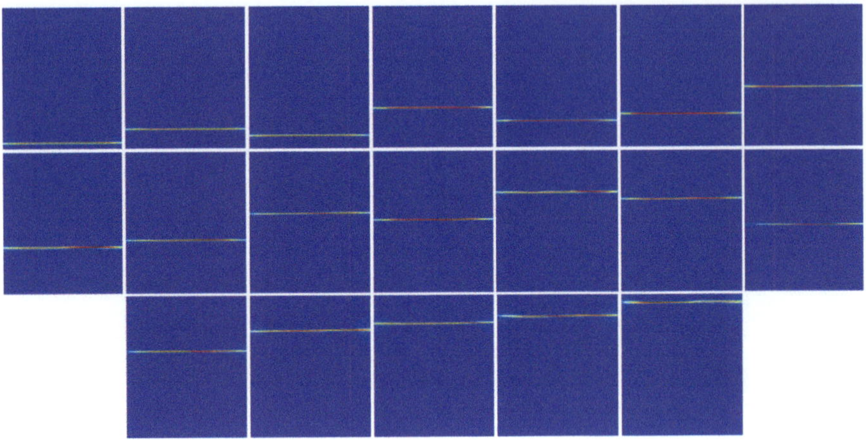

Fig. 2.22 Separated components of violin signal

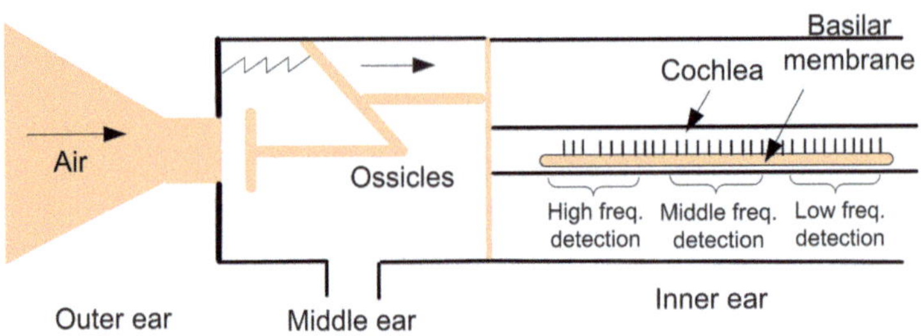


Fig. 2.23 Illustration of human auditory system

auditory system up to the inner ear can be simply represented as a combination of a horn and open pipes.

Sound waves, collected by the ear shell, are forwarded over the ear channel. In the inner ear there is the organ of Corti, which contains the fibrous elements with different lengths and resonant frequencies. These elements are connected to the auditory nerve that is used to convey information to the brain. As a consequence of the applied sound wave, the mechanical vibrations are passed through the ossicles to the Cochlea causing the basilar membrane to oscillate. The parts of basilar membrane resonate depending on the frequencies (Fig. 2.23). In the case of high frequencies the resonance is produced in the front part of basilar membrane, while in the case of low frequencies, it occurs in the rear part.

The hearing system works effectively as a filter bank. We devote our attention to a particular sound only after our brain focuses on it.

### 2.7.1 Audio Masking

As discussed earlier, there is a threshold value of SPL below which we cannot hear a sound. However, even the components above this threshold can be non-audible if they are masked by other components. Masking effects can be either in the time and/or in the frequency domain. In the case of frequency masking, tones with greater intensity can mask lower intensity tones at neighboring frequencies. Therefore, if we know a value of the threshold below which the adjacent frequencies become non-audible, then we can ignore those frequencies without sacrificing the quality of the sound, as shown in Fig. 2.24. This is particularly important when applied to each of the critical frequency bands, where we can say that the ear is equally sensitive. The sensitivity is different for different critical bands.

It should be mentioned that the width of the critical frequency bands varies from a few hundred Hz at lower frequencies to several KHz at higher frequencies. An overview of the 25 experimentally determined critical bands is given in the following section.

A masking curve is illustrated in Fig. 2.25. Note that the samples below the masking curve are dismissed and only the samples that are not masked are considered for encoding and transmission. In addition to frequency masking, we can use

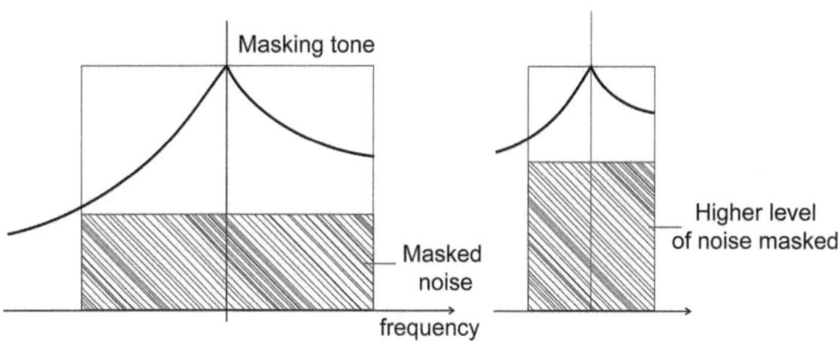


**Fig. 2.24** Masking noise

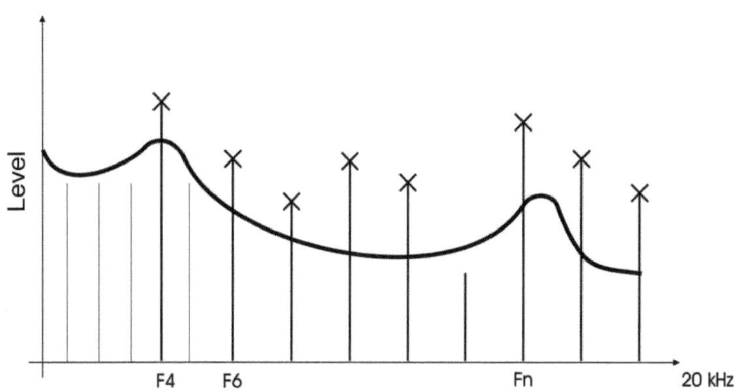


**Fig. 2.25** An illustration of audio masking

time masking where the threshold is defined as a function of time. For example, let us assume that we have a signal with a dominant frequency $f$ at time $t$. Then, it is possible to determine the masking threshold for the interval $(t, t+\Delta t)$ for which the signal becomes non-audible at the given frequency $f$ or adjacent frequencies.

## 2.8 Audio Compression

Based on the aforementioned characteristics of the audio signal, we can conclude that storing high quality digital audio signals requires a large memory space. Therefore, the transmission of such signals also requires a network with large bandwidth. The reduction of the required bandwidth and memory space, while maintaining high audio quality, can be achieved by compression algorithms. Recent advances in computer technology have prompted significant improvements in compression algorithms. Also, there is a growing need to transfer large amount of data over the network. Hence, the compression algorithms have a significant economic impact related to various storage media or better utilization of network connections.

Data compression is performed by a circuit called the encoder. After transmission over a communication channel, the data are restored back into its basic form by using decoders. The encoder is generally much more complex and expensive than the decoder. However, a single encoder can be used to provide data to a large number of decoders.

A compression ratio is the ratio of the compressed signal size versus the original signal size. This ratio is often referred to as a coding gain. The compression is especially important in the Internet-based communications and applications. The need for efficient compression algorithms is also growing in radio broadcasting, as we are trying to use the available bandwidth more efficiently.

### *2.8.1 Lossless Compressions*

Compression in general can be divided into lossless and lossy compression. In lossless compression, the information before and after compression must be identical. To achieve lossless compression, we use algorithms such as Huffman coding and LZW coding. Lossless compression algorithms have limited compression abilities. If the audio signal is compressed by using lossless compression techniques, than we refer to it as heavy due to a low compression ratio.

Figure 2.26 illustrates the concept of entropy as the information content without redundancy. Namely, if we transmit the amount of information smaller than the information content or entropy, we actually introduce the artifacts. This is called lossy compression. Otherwise, the compression scheme is lossless when it is possible to recover the signal by uncompressing, i.e., the compressed signal has

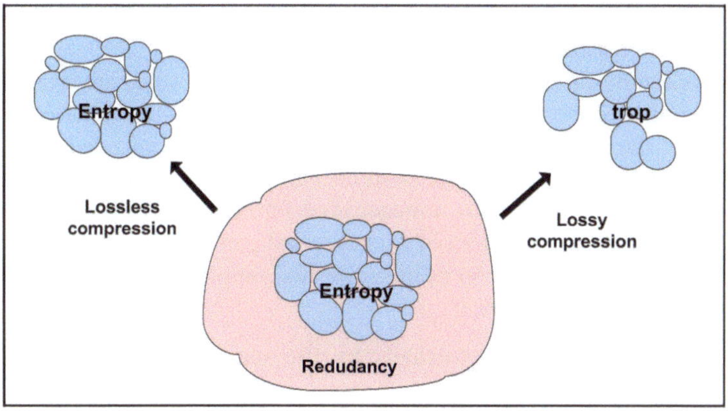


**Fig. 2.26** Lossless and lossy compressions

**Fig. 2.27** Quality of a compressed signal depends on the complexity of the compression system and the compression ratio

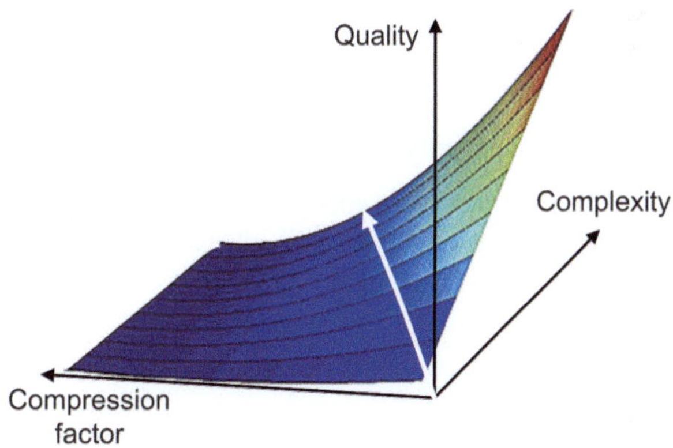


the same entropy as the original one. We can conclude that the redundancy is actually a difference between the information rate and the overall bit rate. An ideal coder should provide the information bit rate defined by the entropy.

The relationship between the compression ratio and the complexity of the compression system is depicted in Fig. 2.27. In order to maintain the quality of signal under high compression ratio, we have to increase the complexity of the system.

#### 2.8.1.1 LZ-77

LZ-77 algorithms achieve compression by replacing repeated occurrences of data with references to a single copy of these existing earlier in the input (uncompressed) data stream. It is especially important to determine the optimal length of the sequence that is encoded. Too short, or too long sequences can cause negative effects on compression.

Pointers can be encoded with 12 bits such that the first 8 bits are used to denote the number of characters we have to go back, and the last 4 bits are used to denote the length of the sequence. In some cases, the pointers are encoded with 18 bits, where the first 12 bits determine the position, and the last 6 bits denote the length of the sequence. Encoding an entire sentence is performed by inserting 1 in front of an uncompressed part and 0 in front of a compressed part.

For the sake of simplicity, let us illustrate this compression principle on the text by using the following sentence:

**she_sells_seashells_by_the_seashore**

The letters *_se* (from the word seashells) are found in the word *_sells* and they are replaced by the pointer (6,3) meaning that we go back six characters and take the following three characters. The sequence *she* from *seashells* is found in the word *she* and can be replaced by a pointer (13,3), meaning that we go back 13 characters and take the next 3 characters. The procedure continues until we reach the end of the sentence, which we are encoding. The sentence can be then encoded as follows:

she_sells<6,3>a<13,3><10,4>by_t<23,5><17,3>ore

Since the pointers are encoded with 12 bits, in this short example we can reduce the amount of information by 76 bits (out of 280).

#### 2.8.1.2 LZW Coding

LZW coding is a generalization of the LZ-77 coding, and it is based on defining a code book (dictionary) of words and strings found in the text. Strings are placed in the dictionary. Since the first 255 entries found in the dictionary are assigned to single characters, the first available index in the dictionary is actually 256. The dictionary is formed by initially indexing any two-character string found in the message. Then, we continue with three-character string, and so on. For example, let us consider the previous example:

```
          she_sells_seashells_by_the_seashore

256 → sh  <sh> e_sells_ seashells_by_the_seashore
257 → he  s < he >_sells_ seashells_by_the_seashore
258 → e_  sh < e_ >sells_ seashells_by_the_seashore
259 → _s  she < _s >ells_ seashells_by_the_seashore
260 → se  she_ < se >lls_ seashells_by_the_seashore
261 → el  she_s < el >ls_ seashells_by_the_seashore
```

**262 → ll she_se < ll >s_ seashells_by_the_seashore**
**263 → ls she_sel < ls >_ seashells_by_the_seashore**
**264 → s_ she_sell< s_ > seashells_by_the_seashore**

*The next two characters are "_s", but they already exist in the dictionary under the number 259. This means that we can now place the three characters "_se" as a new entry in the dictionary and then continue with the strings of two characters:*

**265 → _se she_sells< _ se >ashells_by_the_seashore**
**266 → ea she_sells_s<ea > shells_by_the_seashore**
**267 → as she_sells_se<as > hells_by_the_seashore**

*The next two characters "sh" are already indexed in the dictionary under 256. Therefore, we add a new three-character string "she":*

**268 →she she_sells_sea<she > lls_by_the_seashore**

*The string "el" is already in the dictionary with the label (261), and therefore we add "ell":*

**269 → ell she_sells_seash< ell > s_by_the_seashore**

*The string "ls" is already in the dictionary with the label (263), and we add "ls_", and then continue with the string of two characters:*

**270 → ls_ she_sells_seashel<ls_ >by_the_seashore**
**271 → _b she_sells_seashells<_b >y_the_seashore**
**272 → by she_sells_seashells_<by > _the_seashore**
**273 → y_ she_sells_seashells_b< y_ >the_seashore**
**274 → _t she_sells_seashells_by<_t >he_seashore**
**275 → th she_sells_seashells_by_<th > e_seashore**

*As the string "he" is already in the dictionary with the label (257), "he_" is added:*

**276 → he_ she_sells_seashells_by_t<he_ > seashore**

*String "_s" is already in the dictionary with the label (259), as well as the string "_se" with the label (265). Thus, we add a new string with four characters "_sea":*

**277 → _sea she_sells_seashells_by_the<_ sea > shore**
**278 → ash she_sells_seashells_by_the_se<ash>ore**
**279 → ho she_sells_seashells_by_the_seas<ho>re**
**280 → or she_sells_seashells_by_the_seash<or>e**
**281 → re she_sells_seashells_by_the_seasho<re>**

*Finally, the dictionary will contain the following strings:*

| 256 → sh | 269 → ell |
|---|---|
| 257 → he | 270 → ls_ |
| 258 → e_ | 271 → _b |
| 259 → _s | 272 → by |
| 260 → se | 273 → y_ |
| 261 → el | 274 → _t |
| 262 → ll | 275 → th |
| 263 → ls | 276 → he_ |
| 264 → s_ | 277 → _sea |
| 265 → _se | 278 → ash |
| 266 → ea | 279 → ho |
| 267 → as | 280 → or |
| 268 →she | 281 → re |

In parallel to forming the dictionary, the encoder continuously transmits characters until it encounters the string that is in the dictionary. Then, instead of sending the string, the index from the dictionary is sent. This process is repeated until the whole message is transmitted. It means that the compressed messages in our case will be:

```
she_sells<259>ea<256><261><263>_by_t<257><265><267>hore
```

Note that it is not necessary to send to the decoder the dictionary created by the encoder. While reading and decoding the message, the decoder creates the dictionary in the same way as the encoder.

Let us consider another example:

*Thomas_threw_three_free_throws*

```
256 → th  < Th >omas_threw_three_free_throws
257 → ho  T <ho>mas_threw_three_free_throws
258 → om  Th <om>as_threw_three_free_throws
259 → ma  Tho<ma>s_threw_three_free_throws
260 → as  Thom<as>_threw_three_free_throws
261 → s_  Thoma<s_>threw_three_free_throws
262 → _t  Thomas <_t>hrew_three_free_throws
263 → thr Thomas_<thr>ew_three_free_throws
264 → re  Thomas_th<re>w_three_free_throws
265 → ew  Thomas_thr<ew>_three_free_throws
266 → w_  Thomas_thre<w_>three_free_throws
267 → _th Thomas _threw<_th>ree_free_throws
268 → hr  Thomas _threw_t<hr>ee_free_throws
269 → ree Thomas _threw_th<ree>_free_throws
```

**270 → e_ Thomas _threw_thre<e_>free_throws**
**271 →_f Thomas _threw_three<_f>ree_throws**
**272 → fr Thomas _threw_three_<fr>ee_throws**
**273 → ree_ Thomas _threw_three_f<ree_>throws**
**274 →_thr Thomas _threw_three_free<_thr>ows**
**275 → ro Thomas _threw_three_free_th<ro>ws**
**276 → ow Thomas _threw_three_free_thr<ow>s**
**277 → ws Thomas _threw_three_free_thro<ws>**

*The dictionary is formed as follows:*

| **256 → th** | **267 → _th** |
|---|---|
| **257 → ho** | **268 → hr** |
| **258 → om** | **269 → ree** |
| **259 → ma** | **270 → e_** |
| **260 → as** | **271 → _f** |
| **261 → s_** | **272 → fr** |
| **262 → _t** | **273 → ree_** |
| **263 → thr** | **274 → _thr** |
| **264 → re** | **275 → ro** |
| **265 → ew** | **276 → ow** |
| **266 → w_** | **277 → w** |

*while the coded message is:*

Thomas_<256>rew<262>h<264>e_f<269><267> rows

#### 2.8.1.3 Huffman Coding

The idea behind the Huffman coding is to encode each character with a code word whose length is inversely proportional to the probability of occurrence of that character. In other words, if a character appears more frequently, it should be encoded with the shortest possible code word.

The characters are firstly sorted according to the number of occurrences (NO). Then, we observe a pair of characters with the lowest NO. If the logical value of "1" is assigned to the character with a higher NO, then "0" is assigned to the character with a lower NO. The cumulative NO for the two characters is calculated and it replaces this pair in the next iterations. The next character is used in the new iteration and its NO is compared with the smaller between: NO for another character and cumulative NO from the previous iteration. Again, "1" is assigned to the higher NO, while "0" is assigned to lower NO. The procedure is repeated until we get the entire tree. Each branch within the tree corresponds to one character, and it is uniquely determined by the resulting sequence of logical values "1" and "0".

Consider an example with the following characters A, M, R, C, D and U. Assume that the NOs of characters in a text are: A = 60, M = 38, R = 21, C = 11, D = 34, U = 51. For Huffman coding we form the following tree:

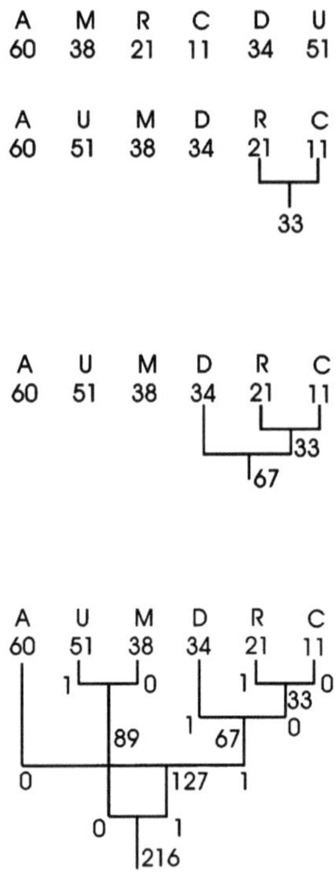


Therefore, the binary combinations denoting each of the characters are given as:

| *A* | *U* | *M* | *D* | *R* | *C* |
|---|---|---|---|---|---|
| 10 | 01 | 00 | 111 | 1101 | 1100 |

## 2.8.2 Lossy Compressions

The idea of lossy compression is based on the perceptual characteristics. Namely, the information that is the least important, from a perceptual point of view, is omitted. For lossy compressions we utilize our understanding of psychoacoustics (e.g., the auditory system responds differently to different frequencies and some sounds may be masked by the others). Therefore, this coding method is often referred to as the perceptual coding. MPEG (Moving Picture Experts Group) compression algorithms represent the important and widely used cases of lossy compression.

The amount of compressed data depends on the signal nature (i.e., the encoding may have a variable compression factor), which causes a variable bit rate through

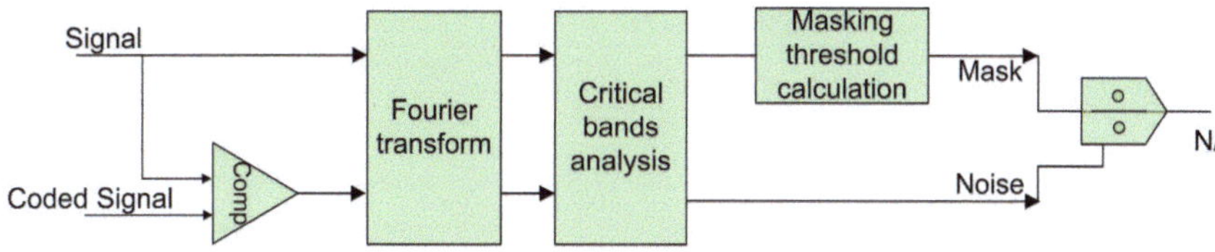


**Fig. 2.28** The system for measuring the noise/masking ratio

the channel. In practice, it is often required that coders have a constant compression factor in order to transmit at a constant rate.

In order to use the perceptual coding, it is important to adjust and calibrate correctly the microphone gain and reproduction system volume control. The overall gains should be adjusted to the human hearing system such that the coder uses the SPL which is actually heard. Otherwise, we might have a situation that the low gain from the microphone is interpreted as low SPL, which further causes inappropriate masking of the coded signal. Thus, the compression systems must include the calibration model based on human hearing system. In addition to calibration, an important role in perceptual coding has a masking model. The accuracy of the model used for the separation of relevant and irrelevant components is of particular importance. Based on this model, we decide to ignore a certain amount of information that will not affect the signal quality. The most reliable approach for assessing the quality of the masking is listening. Such methods are usually expensive to carry out. Therefore, systems have been developed to measure the quality of sound masking. A system based on noise measurements is shown in Fig. 2.28.

The system compares the original and coded signals and determines the noise introduced by encoder. The lower branch performs the noise analysis and provides the critical band spectrum of the noise. The blocks in the upper branch of the system are used to calculate the masking threshold of the input signal. The noise to masking ratio (N/M or NMR) is obtained at the output of the observed system (Fig. 2.28). This ratio is a quality measure of masking. Smaller values denote more accurate masking models.

#### 2.8.2.1 Critical Subbands and Perceptual Coding

The spectrum of the audio signal can be divided into the subbands (critical bands) within which is assumed that the human hearing system has equal sensitivity for all frequencies. Table 2.1 provides the lower ($F_l$) and upper ($F_u$) limit frequencies, the center frequency ($F_c$) and the bandwidth for each critical band.

Thus, the auditory system can be approximately modelled as a filter bank. However, implementing selected critical bands would be a demanding task. Hence, we can obtain a simpler system with some approximations as shown in Fig. 2.29.

**Table 2.1** Critical bands

| Subband | $F_l$ | $F_c$ | $F_u$ | Bandwidth (Hz) |
|---|---|---|---|---|
| 1 | 0 | 50 | 100 | 100 |
| 2 | 100 | 150 | 200 | 100 |
| 3 | 200 | 250 | 300 | 100 |
| 4 | 300 | 350 | 400 | 100 |
| 5 | 400 | 450 | 510 | 110 |
| 6 | 510 | 570 | 630 | 120 |
| 7 | 630 | 700 | 770 | 140 |
| 8 | 770 | 840 | 920 | 150 |
| 9 | 920 | 1000 | 1080 | 160 |
| 10 | 1080 | 1170 | 1270 | 190 |
| 11 | 1270 | 1370 | 1480 | 210 |
| 12 | 1480 | 1600 | 1720 | 240 |
| 13 | 1720 | 1850 | 2000 | 280 |
| 14 | 2000 | 2150 | 2320 | 320 |
| 15 | 2320 | 2500 | 2700 | 380 |
| 16 | 2700 | 2900 | 3150 | 450 |
| 17 | 3150 | 3400 | 3700 | 550 |
| 18 | 3700 | 4000 | 4400 | 700 |
| 19 | 4400 | 4800 | 5300 | 900 |
| 20 | 5300 | 5800 | 6400 | 1100 |
| 21 | 6400 | 7000 | 7700 | 1300 |
| 22 | 7700 | 8500 | 9500 | 1800 |
| 23 | 9500 | 10500 | 12000 | 2500 |
| 24 | 12000 | 13500 | 15500 | 3500 |
| 25 | 15500 | 18775 | 22050 | 6550 |

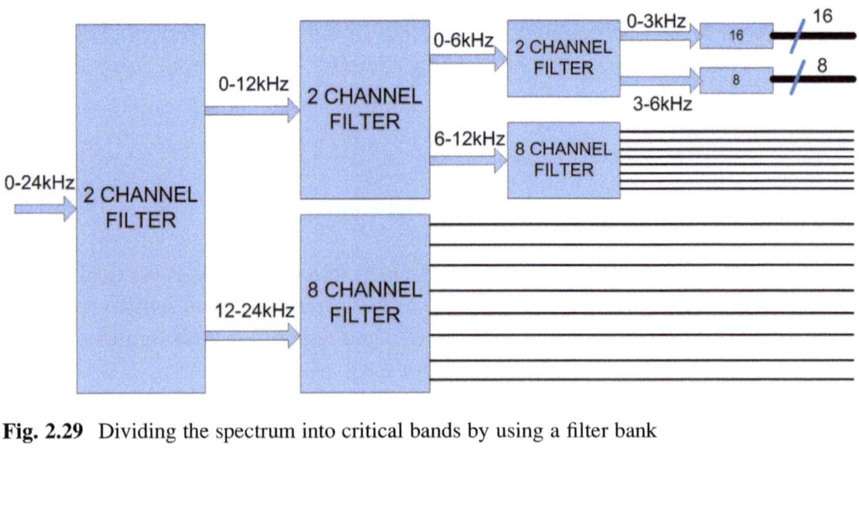


**Fig. 2.29** Dividing the spectrum into critical bands by using a filter bank

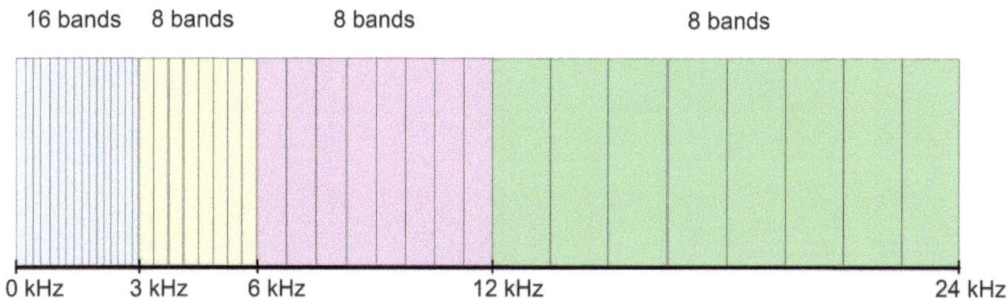


**Fig. 2.30** An illustration of the critical bands

At each filtering stage, the signal bandwidth is halved, allowing us to decrease the sampling frequency[1] by 2. The high-frequency part of the spectrum is obtained as a difference between the input and the filtered spectrum (low-frequency part). In this way, a ladder scheme of the spectrum partition into critical bands is obtained. The frequency subbands are illustrated in Fig. 2.30.

The scale used to number these critical bands is known as the Bark scale named after the German scientist Barkhausen. The scale depends on the frequencies (expressed in Hz) and can be approximately given by:

$$B\,(Bark) = \begin{cases} \dfrac{f}{100} & \text{for } f < 500Hz, \\ 9 + 4\log_2\left(\dfrac{f}{1000}\right) & \text{for } f \geq 500Hz, \end{cases} \tag{2.39}$$

where $B$ is the number of the critical band. It is often used by the following approximate relation:

$$\mathrm{B(Bark)} = 13\arctan(0.76(f(Hz)/1000)) + 3.5\arctan\left((f(Hz)/7500)^2\right).$$

For example, the frequency of 200 Hz can be represented by 2 from the Bark scale, while the frequency of 2000 Hz can be represented by 13 from the Bark scale.

To obtain the frequency in Hz from the Bark scale, we can use the following relationship:

$$f(Hz) = 1000\left\{((\exp(0.219 \cdot B)/352) + 0.1) \cdot B - 0.032 \cdot \exp\left(-0.15 \cdot (B-5)^2\right)\right\}$$

Figure 2.31 shows the masking effects versus the frequency expressed in KHz and the Bark scale. In both cases, the dotted line shows the curve representing a hearing threshold in quiet. Figure 2.31a also depicts masking curves for samples at

[1] For the signal with spectrum bandwidth B, the sampling frequency is $f_s$ =2B if $(2f_c + \mathrm{B})/2\mathrm{B}$ is an integer ($f_c$ is the central frequency)

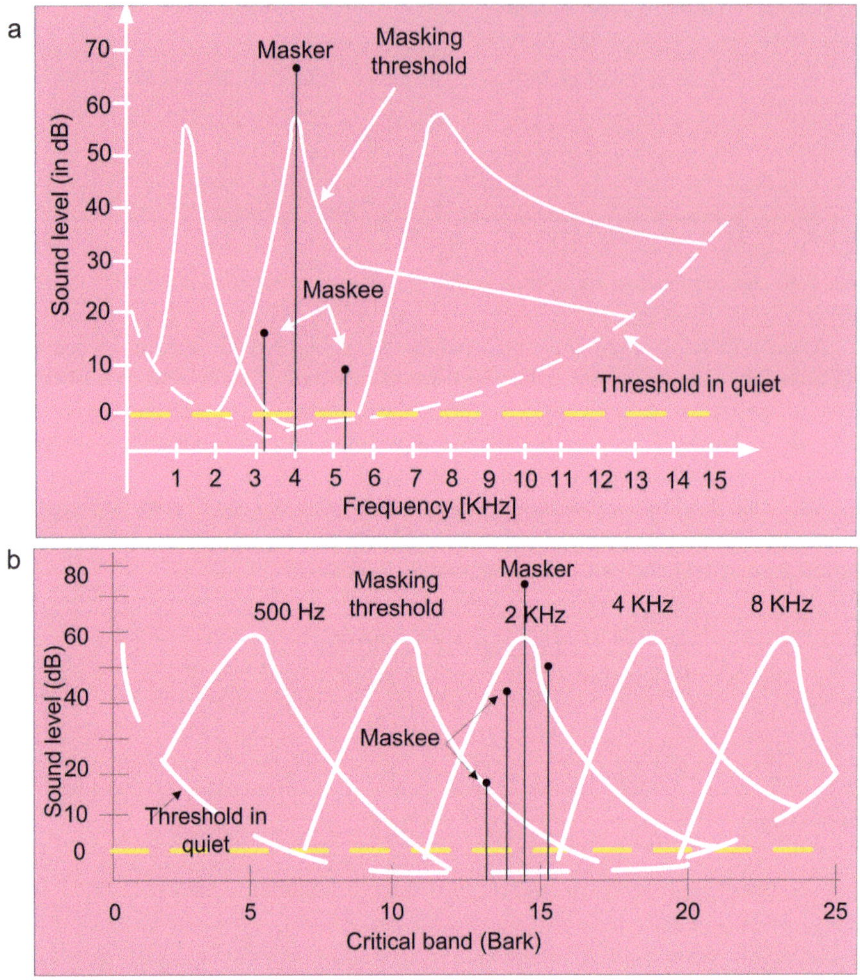


**Fig. 2.31** An illustration of the effects of masking tones: (**a**) masking in frequency, (**b**) masking of the bandwidth range

frequencies 1 KHz, 4 KHz and 8 KHz, respectively. Similarly, Fig. 2.31b shows the masking curve for different ranges on the Bark scale.

Consider the following example, where the amplitude levels in certain frequency bands are provided in the Table 2.2.

Note that the amplitude level in the 12th band is 43 dB. Suppose that it masks all components below 15 dB in the 11th band and the components below 17 dB in the 13th band.

- The signal level in the 11th band is 25 dB (>15 dB) and this band should be encoded for transmission. However, the quantization noise of 12 dB will be masked, and therefore, we can use 2 bits less to represent the samples in this band.

**Table 2.2** An example of amplitude levels in different frequency bands

| Band | 1 | 2 | 3 | 4 | 5 | 6 | 7 | 8 | 9 | 10 | 11 | 12 | 13 | 14 | 15 | 16 |
|---|---|---|---|---|---|---|---|---|---|---|---|---|---|---|---|---|
| Level (dB) | 12 | 5 | 3 | 1 | 2 | 12 | 8 | 28 | 19 | 10 | 25 | 43 | 14 | 2 | 6 | 35 |

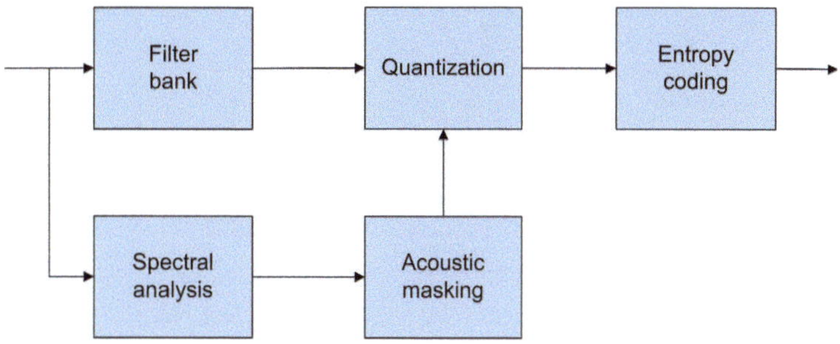


**Fig. 2.32** Block diagram of MUSICAM based coder

- The signal level in the 13th band is 14 dB (<17 dB). Hence, the components in the 13th band are masked and there is no need to transmit this band.

### *2.8.3 MPEG Compression*

In 1988, the ISO (International Standards Organization) and IEC (International Electrotechnical Commission) have begun to establish international standards for audio compression. As a result, they established guidelines for MPEG audio compression, which is currently used for the audio coding in DAB (digital audio broadcasting). Algorithms for MPEG audio compression were derived from MUSICAM (Masking-pattern Universal Subband Integrated Coding And Multiplexing) algorithm. A block diagram for an audio compression coder based on the MUSICAM is shown in Fig. 2.32.

MUSICAM compresses audio data such that the optimal bit rate is approximately 700 Kb/s. In parallel to the MUSICAM, a compression algorithm known as ASPEC (Adaptive Spectral Perceptual Entropy Coding) was developed. Its main goal was to achieve high compression factors in order to facilitate transmission of audio signals over the ISDN lines. By combining MUSICAM and ASPEC, MP3 (MPEG Layer III) algorithm was created. That is, while the MPEG Layer I and MPEG Layer II represent simplified versions of MUSICAM, MP3 combines the best features of MUSICAM and ASPEC. The layers of MPEG audio coding deal with signals having maximal frequencies: 16, 22.05, and 24 KHz and support different bit rates such as: 32, 48, 56, 64, 96, 112, 128, 192, 256, 320, and 384 Kb/s. MPEG Layer I is based on two channels (i.e., a stereo signal), while MPEG Layer II can handle a five-channel audio signal. MPEG Layer II can also

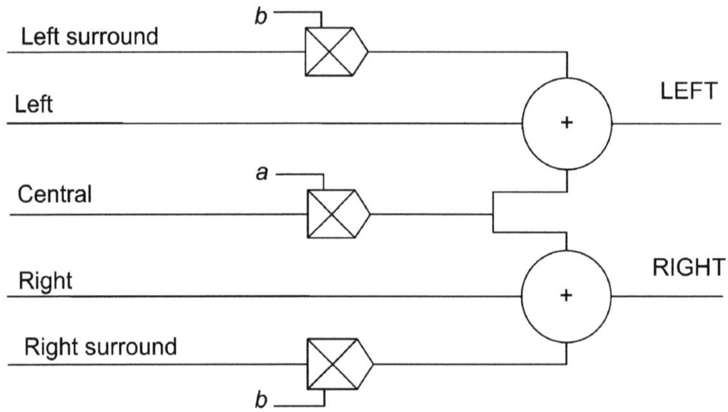


**Fig. 2.33** A scheme for converting from five channels into two channels

convert a five-channel signal into a two-channel signal and such a system is illustrated in Fig. 2.33.

The compression algorithm known as AC-3, developed by Dolby Laboratories, is also used in North America. At the beginning, the AC-3 was developed as a compression scheme that provides the surround sound for the theater and cinema. Nowadays, it is usually referred as Dolby Digital and can be found in the HDTV, home theaters, DVD players, some TV receivers, etc.

#### 2.8.3.1 MPEG Layer I

As already noted, the MPEG Layer I is a simplified version of the MUSICAM algorithm. According to the MPEG Layer I algorithm, an audio signal is divided into 32 subbands. All 32 subbands are of the same width, which is one of the drawbacks of this compression scheme, since the bandwidths of the critical bands are frequency dependent. Thus, subbands can be either too wide at lower frequencies or too narrow at higher frequencies. In order to compensate the imprecision caused by the uniform subbands width, audio masking is used. The Fourier transform has an important role in the audio masking (it is computed by the FFT algorithm). A block scheme for MPEG layer I compression is given in Fig. 2.34.

The signal compression is carried out in blocks of 384 samples (see Appendix). After coding, we obtain 32 blocks with 12 samples corresponding to the width of 8 ms at the sampling frequency of 48 KHz. The FFT is calculated for 512 points in order to obtain higher resolution. This provides a more accurate model of masking. The data in each block are encoded according to the maximum signal value in that block. A 6-bit scale factor is assigned to each block and it is applied to all 12 block samples. The gain step between two successive 6-bit combinations is 2 dB, thus providing a 128 dB of dynamic range. Having in mind the nature of audio signals, the number of bits reserved for samples will vary for the 32 different blocks, but the

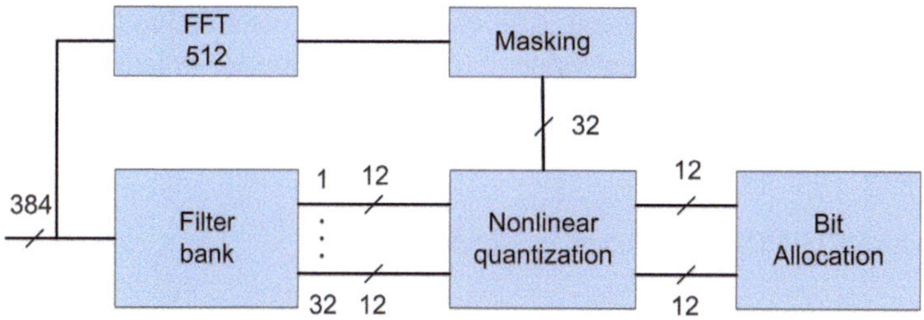


**Fig. 2.34** A block diagram of MPEG layer I compression

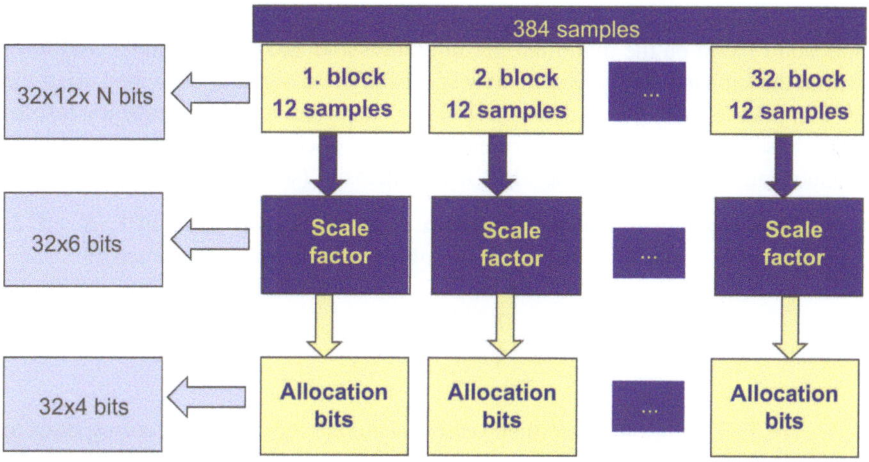


**Fig. 2.35** A part of the bits packing structure in MPEG Layer I (*N* can be between 1 and 15)

total length of 32 blocks has to be equal for each coded block (the size of the output block with 384 samples is fixed for a certain bit rate).

The bit allocation is used to determine the structure of binary code words for the appropriate subband. Namely, four bits are used to describe the samples code length. The length can range between 0 and 15 bits (i.e., from the combination 0000 to 1110), excluding the allocation of 1 bit due to the nature of midtread quantizer (it has 0 as one of its quantized values, and generally an odd number of decision intervals). 0000 denotes that zero bits is allocated for samples within the block, while 1110 denotes that we need 15 bits for each sample in the block. 1111 is not used in order to avoid possible conflict with the synchronization code. There is also a special code if all samples in the block are equal to zero. Hence, for each block it is necessary to send 4 allocation bits and 6 bits that define the amplification factor (Fig. 2.35).

Note that the block length of 8 ms is quite long to avoid pre-masking effects that may appear due to the abrupt changes in signal followed by silence at the transition between two blocks. This phenomenon can be avoided by comparing the values in neighboring blocks. A significant difference between consecutive blocks indicates transient in the signal. A typical value of the compression factor in MPEG Layer I is 1:4, and the bit rate is 384 Kb/s.

### Calculating SMR Using Psychoacoustic Model

In order to perform the bit allocation procedure and quantization of the subband values, the psychoacoustic model is used to calculate the signal to mask ratio (SMR), which is a basis for the bit allocation process. The SMR determines the dynamic range within the subband that needs to be quantized in a way to keep the quantization noise below the masking threshold. The *SMR*($i$) is calculated for each subband $i$ based on the minimum masking threshold and the maximal signal level within the subband. The power density spectrum of the signal is estimated using the FFT in parallel with the analysis filter bank. Therefore, a higher frequency resolution is used for estimating power density spectrum compared to the resolution of 32 subband analysis filter. The SMR is calculates as (see Appendix):

1. **Calculating power density spectrum** using block of $N$-point FFT ($N=512$) and $w(n)$ is a window function (Hanning window is usually assumed):

$$X(k) = 10\log_{10}\left|\frac{1}{N}\sum_{n=0}^{N-1} w(n)x(n)e^{-j2\pi nk/N}\right|^2 \ [dB]$$

   The FFT values in X are then scalled such that the maximum corresponds to 96dB. The window $w(n)$ is later shifted for 384 samples in order to process the next set of samples.

2. **Calculating the SPL** in each subband $i$=1,...,32:

$$\begin{aligned} SPL(i) &= max\{\overline{X}(i),\ 20\log_{10}(SCF(i)*32768) - 10\}\ [dB], \\ where\ \ \overline{X}(i) &= max\{X((i-1)\cdot 8+1), ..., X((i-1)\cdot 8+8)\},\ i = 1, ..., 32 \end{aligned}$$

   where scaling factor $SCF_{\max}(i)$ is selected from the predefined lookup table (specified by the standard) on the basis of the absolute value of the largest among 12 samples in the subband. Note that the SPL is calculated as the larger value between the maximum amplitude FFT spectral line and lowest level determined by the maximal scaling factor in the $i$-th subband. Since the scaling factors defined by the MPEG-1 layer 1 standard range from very small number up to value 2, the multiplication by 32768 ($2^{15}$) is used for normalization of scaling factors, so that the largest value after normalization corresponds to 96 dB.

3. **Determining the absolute threshold** which is defined by the MPEG standard for different sampling rates.
4. **Calculating tonal and non-total masking components**, determining the relevant maskers and calculating the individual thresholds.
5. **Calculating global masking threshold** as a sum of all contributions from masking components.

6. **Determining the minimum masking threshold** in every subband $T_{\min}(i)$ as a minimum of the global masking threshold
7. **Calculating *SMR*(*i*)** per subband as: $SMR(i) = SPL(i) - T_{\min}(i)$.

### Bit Allocation

The MPEG audio compression is based on the principle of quantization, where the quantized values are not the audio samples but so-called signals taken from the frequency domain representation. Generally, the desired bit rate and consequently the compression ratio are known to the encoder, and thus the adaptive (dynamic) bit allocation is applied to quantized signals until the desired rate is achieved. In other words, the MPEG algorithm uses the known bit rate and the frequency spectrum of the most recent audio samples to quantize the signals in a way that allows inaudible quantization noise (quantization noise should be below the masking threshold).

The bit allocation process is performed as an iterative procedure used to allocate bits for each subband. We saw that the *SMR*($i$) is the result of the psychoacoustic model, while the *SNR*($i$) is defined by a Table 2.3 [ISO92], where every number of bits has specified a corresponding SNR. The new bits are allocated as long as the mask-to-noise ratio (MNR) is less than zero in dB.

If we assume that $R$ is a required bit rate (in Kb/s), $f_s$ is a signal sampling rate while the number of samples within the frame is 32×12, then the available number of bits per frame is calculated as:

$$B_{avail} = R \times (32 \times 12) \times \frac{1}{f_s}. \tag{2.40}$$

**Table 2.3** Amplitude levels in different frequency bands

| Bit allocation | Code | Number of levels | SNR |
|---|---|---|---|
| 0 | 0000 | 0 | 0.00 |
| 2 | 0001 | 3 | 7.00 |
| 3 | 0010 | 7 | 16.00 |
| 4 | 0011 | 15 | 25.28 |
| 5 | 0100 | 31 | 31.59 |
| 6 | 0101 | 63 | 37.75 |
| 7 | 0110 | 127 | 43.84 |
| 8 | 0111 | 255 | 49.89 |
| 9 | 1000 | 511 | 55.93 |
| 10 | 1001 | 1023 | 61.96 |
| 11 | 1010 | 2047 | 67.98 |
| 12 | 1011 | 4095 | 74.01 |
| 13 | 1100 | 8191 | 80.03 |
| 14 | 1101 | 16383 | 86.05 |
| 15 | 1110 | 32767 | 92.01 |

*Note*: Code 1111 is not used

The total number of bits that are required for each frame is:

$$B_{total} = B_{header} + \sum_{i=0}^{31} (12 \times B_{data}(i) + B_{alok} + B_{\exp}(i)), \tag{2.41}$$

where $B_{header} = 32$ is the number of bits reserved for the frame header, $B_{data}(i)$ is the number of bits per sample in the $i$-th subband, with allowed range of bits: $B_{data}(i) \in \{0,2,3,4, \ldots, 15\}$. $B_{alok}$ denotes the allocation bits, and this number is fixed and equal to 4 bits. Recall that $B_{data}(i)$ is not allowed to take value 1 or values >15. The number of bits for exponent (used to encode the scaling factor) for the $i$-th subband can be either $B_{exp} = 6$ if $B_{data} > 0$ or $B_{exp} = 0$ if $B_{data} = 0$. Thus, the scaling factor is not coded for the subband whose data are not coded.

According to the psychoacoustic model, the quantization noise will not be perceptible as long as the MNR is greater than zero in dB. It means that the MNR need to be positive (higher than 0) on a dB scale. The MNR is defined as follows:

$$MNR(i) = SNR_q(B_{data}(i)) - SMR(i), \tag{2.42}$$

where $SNR_q(B)$ is the signal power to quantization noise power ratio for a $B$-bit quantizer. The values of $SNR_q$ are given in the Table 2.3.

The procedure for dynamic bit allocation can be briefly summarized as follows:

1. ***Input data***:
   (a) Set initial value for $B_{data}(i) = 0$ for each subband $i$ within the frame of 384 samples
   (b) Calculate $SMR(i)$ for each subband $i$=1,...,32
2. ***Calculate*** $MNR(i) = SNR_q(i)\text{-}SMR(i)$, $i$=1,...,32
3. ***Find*** $i_m = \arg\min_i \{MNR(i)\}, i = 1, \ldots, 32 \text{ and } B_{data}(i) < 15$
4. ***If*** $B_{total} \leq B_{aval}$

      ***If*** $B_{data}(i_m) = 0$,

         ***Set*** $B_{data}(i_m) = 2$ and $B_{exp}(i_m) = 6$

      ***else***

         ***Set*** $B_{data}(i_m) = B_{data}(i_m) + 1$

      ***end***

   ***end***
5. ***Go to*** Step 2

#### 2.8.3.2 MPEG Layer II

The MPEG Layer II is an improved version of the MPEG Layer I algorithm which almost completely utilizes the MUSICAM algorithm. The scheme with 32 blocks is also used for this compression. However, the total frequency range is divided into three parts: low, medium and high. Given the different sensitivities of the auditory system to these three parts, the number of bits used for encoding will be different in each part. Namely, the low-frequency range uses up to 15 bits, the mid-frequency range uses up to 7 bits, and the high-frequency range uses up to 3 bits. In addition, 4 bits are needed for bit allocation in the low-frequency band, while the middle and high-frequency ranges use 3 and 2 allocation bits, respectively. The input blocks contain 1152 samples, and since they split into three new blocks, each of them will contain 384 samples. In such a way, we get a structure that corresponds to the previously described code scheme for the MPEG Layer I. The masking procedure is done by using the FFT algorithm with 1024 samples. The compression ratio of the MPEG Layer II is approximately equal to six to eight times (Fig. 2.36).

#### 2.8.3.3 MPEG Layer III (MP3)

Unlike the prior two compression algorithms, MP3 is based on ASPEC and MUSICAM. Namely, compression is carried out using samples in the transform domain, and the structure of the blocks resembles the previous algorithms. MP3 uses the blocks containing 1152 samples divided into 32 subbands. The transformation from the time to the frequency domain is performed using the modified discrete cosine transform (MDCT). It is important to note that the MP3 algorithm does not use the fixed-length windows, but they are either 24 ms or 8 ms long. The windows of short duration are used when there are sudden signal changes, since shorter windows ensure a good time resolution. Wider windows are used for slowly varying signals. Figure 2.37 shows window forms used in the MP3 compression.

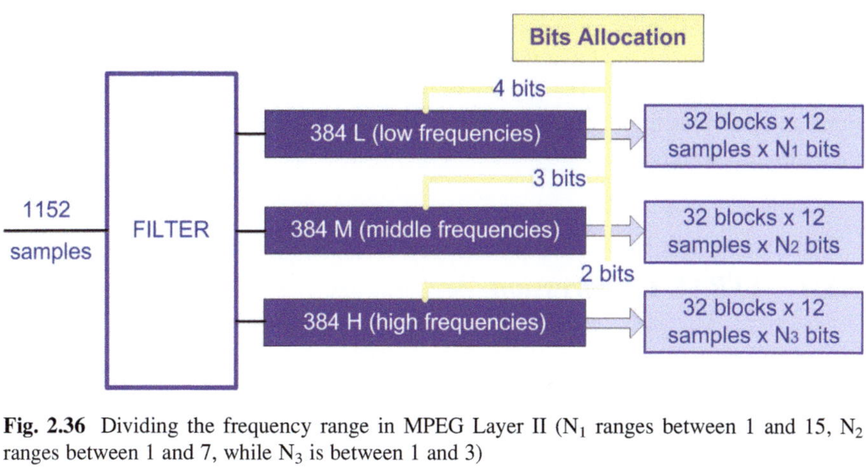


**Fig. 2.36** Dividing the frequency range in MPEG Layer II ($N_1$ ranges between 1 and 15, $N_2$ ranges between 1 and 7, while $N_3$ is between 1 and 3)

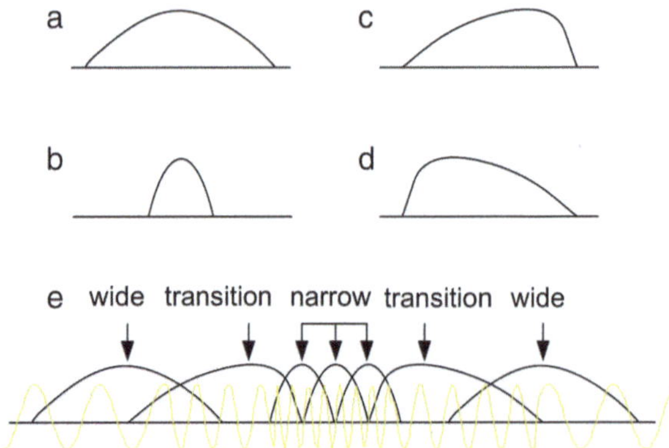


**Fig. 2.37** Windows used by MP3 algorithm: (**a**) Wide window, (**b**) Narrow window, (**c**) and (**d**) Transition windows, (**e**) Shifting from wide to narrow window and vice versa

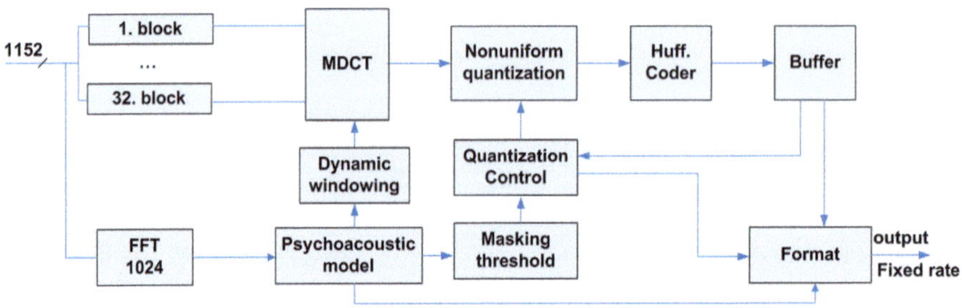


**Fig. 2.38** A block diagram of the system for MP3 compression

The algorithm based on variable window widths provides a better quality of the compressed signals. However, it should be noted that a choice of an appropriate window function depends on a more complex psychoacoustic model than the models used in the MPEG Layer I and Layer II algorithms. Namely, the complexity of the psychoacoustic model is increased due to the use of the MDCT. A block diagram of the MP3 compression is shown in Fig. 2.38.

It should be mentioned that the MP3 coding also uses blocks for entropy coding based on Huffman code. The MP3 was developed primarily for Internet applications and provides high compression ratio (about 12 times) with a good quality of the reproduced signal.

### *2.8.4 ATRAC Compression*

The ATRAC compression algorithm is used for mini-discs in order to store the same amount of audio signals and with same quality as in the case of the CD, but on the significantly smaller disc area. ATRAC stands for Adaptive Transform Acoustic

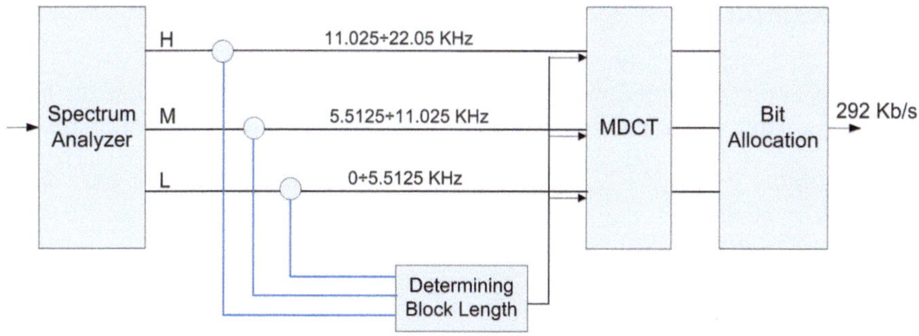


**Fig. 2.39** A block scheme of ATRAC compression system

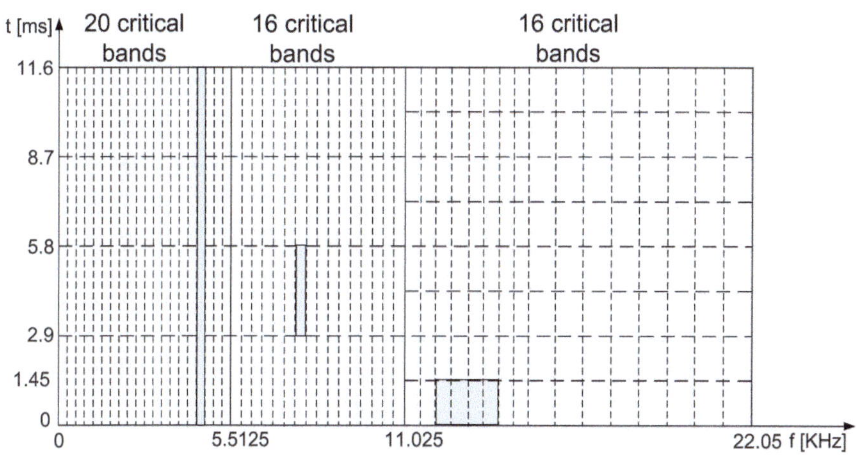


**Fig. 2.40** A division of the time-frequency plane in the ATRAC algorithm

Coder. Using filters, the range of the input signal is divided into three subband (0–5.5125 KHz, 5.5125–11.025 KHz and 11.025–22.05 KHz). Each subband is passed to the MDCT processors. The first subband has 20 blocks, while the other two contain 16 blocks each. Such a resolution corresponds better to the sensitivity of the auditory system. The time slot for the analysis can vary from 1.45 ms to 11.6 ms by using the increments of 1.45 ms. In this way, the time-frequency plane of the signal is divided into a number of different areas which enable successful compression, taking into account the difference in sensitivity of the auditory system in different parts of the time-frequency plane. The ATRAC compression reduces the bit rate from 1.4 Mb/s to 292 Kb/s. A block scheme of ATRAC compression system is shown in Fig. 2.39. Figure 2.40 demonstrates the division of the time-frequency plane as required by the ATRAC compression algorithm.

## 2.9 Examples

2.1. The Sound Pressure Level for a signal is SPL = 20 dB. If the reference level of pressure is $Po = 20$ μPa, calculate the value of the pressure $P$ in Pascals.

Solution:
SPL = 20 dB
$P_o = 20$ μPa

$\text{SPL} = 20 \cdot \log_{10} (P/P_o)$
$20 = 20 \cdot \log_{10}(P/P_o) => \log(P/P_o) = 1 => P/P_o = 10$
$P = P_o \cdot 10 = 20 \cdot 10^{-6} \cdot 10\ \text{Pa} = 2 \cdot 10^{-4}\text{Pa} = 0.2\ \text{mPa}$

2.2. If the signal to quantization noise is $S/N = 61.76$ dB, determine the number of bits used for signal representation?

Solution:
$S/N = 1.76 + 6 \cdot n$ $n$- number of bits used to represent signal
$6 \cdot n = 60 =>$ $n = 10$ bits

2.3. A 13-bit signal is obtained at the output of the floating-point converter, with the signal to noise ratio $S/N = 61.76$ dB. Determine the number of bits used to represent mantissa, and the number of bits used for exponent?

Solution:
$S/N = 61.76$ dB
$6 \cdot n = 60 => n = 10$ bits for mantissa
$m = 13–10 = 3$ bits for exponent

2.4. The communication channel consists of three sections. The average level of transmission power is 400 mW. The first section introduces 16 dB attenuation compared to the average power level, the second introduces 20 dB gain compared to the first section, while the third introduces attenuation of 10 dB compared to the second section. Determine the signal power at the output of each channel section.

Solution:
$P_0 = 400$ mW

First section: $16\,dB = 10\log\left(\frac{P_0}{P_1}\right) = 10\log\left(\frac{400}{P_1}\right) \Rightarrow \text{P}_1 = 10.0475\ \text{mW}$

Second section:

$$20\,dB = 10\log\left(\frac{P_2}{P_1}\right) = 10\log\left(\frac{P_2}{10.0475}\right) \Rightarrow \text{P}_2 = 1004.75\,\text{mW}$$

Third section:

$$10\,dB = 10\log\left(\frac{P_2}{P_3}\right) = 10\log\left(\frac{1004.75}{P_3}\right) \Rightarrow \text{P}_3 = 100.475\ \text{mW}$$

**Fig. 2.41** Filter function

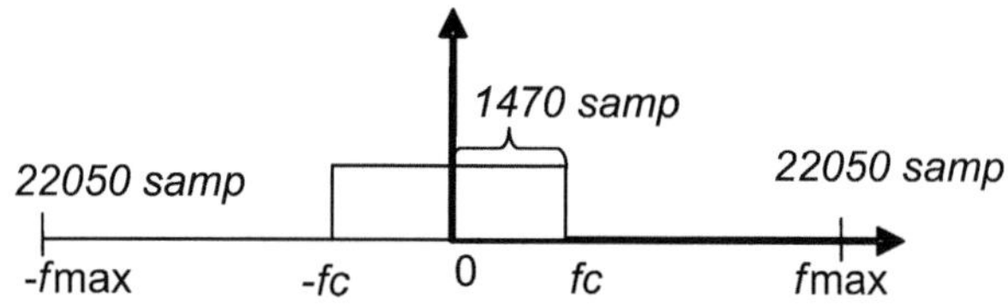


2.5. Load the signal *speech_dft.wav* in Matlab. Make a new signal *y* that will contain 2 seconds of the original speech signal, and listen to the resulting signal.

Solution:

```
[y,fs]=wavread('speech_dft.wav');
length(y)
      ans=110033
fs=22050
y=y(1:2*fs);
soundsc(y,fs)
```

2.6. For a signal obtained in the previous example, design a low-pass filter in Matlab, with the cutoff frequency $f_c = 735$ Hz.

Solution:
The sampling frequency of the considered signal is $f_s = 22050$ Hz. The total length of the signal is 44100 samples. Hence, the Fourier transform will produce 44100 samples in the frequency domain, from which 22050 samples are related to positive and 22050 to negative frequencies (Fig. 2.41).

$f_{max} = f_s/2 = 11025$ Hz.

In the frequency range between zero and the cutoff frequency $f_c = 735$ Hz, we have: $22050 \cdot (735/11025) = 1470$ samples

The filtering operation can be done by using Matlab as follows:

```
F=fftshift(fft(y));    % Fourier transform of the signal
figure(1), plot((abs(F)))
% Filter transfer function
H=[zeros(1,20580) ones(1,2940) zeros(1,20580)];
G=F.*H'; % Signal filtering in the frequency domain
figure(2), plot(abs(G));
% The filtered signal is obtained by applying the inverse Fourier
  transform
  yg=ifft(fftshift(G));
  soundsc(real(yg),fs)
```

The Fourier transform of the original and filtered signal are shown in Fig. 2.42.

2.7. For the speech signal used in previous examples, design the band-pass filter with the band frequencies defines by 1102.5 Hz and 2205 Hz.

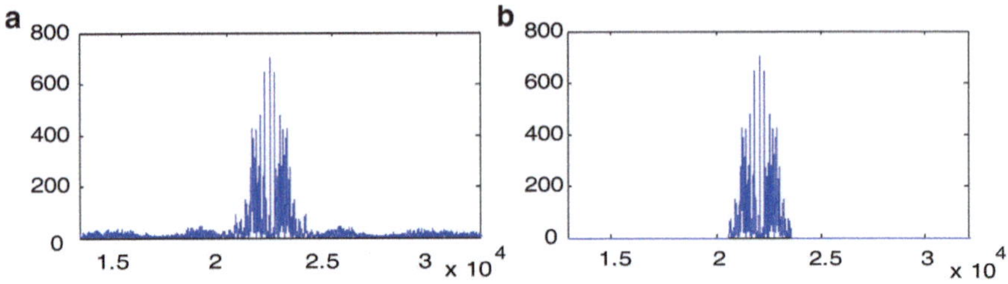


**Fig. 2.42** (**a**) The Fourier transform of the original signal, (**b**) the Fourier transform of the filtered signal

**Fig. 2.43** Parameters of band-pass filter function

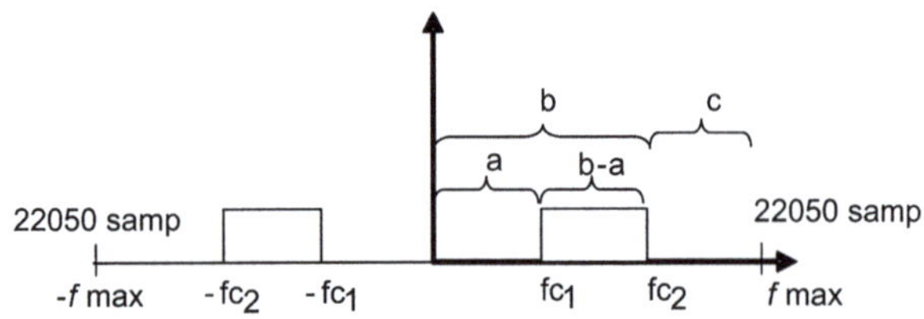


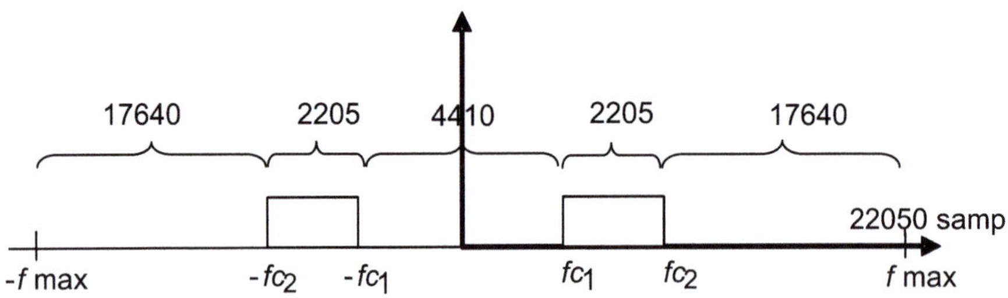


**Fig. 2.44** Filter transfer function

Solution:

The cutoff frequencies of the band-pass filter are: $fc_1 = 1102.5$ Hz and $fc_2 = 2205$ Hz, while the maximal signal frequency is $f_{\max} = 11025$ Hz (Fig. 2.43).

Hence, we made the proportions as:

$fc_1 : \mathrm{a} = f_{\max} : 22050 \Rightarrow \mathrm{a} = 2205$

$fc_2 : \mathrm{b} = f_{\max} : 22050 \Rightarrow \mathrm{b} = 4410$

The number of samples passing unchanged through the filter is

b-a = 2205.

Note that the length between the cutoff frequency $fc_2$ and the maximal signal frequency $f_{\max}$ is:

c = 22050-b = 22050–4410 = 17640 samples.

The filter transfer function in Matlab is given by (Fig. 2.44):

```
>>H=[zeros(1,17640) ones(1,2205) zeros(1,4410) ones(1,2205)
zeros(1,17640)];
```

Finally, we can perform signal filtering in the frequency domain by using the filter transfer function *H*:

```
>>G=F.* H';
```

The filtered signal is obtained by applying the inverse Fourier transform to the filtered signal spectrum:

```
>>yg=ifft(fftshift(G));
>>soundsc(real(yg),fs)
```

2.8. By using the speech signal "*speech_dft.wav*" in Matlab, realize the echo by using a 0.2 s delay, while the echo amplitude is decreased for 50%. Listen to the achieved echo effect.

Solution:

Echo effect can be realized in a way that we make two versions of the original signal: one is obtained by adding a zero sequence at the beginning of the original signal, while the other is obtained by adding zeros at the end of the considered signal. The signal with echo effect is obtained as a sum of two modified signal versions.

The length of the zero sequence is defined by the delay which is equal to 0.2 s. Since the sampling frequency for the observed speech signal is 22050 Hz, the delay 0.2 s corresponds to 4410 samples. The echo realization in Matlab can be done as follows:

```
[y,fs]=wavread('speech_dft.wav');
y1=[zeros(1,4410) y'];
y2=[y' zeros(1,4410)];
echo=0.5*y1+y2;
soundsc(echo,fs)
```

2.9. By using the linear prediction coefficients given by vector *a*, and the set of 20 signal samples (vector *f*), determine the 14th signal sample and the prediction error.

$$a = [-1.7321\ \ 0.9472\ \ -0.3083\ \ 0.0748\ \ -0.0812\ \ 0.1260\ \ 0.2962\ \ldots$$
$$\ldots -0.3123\ \ 0.0005\ 0.0216\ \ -0.1595\ \ 0.2126\ \ -0.0496]$$

$$f = [\text{-}2696\ \ \text{-}2558\ \ \text{-}2096\ \ \text{-}1749\ \ \text{-}1865\ \ \text{-}2563\ \ \text{-}2280\ \ \text{-}1054\ \ \text{-}635\ \ \text{-}41\ \ldots$$
$$\ldots\ 1695\ \ 3645\ \ 5150\ \ 6188\ \ 5930\ \ 4730\ \ 3704\ \ 3039\ \ 2265\ \ 1159]$$

Solution:

Based on the linear prediction analysis, the estimated value of the 14th sample is calculated according to:

$$\hat{f}(n) = -\sum_{i=1}^{L} a_i f(n-i),$$

For $n = 14$, $L = 13$, we have: $\hat{f}(14) = -\sum_{i=1}^{13} a_i f(14 - i) = 6064.$

The prediction error is: $e(14) = f(14) - \hat{f}(14) = 6188 - 6064 = 124.$

2.10. For a given set $f$ of ten signal samples and the corresponding prediction errors (given by vector $e$) calculated as in the previous example, determine the value of prediction gain.

$$e = 10^3 \cdot [-0.0095 \quad -0.7917 \quad -1.1271 \quad -0.3273 \quad 0.0907 \quad -0.1379 \quad -0.1106 \quad \ldots \quad 0.1444 \quad -0.1762 \quad 0.5057]$$

$$f = [6188 \quad 5930 \quad 4730 \quad 3704 \quad 3039 \quad 2265 \quad 1159 \quad 168 \quad -434 \quad 120]$$

Solution:
The prediction gain for the observed set of samples given in $f$ can be calculated as:

$$PG = 10\log_{10}\left(\frac{\sum_{k=1}^{10} f^2(k)}{\sum_{k=1}^{10} e^2(k)}\right) = 17.27dB.$$

2.11. For a set of ten samples (given below), calculate the value of energy-entropy feature EEF.

$$f = [-43 \quad 7 \quad -97 \quad -3 - 163 \quad 182 \quad 143 \quad 225 - 242 - 262].$$

Solution:
Firstly, we calculate the energy $E$ of the frame:

$$E = \sum_{k=1}^{10} f_k^{\ 2} = 269291.$$

The Fourier transform coefficients of the signal $f$ are:

$$F(\omega) = [-2.5300 \quad -5.8848 + 1.0069i \quad 1.4868 - 7.6610i \quad 3.0148 - 0.1360i \quad 3.2532 + 0.1677i \quad -5.5100 \quad 3.2532 - 0.1677i \quad 3.0148 + 0.1360i \quad 1.4868 + 7.6610i \quad -5.8848 - 1.0069i].$$

The probability density function is calculated as:

$$p = F(\omega) / \sum_{k=1}^{10} F(\omega),$$

and the corresponding vector $p$ is obtained:

$$p = [0.0526\ \ 0.1240\ \ 0.1621\ \ 0.0627\ \ 0.0677\ \ 0.1145\ \ 0.0677\ \ 0.0627\ \ 0.1621\ \ 0.1240].$$

The entropy of the observed frame is calculated as:

$$H = \sum_{k=1}^{10} p_k \log p_k = -0.9651.$$

Finally, the energy-entropy feature can be calculated as:

$$EEF = (1 + |E \cdot H|)^{1/2} = 509.8067.$$

2.12. Write the Matlab code for the word endpoints detector based on the energy-entropy feature.

Solution:

```
%% load test speech signal in vector f
k=1;
  for i=1:64: round(length(f)/64)*64
    E(k)=sum(f(i:i+63).^2);
    X=fft(f(i:i+63));
    p=(abs(X)./sum(abs(X)));
    H(k)=sum(p.*log10(p));
    EEF(k)=sqrt(1+abs(E(k).*H(k)));
    k=k+1;
  end
  for i=0:length(EEF)-1
  s(1+i*64:i*64+64)=EEF(i+1);
  end
figure(1),plot(real(s)./max(real(s)))
```

2.13. In this example a short Matlab code for the time-frequency based eigenvalue decomposition is provided. We assume that the S-method is calculated in advance (Chap. 1).

Solution:

```
%Sm is the S-method matrix
%% Calculation of the auto-correlation matrix
R=zeros(N+1);
  for n=1:N+1;
    v=N+n;
    k=n;
    for m=1:N+1;
      R(n,m)=Sm(v,k);
      v=v-1;k=k+1;
    end
  end
% Eigenvalues matrix D and eigenvectors V
[V,D]=eigs(R,Nc,'lm',opt); %columns of V are eigenvectors
D=abs(diag(D));
```

2.14. For the given subband samples, determine the number of bits that will be transmitted, if we know that the samples below 13 dB are masked by the neighboring subband (as shown in Fig. 2.45). Assume that the signal samples are originally represented by 8 bits.

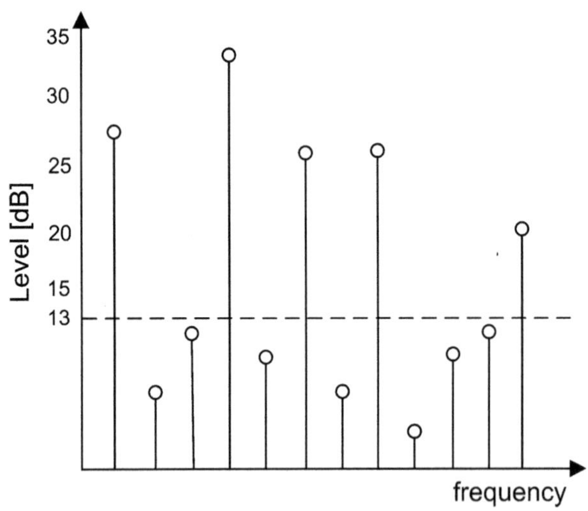


**Fig. 2.45** The subband samples and masking level

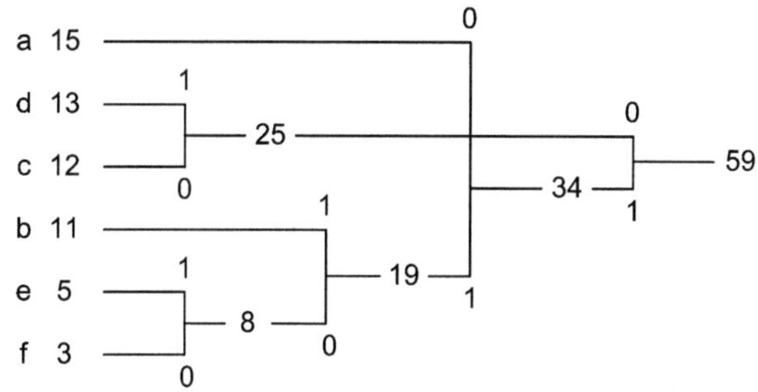


**Fig. 2.46** An example of Huffman coding

Solution:

Due to the audio masking effects, only the samples that are above the masking level will be transmitted. Due to the masking of tones below 13 dB, the quantization noise of 12 dB is masked as well. Therefore, we use 2 bits less to represent the samples, and the total number of transmitted bits is:

5 samples · (8-2) b = 30 b

2.15. Perform the Huffman coding algorithm, for the symbols whose numbers of occurrences within a certain sequence are given below.

| Number of occurrences: | a → 15 |
|---|---|
| | b → 11 |
| | c → 12 |
| | d → 13 |
| | e → 5 |
| | f → 3 |

Solution:

In order to perform Huffman coding, the numbers of occurrences for symbols a, b, c, d, e, and f are firstly sorted in decreasing order. Then the coding is performed according to the scheme in Fig. 2.46.

Thus the symbols are coded as follows:

a → 10 d → 01 c → 00 b →111 e → 1101 f → 1100

2.16. Consider the sequence *this_image_is_damaged.* Code the sequence by using the LZ-77 code. Determine the number of bits that can be saved by applying this coding algorithm. Assume that the pointers are represented by 12 bits.

Solution:

The sequences can be coded as follows:

```
this_image_is_damaged
this_image_(9,3)da(10,4)d
```

(9,3) → 00001001 0011
(10,4) → 00001010 0100
Before LZ-77: 21·8 b = 168 b (21 characters including spaces)
After LZ-77: 14·8 b + 24 b = 136 b
168–136 = 32 (19%)

2.17. Perform the LZW coding of the sequence: *strange strategic statistics.*
Solution:

```
strange_strategic_statistics
256 → st < st >range_strategic_statistics
257 → tr s< tr >ange_strategic_statistics
258 → ra st< ra >nge_strategic_statistics
259 → an str< an >ge_strategic_statistics
260 → ng stra< ng >e_strategic_statistics
261 → ge stran< ge >_strategic_statistics
262 → e_ strang< e_ > strategic_statistics
263 → _s strange <_s > trategic_statistics
264 → str strange _< str >ategic_statistics
265 → rat strange_st< rat > egic_statistics
266 → te  strange_stra< te > gic_statistics
267 → eg strange_strat< eg > ic_statistics
268 → gi  strange_strate< gi > c_statistics
269 → ic  strange_strateg< ic > _statistics
270 → c_ strange_strategi< c_ > statistics
271 → _st strange_strategic< _st > atistics
272 → ta  strange_strategic_s< ta > tistics
273 → at  strange_strategic_st< at > istics
274 → ti  strange_strategic_sta< ti > stics
275 → is  strange_strategic_stat< is > tics
276 → sti  strange_strategic_stati< sti > cs
277 → ics strange_strategic_statist< ics >
```

Coded sequence:
strange_ < 256 > < 258 > tegic < 263 > tati < 256 > <269 > s

2.18. Determine the bit rate (in Kb/s) for the following cases:

(a) Speech signal with the maximal frequency 10 KHz, while the samples are coded by using 12 b/sample.

(b) Musical signal with the maximal frequency 20 KHz, coded using 16 b/sample. How much memory is required to store 10 min of this stereo music?

The speech and musical signals are sampled according to the sampling theorem.

Solution:
(a) Speech signal:
$f_{max} = 10$ KHz $=> f_s \geq 2{\cdot}f_{max} = 20$ KHz . Let us consider $f_s = 20$ KHz.
Therefore, we have:
(20000 samples/s) · (12 b/sample) = 240 Kb/s

(b) Musical signal:
$f_{max} = 20$ KHz $=> f_s \geq 2{\cdot}f_{max} = 40$ KHz

mono signal: (40000 samples/s) · (16 b/sample) = 640 Kb/s
stereo signal: 2·640 Kb/s = 1280 Kb/s

Memory requirements:
1280 Kb/s · 10 min = 1280 Kb/s · 600s = 768000 Kb
768000 Kb / 8 = 93750 KB

2.19. Consider a stereo signal, sampled at 44.1 KHz, and coded by using 16 b/sample. Calculate the memory requirements for storing 1 min of this audio format? What time is required to download 1 min of audio content from the Internet if the connection speed is 50 Kb/s?

Solution:
The sampling rate for the considered signal is 44100 samples per second. This number is multiplied by 2 due to stereo format, so that we have 88200 samples per second. Since each sample is coded with 16 bits, the total number of bits used to represent 1 s of this audio format is:

$$88200{\cdot}16 = 1411200 \text{ b/s}$$

Furthermore, 60 s of audio contains:
1411200b/s·60 s = 84672000 b, or equivalently,

$$\frac{84672000}{8} = 10584000 \text{ B} = 10336 \text{ KB} = 10 \text{ MB}.$$

The time required for a download of 1 min long audio content is:

$$\frac{84672000 \text{ b}}{50000 \ \frac{\text{b}}{\text{s}}} = 1693.44 \text{ s} = 28.22\text{min}.$$

2.20. If the sampling frequency of a signal is $f_s = 32000$ Hz, determine the frequency bandwidth of each subband in the case of the MPEG Layer I compression algorithm.

Solution:

$f_s = 32$ KHz
$f_{max} = f_s/2 = 16$ KHz

In the MPEG Layer I compression algorithm the total frequency bandwidth is divided into 32 subbands. Hence, each of the subbands has the following width:

16000 / 32 = 500 Hz.

2.21. Calculate the bit rate of the compressed 16-bit stereo signal if the sampling frequency is:

(a) 32 KHz, (b) 44.1 KHz, (c) 48 KHz.

Assume that the MPEG Layer I compression factor is 1:4.

Solution:

(a) $$\frac{16\text{b}\cdot 2\cdot 32000\frac{1}{\text{s}}}{4} = 256000\text{b/s} = 256\text{Kb/s}.$$

(b) $$\frac{16\text{b}\cdot 2\cdot 44100\frac{1}{\text{s}}}{4} = 352800\text{b/s} = 352.8\text{Kb/s}.$$

(c) $$\frac{16\text{b}\cdot 2\cdot 48000\frac{1}{\text{s}}}{4} = 384000\text{b/s} = 384\text{Kb/s}.$$

2.22. Consider 1152 signal samples and show that MPEG Layer II compression algorithm provides considerable savings compared to the MPEG Layer I algorithm, even in the case when the samples are coded with the maximal number of bits in each subband.

Solution:

*MPEG Layer I algorithm:*

1152 samples = 3 block × 384 samples
384 samples = 32 block × 12 samples
Four allocation bits are assigned to each block
Maximal number of bits that is available for coding of samples is 15
6 bits that corresponds to scale factor is assigned to each block
3·32·4 b + 3·32·6 b + 3·32·12·15 b = 18240 b

*MPEG Layer II algorithm:*

The signal with 1152 samples is divided into three parts: 384 samples belonging to low frequencies, 384 middle frequency samples and 384 samples corresponding to high frequencies.

We assign 4 allocation bits, for each low-frequency block and consequently we have 15 b/sample at most;

Three allocation bits are assigned to each of 32 middle frequency blocks, meaning that at most 7 b/samples are available;

Finally, high-frequency blocks get 2 allocation bits each, and this means at most 3 b/sample;

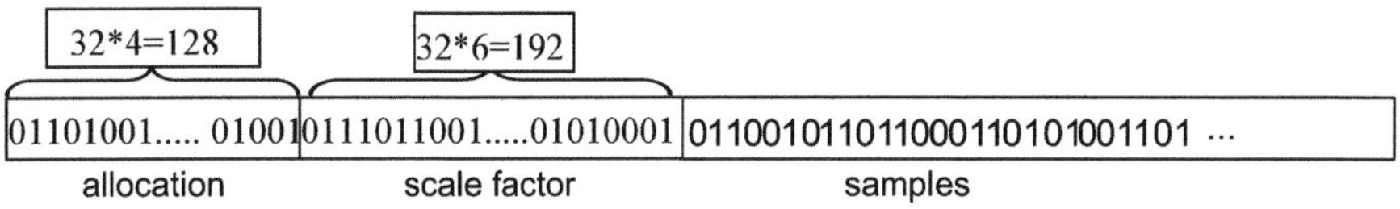


**Fig. 2.47** An illustration of MPEG Layer I sequence part

The scale factor requires 6 bits per block.

Therefore, the total number of bits that is required for coding the set of 1152 samples is:

$$32{\cdot}4 + 32{\cdot}3 + 32{\cdot}2 + 3{\cdot}32{\cdot}6 + 32{\cdot}12{\cdot}15 + 32{\cdot}12{\cdot}7 + 32{\cdot}12{\cdot}3 = 10464 \text{ b}$$

The savings can be calculated as a difference between the number of required bits: 18240–10464 = 7776 b.

2.23. Consider a simplified part of the sequence obtained by using the MPEG Layer I algorithm and determine the value of the third sample in the first block (from 32 blocks)? (Fig. 2.47)

Solution:
First 4 allocations bits—0110—correspond to the first block.

The sequence 0110 determines the samples within the considered block are coded by using 6 + 1 = 7 b/sample. Hence, we have:

I sample: 0110010
II sample: 1101100
III sample: 0110101
The value of the third signal sample is 53.

The scale factor is defined by the sequence 011101, i.e., the scaling factor is 29·2 dB = 58 dB.

2.24. The signal with maximal frequency 24 KHz is coded by using the MPEG Layer II algorithm and the achieved bit rate is 192 Kb/s. Calculate the number of bits required for representation of the constant-length block used as a coding unit.

Solution:
$f_{max} = 24$ KHz $\Rightarrow f_s = 48$ KHz, or in other words 1 s of the signal consists of 48000 samples.

The total number of bits for the coding block within the MPEG Layer II algorithm is:

$$n = \frac{1152 \text{ samples} \cdot 192000 \ \text{b/s}}{48000 \ \text{samples/s}} = 4608 \ \text{b}.$$

2.25 Consider a MPEG-1 layer 1 audio frame of 384 samples divided into 32 subbands. Given are the $SMR(i)$ for each subband $i = 1, \ldots, 32$, the signal sampling frequency 44.1 KHz and the required bit rate $R = 320$

**Table 2.4** *SMR*($i$) for subband $i = 1, \ldots, 32$

| **Subband** | **1** | **2** | **3** | **4** | **5** | **6** | **7** | **8** | **9** | **10** | **11** |
|---|---|---|---|---|---|---|---|---|---|---|---|
| SMR/dB | 24 | 18 | 14 | 14 | 14 | 18 | 18 | 18 | 20 | 18 | 14 |
| **Subband** | **12** | **13** | **14** | **15** | **16** | **17** | **18** | **19** | **20** | **21** | **22** |
| SMR/dB | 6 | 6 | 20 | 20 | 40 | 40 | 20 | 24 | 24 | 40 | 40 |
| **Subband** | **23** | **24** | **25** | **26** | **27** | **28** | **29** | **30** | **31** | **32** | |
| SMR/dB | 40 | 60 | 60 | 60 | 60 | 60 | 64 | 85 | 85 | 85 | |

Kb/s. Determine the number of bits that can be allocated per each subband taking into consideration the required bit rate and $MNR(i) = SNR_q(i)\text{-}SMR(i)$. The values of $SNR_q$ are given earlier in Table 2.3 for different number of allocated bits $B_{data}$ (Table 2.4).
Solution:

The total number of available bits for frame can be calculated as:

$$B_{avail} = R \times 1000 \times (32 \times 12) \times \frac{1}{f_s} = \\ = 320 \times 1000 \times (32 \times 12) \times \frac{1}{f_s} = 2786 \text{ b.}$$

Also observe that each subband has $SMR(i) > 0$ dB which means that each subband will require 6 bits for the exponent and 4 allocation bits, which results in:

$$32(\text{header}) + 32 \times 4(\text{allocation bits}) + 32 \times 6(\text{exponent}) = 352 \text{ b.}$$

The total number of bits that will be reserved for the frame is calculated as:

$$B_{total} = 352 \text{ b} + \sum_{i=0}^{31} (12 \times B_{data}(i)),$$

where $B_{data}(i)$ will be increased though the iteration until $MNR(i) > 0$ for all $i$, or until we reach the available number of bits $B_{avail}$.

In the first iteration we allocated $B_{data}(i) = 2$ b bits per sample in each subband, which is equivalent to $SNR_q = 7$ dB. Then we calculate $MNR(i)$ for each subband $i$ as given in Table 2.5.

Now the total number of used bits is: $B_{total} = 352 \text{ b} + \sum_{i=0}^{31} (12 \times 2) = 1120.$

As long as there are still bits available, we increase the number of $B_{data}(i)$ as follows: $B_{data}(i) = B_{data}(i) + 1$, starting from the subbands $i$ with the lowest $MNR(i)$. As we increase $B_{data}(i)$ for a specific subband $i$, the $SNR_q(B_{data}(i))$ increase according to the Table 2.3, and the $MNR(i)$ needs to be recalculated through iterations.

**Table 2.5** $MNR(i)$ for $i = 1, \ldots, 32$ and $B_{data}(i) = 2$ ($SNRq = 7$ dB)

| **Subband** | **1** | **2** | **3** | **4** | **5** | **6** | **7** | **8** | **9** | **10** | **11** |
|---|---|---|---|---|---|---|---|---|---|---|---|
| MNR | −17 | −11 | −1 | −7 | −7 | −11 | −11 | −11 | −13 | −11 | −7 |
| **Subband** | **12** | **13** | **14** | **15** | **16** | **17** | **18** | **19** | **20** | **21** | **22** |
| MNR | 1 | 1 | −13 | −13 | −33 | −33 | −13 | −17 | −17 | −33 | −33 |
| **Subband** | **23** | **24** | **25** | **26** | **27** | **28** | **29** | **30** | **31** | **32** | |
| MNR | −33 | −53 | −53 | −53 | −53 | −53 | −57 | −78 | −78 | −78 | |

**Table 2.6** $B_{data}$ allocated for samples of different subbands

| **Subband** | **1** | **2** | **3** | **4** | **5** | **6** | **7** | **8** | **9** | **10** | **11** |
|---|---|---|---|---|---|---|---|---|---|---|---|
| $B_{data}$ | 4 | 4 | 3 | 3 | 3 | 4 | 4 | 4 | 4 | 4 | 3 |
| **Subband** | **12** | **13** | **14** | **15** | **16** | **17** | **18** | **19** | **20** | **21** | **22** |
| $B_{data}$ | 2 | 2 | 4 | 4 | 7 | 7 | 4 | 4 | 4 | 7 | 7 |
| **Subband** | **23** | **24** | **25** | **26** | **27** | **28** | **29** | **30** | **31** | **32** | |
| $B_{data}$ | 7 | 10 | 10 | 10 | 10 | 10 | 11 | 14 | 14 | 14 | |

**Table 2.7** $SNR_q(B_{data}(i))$ for different subbands

| **Subband** | **1** | **2** | **3** | **4** | **5** | **6** | **7** | **8** | **9** | **10** | **11** |
|---|---|---|---|---|---|---|---|---|---|---|---|
| $SNR_q$ | 25.28 | 25.28 | 16 | 16 | 16 | 25.28 | 25.28 | 25.28 | 25.28 | 25.28 | 16 |
| **Subband** | **12** | **13** | **14** | **15** | **16** | **17** | **18** | **19** | **20** | **21** | **22** |
| $SNR_q$ | 7 | 7 | 25.28 | 25.28 | 43.84 | 43.84 | 25.28 | 25.28 | 25.28 | 43.84 | 43.84 |
| **Subband** | **23** | **24** | **25** | **26** | **27** | **28** | **29** | **30** | **31** | **32** | |
| $SNR_q$ | 43.84 | 61.96 | 61.96 | 61.96 | 61.96 | 61.96 | 67.98 | 86.05 | 86.05 | 86.05 | |

**Table 2.8** $MNR(i)$ for $i = 1, \ldots, 32$ and $B_{data}(i)$ given in Table 2.6

| **Subband** | **1** | **2** | **3** | **4** | **5** | **6** | **7** | **8** | **9** | **10** | **11** |
|---|---|---|---|---|---|---|---|---|---|---|---|
| MNR | 1.28 | 7.28 | 2 | 2 | 2 | 7.28 | 7.28 | 7.28 | 5.28 | 7.28 | 2 |
| **Subband** | **12** | **13** | **14** | **15** | **16** | **17** | **18** | **19** | **20** | **21** | **22** |
| MNR | 1 | 1 | 5.28 | 5.28 | 3.84 | 3.84 | 5.28 | 1.28 | 1.28 | 3.84 | 3.84 |
| **Subband** | **23** | **24** | **25** | **26** | **27** | **28** | **29** | **30** | **31** | **32** | |
| MNR | 3.84 | 1.96 | 1.96 | 1.96 | 1.96 | 1.96 | 3.98 | 1.05 | 1.05 | 1.05 | |

After a certain number of iterations, $B_{data}(i)$ for 32 subbands are given in Table 2.6. The total number of used bits is: $B_{total} = 352\text{ b} + 12 \times 202 = 2776\text{ b}$, which is still below $B_{avail} = 2786$, but we cannot allocate any additional bit to 12 samples of certain subband.

For better readability, $SNR_q(B_{data}(i))$ are rewritten from Table 2.3 in Table 2.7 having in mind $B_{data}$ from the Table 2.6.

The corresponding $MNR(i)$ are given in Table 2.8. We may observe that $MNR(i) > 0$ dB for each subband $i$, which means that the quantization noise is below mask in each subband.

# Appendix

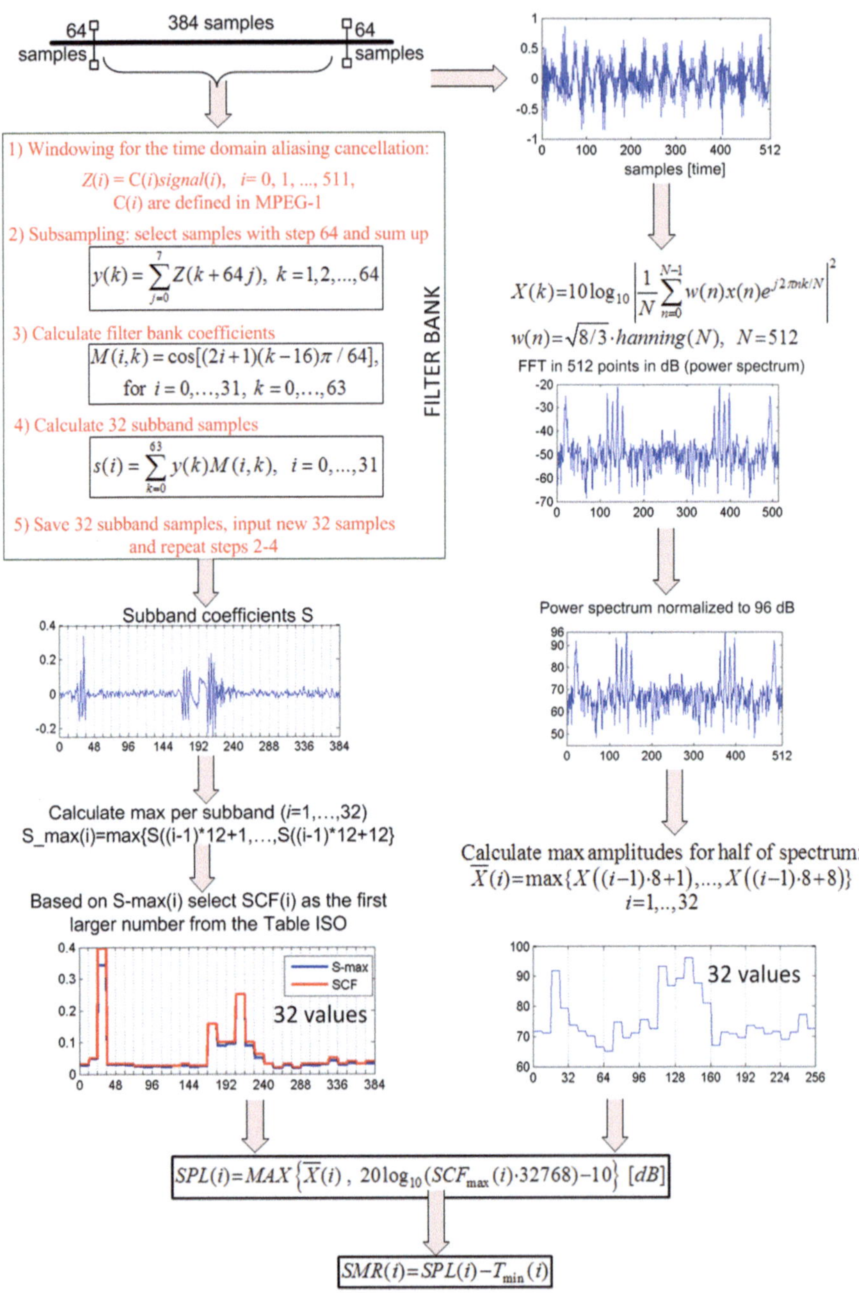


**Fig. 2.48** MPEG layer 1 - SMR calculation

## References

1. Bosi M, Goldberg RE (2003) Introduction to digital audio coding and standards. Springer, New York
2. Chu WC (2003) Speech coding algorithms. Wiley, New York
3. Fuhrt B (2008) Encyclopedia of multimedia, 2nd edn. Springer, New York
4. Gibson J, Berger T, Lookabaugh T, Baker R, Lindbergh D (1998) Digital compression for multimedia: principles & standards. Morgan Kaufmann, San Francisco, CA
5. Hankersson D, Greg AH, Peter DJ (1997) Introduction to information theory and data compression. CRC Press, Boca Raton
6. Hassanpour H, Mesbah M, Boashash B (2004) Time-frequency feature extraction of newborn EEG seizure using SVD-based techniques. EURASIP J Appl Signal Process 16:2544–2554
7. Hoeg W, Lauterbach T (2003) Digital audio broadcasting: principles and applications of Digital Radio. John Wiley and Sons, New York
8. Kaplan R (1997) Intelligent multimedia systems. Willey, New York
9. Kovačević B, Milosavljević M, Veinović M, Marković M (2000) Robustna digitalna obrada signala. Akademska misao, Beograd
10. Maes J, Vercammen M, Baert L (2002) Digital audio technology, 4th edn. Focal Press, Oxford, In association with Sony
11. Mandal M (2003) Multimedia signals and systems. Springer, New York
12. Mataušek M, Batalov V (1980) A new approach to the determination of the glottal waveform. IEEE TransAcoust Speech Signal Process ASSP-28(6):616–622
13. Noll P (1997) MPEG digital audio coding. IEEE Signal Process Mag 59–81
14. Orovic I, Stankovic S, Draganic A (2014) Time-frequency analysis and singular value decomposition applied to the highly multicomponent musical signals. Acta Acust 100(1):93–101
15. Painter T (2000) Perceptual coding of digital audio. Proc IEEE 88(4):451–513
16. Pan D (1995) A tutorial on MPEG/audio compression. IEEE Multimedia 2(2):60–74
17. Pohlmann KC (2005) Principles of digital audio. McGraw-Hill, New York
18. Salomon D, Motta G, Bryant D (2009) Handbook of data compression. Springer, New York
19. Salomon D, Motta G, Bryant D (2006) Data compression: a complete reference. Springer, London
20. Sayood K (2000) Introduction to data compression, 2nd edn. Morgan Kaufmann, San Francisco, CA
21. Smith MT (1999) Audio Engineer's reference book, 2nd edn. Focal Press, Oxford
22. Schniter P (1999) www2.ece.ohiostate.edu/~schniter/ee597/handouts/ homework5.pdf
23. Spanias A, Painter T, Atti V (2007) Audio signal processing and coding. Wiley-Interscience, New York
24. Stanković L (1994) A method for time-frequency signal analysis. IEEE Trans Signal Process 42(1):225–229
25. Stanković S, Orović I (2010) Time-Frequency based Speech Regions Characterization and Eigenvalue Decomposition Applied to Speech Watermarking. EURASIP Journal on Advances in Signal Processing, Special Issue on Time-Frequency Analysis and its Application to Multimedia signals, Article ID 572748, Pages(s) 10 pages
26. Steinmetz R (2000) Multimedia systems. McGrawHill, New York
27. Vetterli M, Kovačević J (1995) Wavelets and subband coding. Prentice-Hall, Englewood Cliffs, NJ
28. Watkinson J (2001) The art of digital audio, 3rd edn. Focal Press, Oxford
29. Watkinson J (2001) The MPEG handbook. Focal Press, Oxford
30. Wong DY, Markel JD, Gray AH (1979) Least squares glottal inverse filtering from the acoustic speech waveform. IEEE Trans Acoust Speech Signal Process ASSP-27(4)
31. Zolcer U (2008) Digital audio signal processing. John Willey & Sons, Ltd., Chichester

# Chapter 3
# Storing and Transmission of Digital Audio Signals

In this chapter we consider the widely used media for storing digital data. A special attention is given to CD, Mini Disc (MD), DVD (concepts of data writing and reading processes are considered), as well as to the coding principles. Different error correction and interleaving algorithms such as cyclic redundancy check, cross-interleaving, Reed–Solomon code, and Eight-to-Fourteen Modulation are presented. Also, the basic concepts of the digital audio broadcasting system are considered.

## 3.1 Compact Disc: CD

The basic characteristics of a CD are provided in Table 3.1.

CD has 20625 tracks, where the distance between tracks is 1.6 μm. The audio storage space is placed between the lead-in and the lead-out area, having diameters of 46 mm and 116 mm, respectively. The lead-in area contains information about the CD content, the length and the starting time of audio sequences. The lead-out area provides the information that the playback is completed. The internal structure of the CD is given in Fig. 3.1, Table 3.2.

On the CD surface, there are pits and flat layers called lands. The pits can have one of nine different lengths, from T3 to T11. The smallest pit size is $0.833 \times 0.5$ μm. However, the pit and land lengths may vary depending on the disc writing (turning) speed while recording. For example, T3 pit length is 833 nm for the writing speed 1.2 m/s, while for the speed 1.4 m/s it is 972 nm.

Laser rays that fall on the land of the CD are reflected with the same path and the same intensities, while the intensities of rays scattered from the bumps are lower. The intensity of reflected beam is detected as one of the logical values (1 or 0). Figure 3.2 illustrates the laser beam reflections from a CD.

It is noteworthy that a CD is not sensitive to an amount of dust, fingerprints, and scratches. One reason that a CD has a good performance in terms of sensitivity to



S. Stanković et al., *Multimedia Signals and Systems*,
DOI 10.1007/978-3-319-23950-7_3

**Table 3.1** Basic features of CD

| Characteristics | Values |
|---|---|
| Frequency range | 20 Hz–20 KHz |
| Dynamic range | ≈96 dB |
| Diameter | 12 cm |
| Playing time | 60 min–74 min |

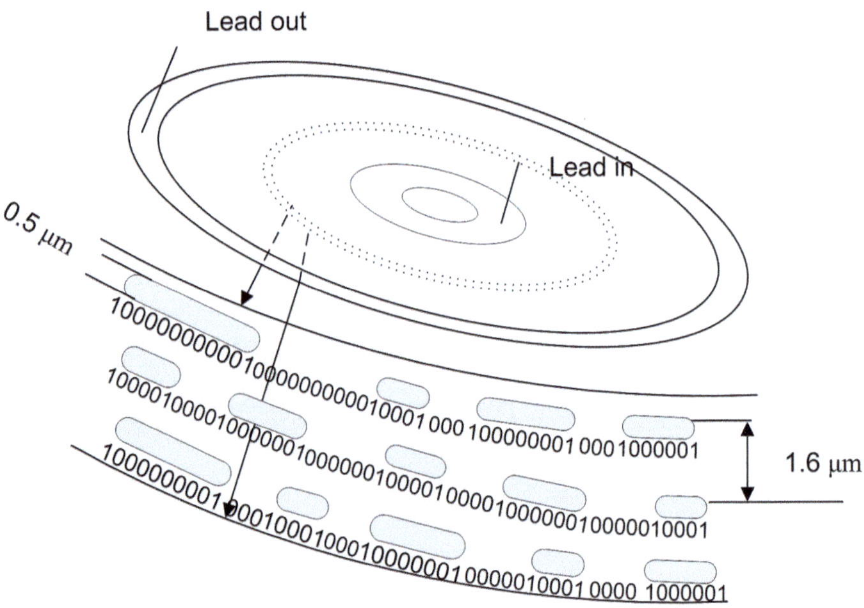


**Fig. 3.1** The structure of CD

**Table 3.2** Lengths of pits at 1.2 m/s

| Pits length | | Size in nm |
|---|---|---|
| T3 = 10001 | | 833 |
| T4 = 100001 | | 1111 |
| T5 = 1000001 | | 1388 |
| ... | ... | ... |
| T11 = 1000000000001 | | 3054 |

dust is a protective layer of 1.2 mm thickness, which completely passes the laser beam through. For example, if there is a dust grain on the protective layer, the laser ray will pass to the focal point without obstacles as long as the dust grain diameter is less than 0.8 mm. The intensity of the reflected ray will correspond to the same logical value as in the case of reflection from the clean surface, Fig. 3.2.

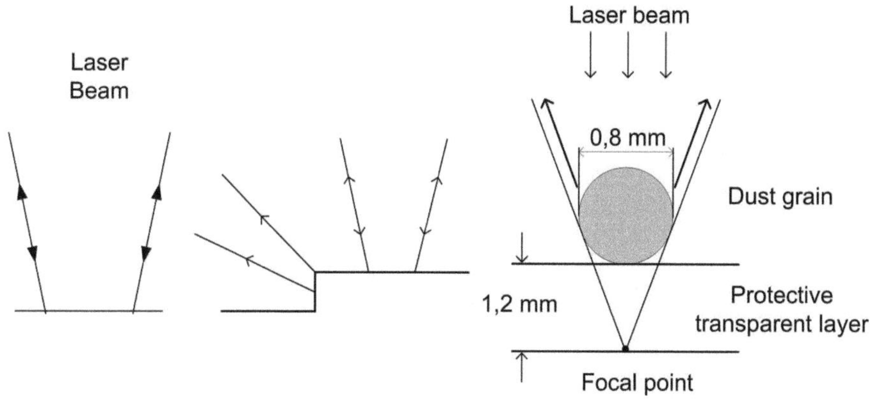


**Fig. 3.2** Reflections from a CD

## 3.1.1 *Encoding CD*

In order to be resistant to scratches and other errors, an efficient coding scheme is applied. Namely, the audio signal stored on a CD is encoded within four steps:

1. Cross-Interleaved Reed–Solomon Coding (CIRC).
2. Generating a control word.
3. EFM encoding.
4. Generating synchronization word.

When passing through the CD encoding system, a bit rate for a 16-bit stereo audio signal changes from $1.4112 \cdot 10^6$ b/s to $4.3218 \cdot 10^6$ b/s. To easily understand the coding schemes used for CD, let us first briefly consider Cyclic Redundancy Check (CRC) and interleaving.

### 3.1.1.1 Cyclic Redundancy Check: CRC

CRC is a general method used for error detection. The method relies on the division of a polynomial corresponding to the original sequence by another predefined polynomial function, resulting in a residue which is actually a CRC. The length of a residue is always smaller than the length of the polynomial. CRC coding can be also done by using the exclusive OR operation (XOR), but both polynomials have to be represented as binary sequences.

Let us assume that the message is 11010011101100, while the divisor sequence is equal to 1011 (as a polynomial it is defined by $x^3 + x + 1$). Now, the XOR operation should be carried out between the original sequence and the divisor sequence (from left to right). It should be mentioned that if the first bit in the original sequence is equal to 0, then we begin the XOR operation on the next bit that has the value 1. The second step is to move the divisor sequence by one position to the right and perform the XOR operation again. This procedure is repeated until the sequence 1011 reaches the end of the original sequence, as illustrated in the example. At the end, a binary residual sequence is obtained, representing the CRC function.

```
11010011101100
1011
______________
01100011101100
 1011
______________
00111011101100
  1011
______________
00010111101100
   1011
______________
00000001101100
       1011
______________
00000000110100
        1011
______________
00000000011000
         1011
______________
00000000001110
          1011
______________
00000000000101 (the remaining 3 bits)
```

Typical polynomials used in the CRC encoding are given in Table 3.3.

#### 3.1.1.2 Interleaving

Interleaving is an approach that arranges the data in noncontiguous order to decrease the effects of errors. This enables us to possibly recover damaged information by an interpolation method. For example, consider a signal with 24 samples and divide it into blocks of 12 samples. A simple interleaving can be obtained by reordering samples as shown in Fig. 3.3. In this case, interleaving is based on

**Table 3.3** Some of the polynomials used for CRC coding

| Code | Polynomial |
|---|---|
| CRC-1 | $x + 1$ |
| CRC-4 ITU | $x^4 + x + 1$ |
| CRC-5 ITU | $x^5 + x^4 + x^2 + x1$ |
| CRC-8-CCITT | $x^8 + x^2 + x + 1$ |
| CRC-10 | $x^{10} + x^9 + x^5 + x^4 + x + 1$ |
| CRC-12 | $x^{12} + x^{11} + x^3 + x^2 + x + 1$ |
| CRC-16 CCIT | $x^{16} + x^{12} + x^5 + 1$ |
| CRC-16 IBM | $x^{16} + x^{15} + x^2 + 1$ |

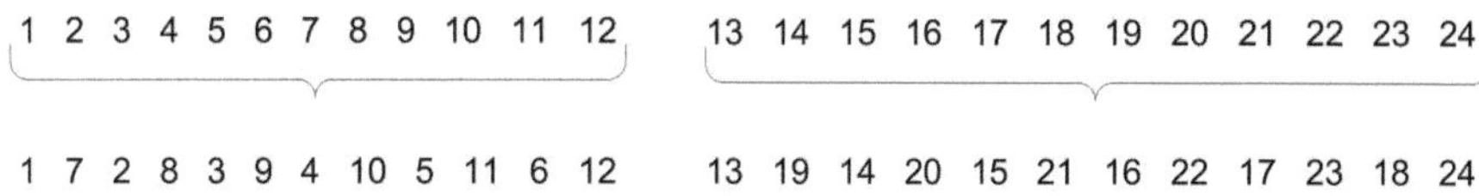


**Fig. 3.3** An example of interleaving

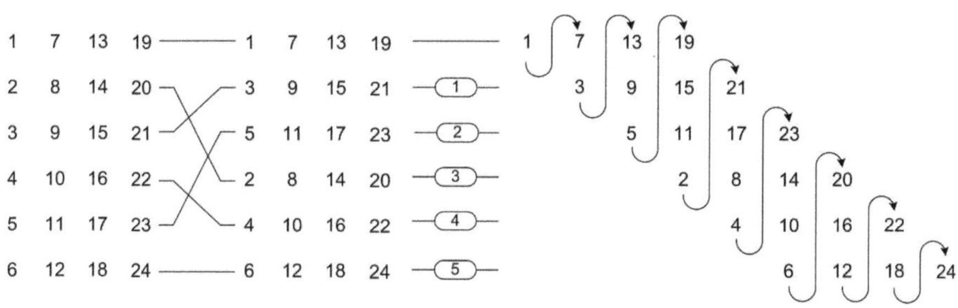

**Fig. 3.4** Interleaving based on the delay lines

moving the first 6 samples within the block to the right by $i$-1, where $i$ is the sample position (e.g., the third sample is moved for 3–1 = 2 positions to the right).

Another simple example of interleaving which is closer to the concept used in the CIRC encoding, is shown in Fig. 3.4. Note that the distance between the consecutive samples is increased. Each row has a different delay (the first row has no delay, the second row has a unit delay, etc.).

#### 3.1.1.3 CIRC Coding

Consider now the interleaving procedure used in the CD coding, which is considerably more complex and it is illustrated in Fig. 3.5.

The structure is based on a group of six samples for the left and six samples for the right channel of stereo audio signals. Each sample is represented by 16 bits.

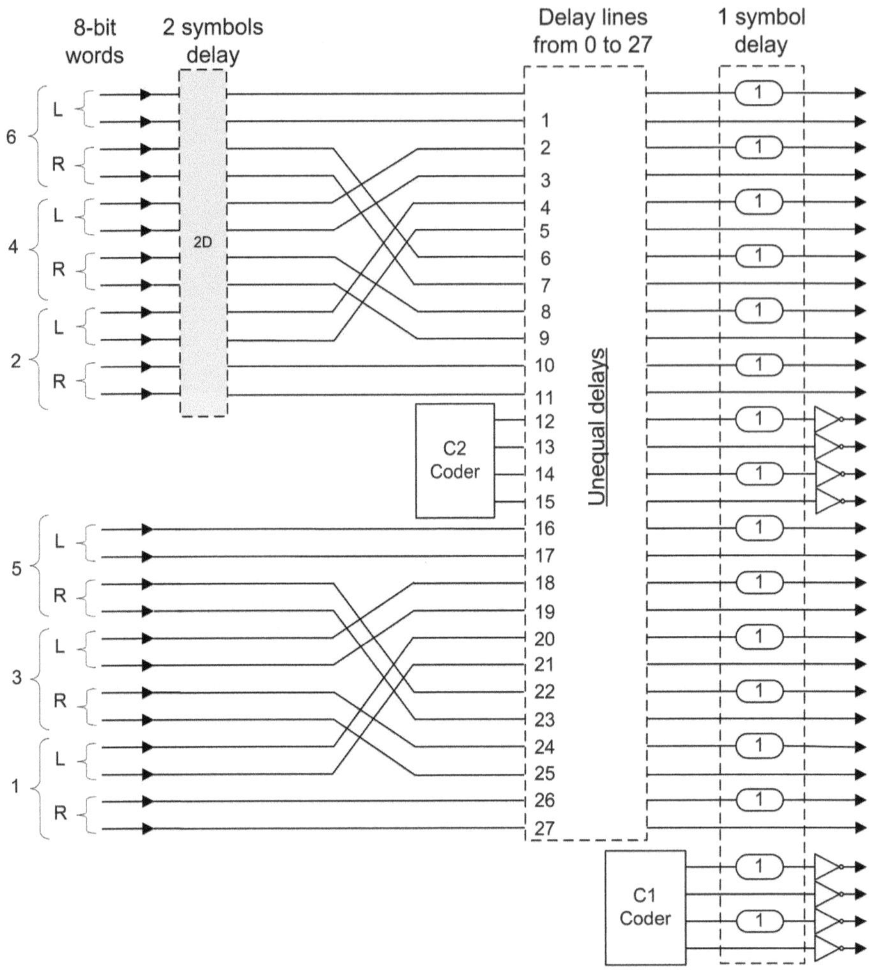


**Fig. 3.5** Interleaving used in CD coding

Odd and even samples are separated. From each sample, two 8-bit words are formed (24 words in total). Then, all even samples are delayed by two symbols.

Figure 3.6 illustrates an example depicting how even the part of the system with a two-symbol delay, can be useful to reconstruct the damaged part of the signal. Labels $L_i$ and $R_i$ represent the left and right *i-th* sample, respectively. Shaded parts denote damaged samples. In the lower part of Fig. 3.6, the delay compensation is performed and the samples are synchronized according to their initial order. Based on the even samples, the damaged odd samples are reconstructed by interpolation, and vice versa.

The C2 encoder, shown in Fig. 3.5, generates four Q words that are 8 bits long. These words represent the parity bytes used to increase the distance between the odd and even samples and to allow the errors detection. Additional interleaving is performed after the C2 encoder, which arranges the order and distances between the

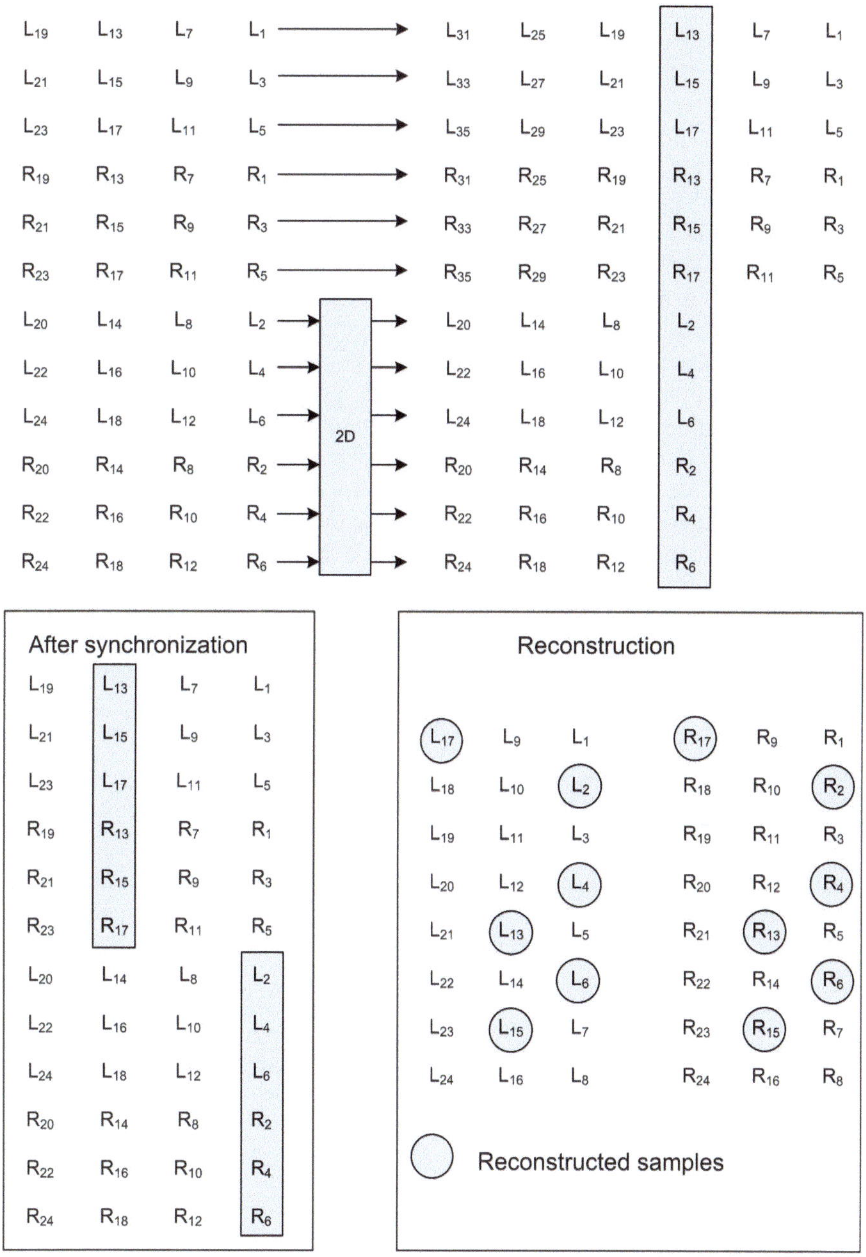


**Fig. 3.6** The reconstruction principle of the damaged signal part

existing 28 words. The introduced delay between the words is used to dissipate the error across distant positions in order to increase the ability to recover as many samples as possible. After the interleaving subsystem, the C1 encoder generates four P words (P parity bytes). Therefore, the CIRC encoder ends up with 32 words from the initial 24 input words, introducing the redundancy of 8 words and increasing the bit rate from $1.4112 \cdot 10^6$ b/s to $1.8816 \cdot 10^6$ b/s. The resultant 32 sequences are included within the unit called frame.

The procedure for determining P and Q parity words is done by using the Reed–Solomon code. It is based on the finite field arithmetic, which is usually referred to as Galois field. A finite field of $q$ elements is denoted as $GF(q)$. The field $GF(q)$ always contains at least one element, called a primitive element, with the order $(q-1)$. If $a$ is a primitive in $GF(q)$, then $(q-1)$ consecutive powers of $a$: $\{1, a, a^2, \ldots, a^{q-2}\}$, must be distinct and they are $(q-1)$ nonzero elements of $GF(q)$. The "exponential representation" allows describing the multiplication operation as an addition: $a^x a^y = a^{x+y}$. A primitive element is a root of a *primitive polynomial* $p(x)$. For example, if we consider the polynomial: $p(x) = x^3 + x + 1$, then $a^3 + a + 1 = 0$. Note that the addition is done as the XOR operation.

The Reed–Solomon code use the Galois field in the form $GF(2^k)$, where the elements of the field are represented by $k$ bits. The 3-bit terms given in Table 3.4 describe a *Galois field* $GF(2^3)$.

In order to understand how to obtain $P$ and $Q$ words, let us consider one simplified, but illustrative example. Suppose that we have five data words labelled as $A$, $B$, $C$, $D$, and $E$ (3-bit words are used). Then, we set the following equations:

$$A \oplus B \oplus C \oplus D \oplus E \oplus P \oplus Q = 0, \tag{3.1}$$

$$a^7 A \oplus a^6 B \oplus a^5 C \oplus a^4 D \oplus a^3 E \oplus a^2 P \oplus aQ = 0, \tag{3.2}$$

where $a^i$ are the above defined constants. By solving the equations simultaneously, the expressions for $P$ and $Q$ words are obtained. Hence, Eq. (3.2) is divided by $a$, and then $Q$ is replaced by $A \oplus B \oplus C \oplus D \oplus E \oplus P$ (since from Eq. (3.1) $Q = A \oplus B \oplus C \oplus D \oplus E \oplus P$ holds):

**Table 3.4** The Galois field $GF(2^3)$

| Exponential | Algebraic | Binary |
|---|---|---|
| 0 | 0 | 000 |
| 1 | 1 | 001 |
| $a$ | $a$ | 010 |
| $a^2$ | $a^2$ | 100 |
| $a^3$ | $a+1$ | 011 |
| $a^4$ | $a \cdot a^3 = a^2 + a$ | 110 |
| $a^5$ | $a^2 \cdot a^3 = a^2 + a + 1$ | 111 |
| $a^6$ | $a \cdot a^5 = a^3 + a^2 + a = a + 1 + a^2 + a = a^2 + 1$ | 101 |
| $a^7$ | $a \cdot a^6 = a \cdot (a^2 + 1) = a + 1 + a = 1$ | 001 |

$$
\begin{aligned}
&a^6A \oplus a^5B \oplus a^4C \oplus a^3D \oplus a^2E \oplus aP \oplus Q \\
&= a^6A \oplus a^5B \oplus a^4C \oplus a^3D \oplus a^2E \oplus aP \oplus A \oplus B \oplus C \oplus D \oplus E \oplus P \\
&\Rightarrow (a^6 \oplus 1)A \oplus (a^5 \oplus 1)B \oplus (a^4 \oplus 1)C \oplus (a^3 \oplus 1)D \oplus (a^2 \oplus 1)E = (a \oplus 1)P.
\end{aligned} \tag{3.3}
$$

By using the binary representation of constants from the Table 3.4, Eq. (3.3) can be simplified as:

$$
\begin{aligned}
&a^2A \oplus a^4B \oplus a^5C \oplus aD \oplus a^6E = a^3P \\
&P = a^6A \oplus aB \oplus a^2C \oplus a^5D \oplus a^3E,
\end{aligned} \tag{3.4}
$$

where $a^2/a^3 = a^{-1} = a^{7-1} = a^6$. Similarly, by multiplying the Eq. (3.1) by $a^2$, we have:

$$
\begin{aligned}
&a^2A \oplus a^2B \oplus a^2C \oplus a^2D \oplus a^2E \oplus a^2P \oplus a^2Q = 0 \;\;\Rightarrow \\
&a^7A \oplus a^6B \oplus a^5C \oplus a^4D \oplus a^3E \oplus (a^2A \oplus a^2B \oplus a^2C \oplus a^2D \oplus a^2E \oplus a^2Q) \oplus aQ = 0 \Rightarrow \\
&(a^7 \oplus a^2)A \oplus (a^6 \oplus a^2)B \oplus (a^5 \oplus a^2)C \oplus (a^4 \oplus a^2)D \oplus (a^3 \oplus a^2)E \oplus (a \oplus a^2)Q = 0
\end{aligned}
$$

Again using the binary representation of constants, the $Q$ word is obtained as:

$$
\begin{aligned}
&a^6A \oplus B \oplus a^3C \oplus aD \oplus a^5E = a^4Q \quad \Rightarrow \\
&Q = a^2A \oplus a^3B \oplus a^6C \oplus a^4D \oplus aE.
\end{aligned} \tag{3.5}
$$

In order to detect errors, two syndromes are considered:

$$
\begin{aligned}
&S_1 = A' \oplus B' \oplus C' \oplus D' \oplus E \oplus P' \oplus Q', \\
&S_2 = a^7A' \oplus a^6B' \oplus a^5C' \oplus a^4D' \oplus a^3E' \oplus a^2P' \oplus aQ',
\end{aligned} \tag{3.6}
$$

where $A'$, $B'$, …, $Q'$ denote received words that may contain an error. Assume that the error occurred in the word $C$ ($C' = C + G$), while the other words are without errors. Then, we obtain:

$$
\begin{aligned}
&S_1 = A \oplus B \oplus (C + G) \oplus D \oplus E \oplus P \oplus Q = G, \\
&S_2 = a^7A \oplus a^6B \oplus a^5(C + G) \oplus a^4D \oplus a^3E \oplus a^2P \oplus aQ = a^5G,
\end{aligned} \tag{3.7}
$$

or $S_2 = a^5S_1$. Therefore, the error is equal to the syndrome $S_1$ and the error location is obtained based on the weighting coefficient. After calculating the coefficient as: $a^x = \frac{S_2}{S_1}$, and concluding that $a^x = a^5$ holds, one may know that an error occurred within the $C$ word, because $C$ is multiplied by $a^5$.

#### 3.1.1.4 Generating Control Word

The next step in the CD coding procedure is a control word generation. The control word is added to each block of 32 words. This word consists of the codes $P, Q, R, S, T, U, V, W$. Note that the choice of $P$ and $Q$ labels is made a bit unadvisedly, since we used them to obtain new code sequences independent of $P$ and $Q$ words generated in CIRC coding. $P$ can have values 0 or 1. From Fig. 3.7, we can observe that $P$ has value 1 between two sequences recorded on CD and value 0 during the sequence duration. Switching from 0 to 1 with frequency equals to 2 Hz in the lead-out area indicates the end of the disc. The $Q$ word specifies the number of audio channels. It should be noted that the total length of these subcodes is 98 bits, which means that it can be read from 98 frames. After adding the control word, the bit rate is increased to:

$$33/32 \cdot 1.8816 \cdot 10^6 \mathrm{b/s} = 1.940410^6 \mathrm{b/s}.$$

An example of the $P$ and $Q$ words is given in Fig. 3.7.

The $Q$ word in the BCD format contains the current track number (01, 02, 03, etc.), the index number, running time, etc. TNO (track number) represents the current track number and ranges from 01 to 99. The TNO within the lead in area has the value 00. The index point is a two-digit number in the BCD format and within the sequences can be up to 99 index points. During a pause, the index point is equal to 00, while the index point at the beginning of each sequence is equal to 01. Also, the index point in the lead out area is equal to 01. Setting up the values for index pointers is a way to divide the sequence into smaller parts. Index pointers are primarily intended for CDs with long sequences (e.g., a classical music CD), since they allow direct access to some parts of the sequence. However, they are rarely used nowadays.

Other subcodes are used for transmitting additional information such as text and information on duration of individual sequences.

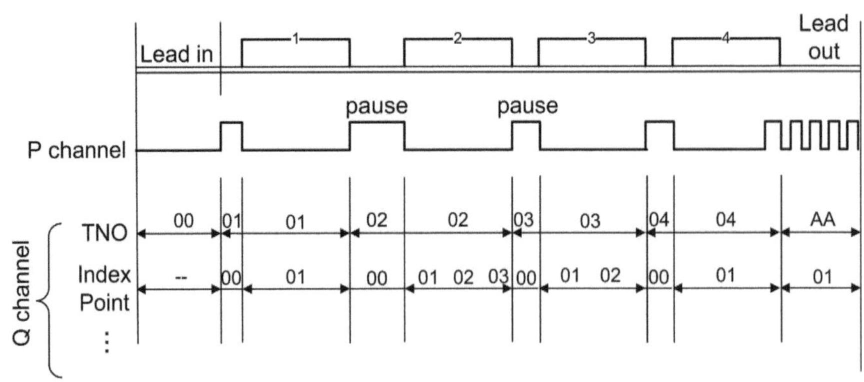


**Fig. 3.7** An illustration of timing diagrams for P and Q channels

**Table 3.5** Examples of extending 8-bit words to 14-bit words

| 8-bit word | 14-bit words |
|---|---|
| 00000011 | 00100100000000 |
| 01001110 | 01000001001000 |
| 10101010 | 10010001000100 |
| 11110010 | 00000010001001 |

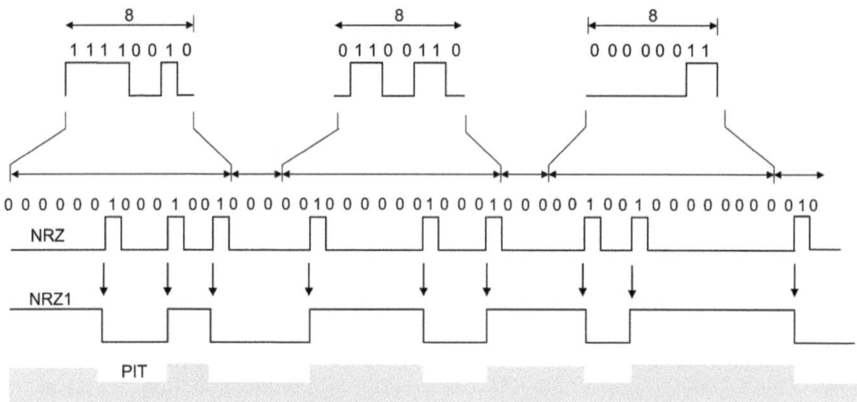


**Fig. 3.8** An example of EFM encoding

After we determine the control word, the EFM (**E**ight to **F**ourteen **M**odulation) is used to convert 8-bit symbols into 14-bit symbols. Observe that with 8 bits we can make 256 combinations, while with 14 bits we can achieve 16384 combinations. The basic idea of EFM coding is to map 8-bit words into 14-bit words such that the number of inversions between consecutive bits is reduced, i.e., the distance between transitions on the disc surface is increased (logical value 1 is used to determine the transitions). An example of an EMF mapping is shown in Table 3.5, while the EFM encoding procedure is illustrated in Fig. 3.8.

Note that the 14-bit signals are separated by using 3 merging bits to additionally increase the distance between consecutive values 1. In other words, the initial 8-bit sequences are extended to 17 bits and represented by the NRZ code. Then the sequence of bits from the NRZ code is transferred into the NRZ1 code, such that each value 1 in the NRZ code makes the transition in NRZ1, as shown in Fig. 3.8. The NRZ1 sequence defines the position of pits when writing data to a CD. The minimum duration of the NRZ1 signals is 3 T (3 clock periods) and the maximum duration is 11 T.

The bit rate after this coding stage is:

$$17/8 \cdot 1.9404 \cdot 10^6 \text{b/s} = 4.12335 \cdot 10^6 \text{b/s}.$$

Finally, the CD encoding process ends with a synchronization (sync) word. This word is added after each frame to indicate the beginning of the frame, but also serves to control the spinning motor speed. The sync word consists of 12 values equal to 1, another 12 values 0, and 3 filter bits, making a total of 27 bits. Hence, from the previously achieved 561 bits per frame, now we get 588 bits within the frame (33 words · 17 bits = 561).

The final bit rate is:

$$4.12335\text{b/s} \cdot 588/561 = 4.3218 \cdot 10^6\text{b/s}.$$

## 3.2 Mini Disc

Mini Disc (MD) has a diameter of 6.4 cm, almost twice smaller than the CD, with the same playing time of 74 min. The sound quality is almost identical to the CD audio quality. The structure of MD is depicted in Fig. 3.9.

Sophisticated compression algorithms are needed to reduce the amount of information that has to be stored in order to retain a high-quality sound on MDs. For this purpose, the MD uses ATRAC compression described in the previous chapter. Note that the sampling frequency used for MDs is the same as for CDs (44.1 KHz), and the track width is 1.6 μm.

Data recording is done through the magnetization performed by the magnetic head. The magnetization is done at the specific temperature, which is above the Curie point (about 185 °C). Note that the materials which are easily magnetized are not used for manufacturing of MDs due to the possibility of data loss in the presence of an external magnetic field. Therefore, even when exposed to an external magnetic field, MDs will not lose its contents, unless the required temperature is achieved. A system for the MD magnetization is illustrated in Fig. 3.10.

When recording the data, the laser heats the magnetic layer up to the Curie temperature. Then the magnetic head, placed on the opposite disc surface, performs the magnetization by producing the correct polarity for each logical value (north or south depending on the bit, 1 or 0). The laser beam is reflected from the magnetic layer while reading the data. The polarization of the laser beam is changed based on the orientation of the magnetic layer (Fig. 3.11). An optical device with a polarizing filter collects reflected polarized signals. When the laser beam passes through the filter, the intensity changes according to the laser beam polarization, and the output signal is generated.

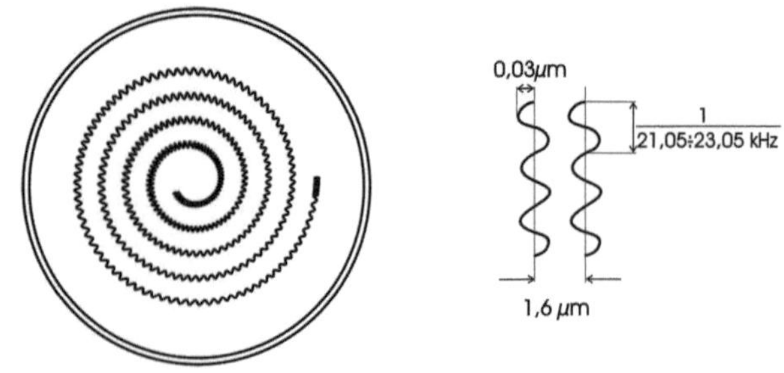


**Fig. 3.9** The structure of MD

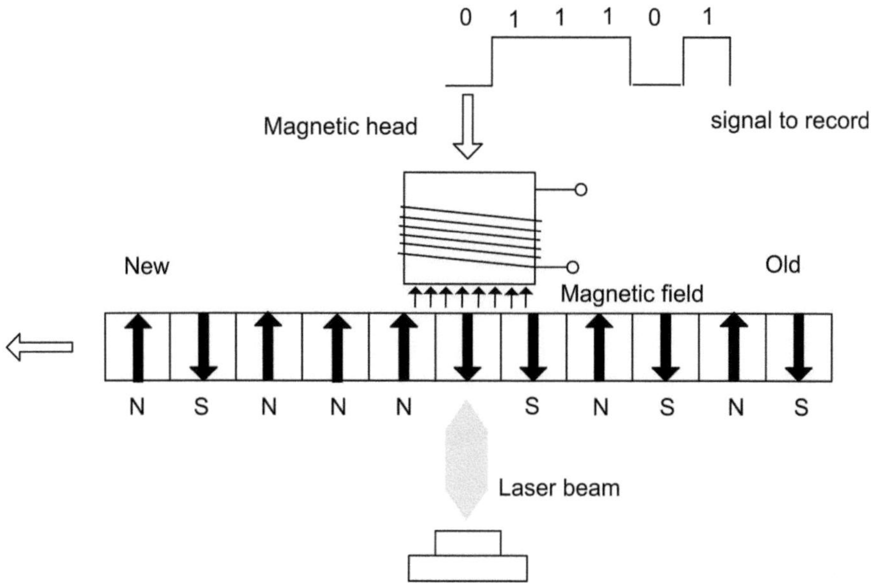


**Fig. 3.10** Recording the data on the MD

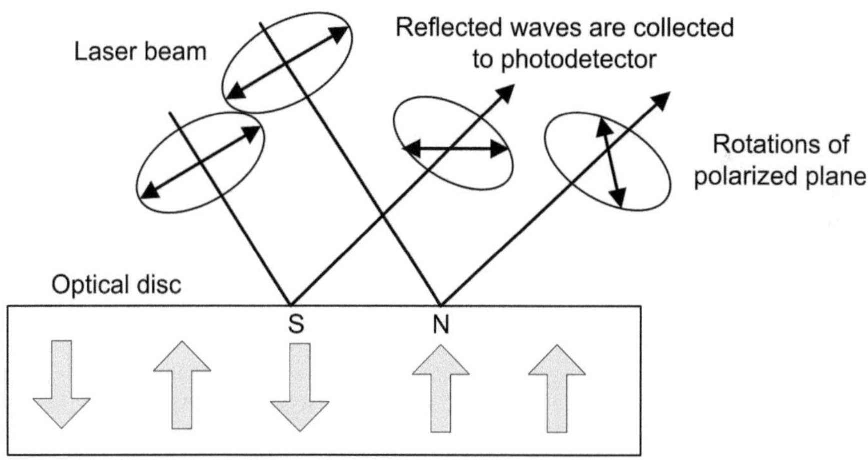


**Fig. 3.11** Reflections of the laser beam in the case of MD

MDs use the ACIRC (Advanced CIRC) for encoding, which is similar to the CIRC encoding used in CDs. It also uses the EFM encoding, along with the ATRAC compression which is not used in CDs.

The antishock system is an important part of MDs as it enables the system to recover from any shocks during playback. This system is based on the RAM, allowing recovery from the shock with duration of several seconds.

A block diagram of the entire MD system is illustrated in Fig. 3.12.

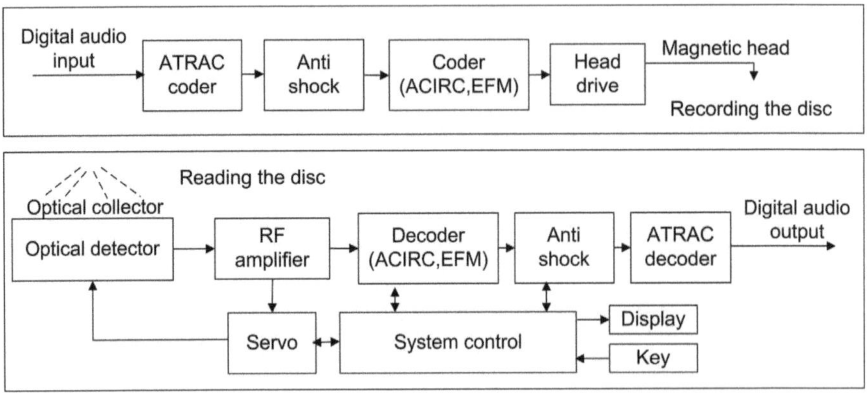


**Fig. 3.12** A block diagram of the MD system

When data are written to the MD (upper part in Fig. 3.12), the digital audio signal is fed to the ATRAC encoder. ATRAC data compression is performed and the data are loaded to the antishock system, and further through the EFM/ACIRC encoder (which includes interleaving, error detection and EFM coding). The signal from the EFM/ACIRC encoder is used to control the magnetic head when recording the data.

Reproduction or reading of the data starts at the unit called the optical signal collector, shown in the lower part of Fig. 3.12. A special device within this unit is called the optical detector. Then, the signal is amplified by the RF (radio frequency) amplifier and fed to the EFM/ACIRC decoder. The data is decompressed using the ATRAC decoder that follows the antishock system. The output of the ATRAC decoder is a digital audio signal.

## 3.3 Super Audio CD (SACD)

SACD provides a high quality sound, with the option of multichannel records. The diameter of SACD is the same as for a CD, while the width of pit lane is less than in the case of CD (the track width is 0.74 μm and the length of a pit is 0.40 μm). The sampling frequency is 2.8224 MHz and the 1-bit DSD encoding is used. The maximum frequency of the reproduced sound is up to 100 KHz and with 120 dB dynamic range. Data are protected with the SACD watermarking techniques.

The memory space required to store 74 min stereo audio recording is $(2 \cdot 74 \cdot 60 \cdot 2.8224 \cdot 10^6)/8$ B = 2.9 GB. Hence, in the case of 6-channel format, the required memory capacities would be certainly much larger. Therefore, SACDs use lossy compression (e.g., AC3) or lossless compression based on the complex algorithms with adaptive prediction and entropy coding.

## 3.4 DVD-Audio

DVD-audio (DVD hereinafter) is also used to record high quality sound with the sampling frequency of 192 KHz and 24 bit data format. DVD allows the signal to noise ratio of S/N = 146 dB. The capacity of a DVD is 4.7 GB and its diameter is 8 or 12 cm. The maximum number of channels is 6. Based on these requirements, a DVD cannot store 74 min of high quality music within 4.7 GB of memory space. Therefore, the data have to be compressed. For this purpose, the lossless compression called Meridian Lossless Packing (or Packed PCM) has been developed. It is based on three lossless techniques: Infinite Impulse Response (IIR) waveform predictor selected from a set of predetermined filters to reduce the intersample correlation, lossless inter-channel decorrelation and Huffman coding. This compression algorithm compresses the original data by 50 %. However, even with this high compression ratio, it is not possible to record six channels with sampling frequency of 192 KHz and 24 bits. Therefore, the channels reflecting the influence of the environment (surround sound) use different sampling frequencies. For example, the direct left, right and center channels are characterized by 24-bit format and the sampling frequency of 192 KHz, while the signals in the remaining three channels (representing the surround effects) have a sampling frequency 96 KHz and they are coded by 16 bits.

## 3.5 Principles of Digital Audio Broadcasting: DAB

Before we consider the main characteristics of DAB systems, let us review some basic facts about the FM systems. In order to receive an FM signal with a stable high quality, the fixed and well-directed antennas are required. For example, it is impossible to achieve this condition with car antennas. Also, due to the multipath propagation, the waves with different delays (i.e., different phases) can cause a significant amplitude decrease and hence the poor reception of such signals (Fig. 3.13).

The DAB system can avoid the aforementioned problems. Consider a DAB system given in Fig. 3.14.

The first block compresses the data, which are then forwarded to the second block. The second block encodes the data in order to become less sensitive to noise. Lastly, the signal is forwarded to a transmitter that broadcasts the data. Figure 3.15 shows the channel interleaving process used to combine data from different channels into one transmission channel.

If interference occurs, it will not damage only the signal in one channel, but will be scattered across all channels. Hence, a significant damage of signal belonging to only one channel is avoided.

**Fig. 3.13** Direct and reflected FM signals with equal strength are nulled when 180° phase difference occur

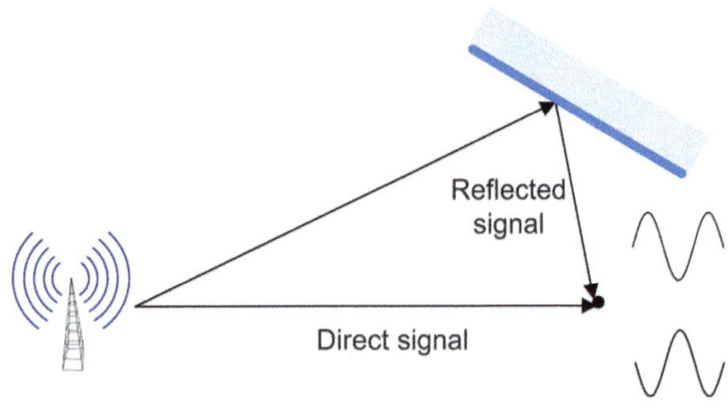


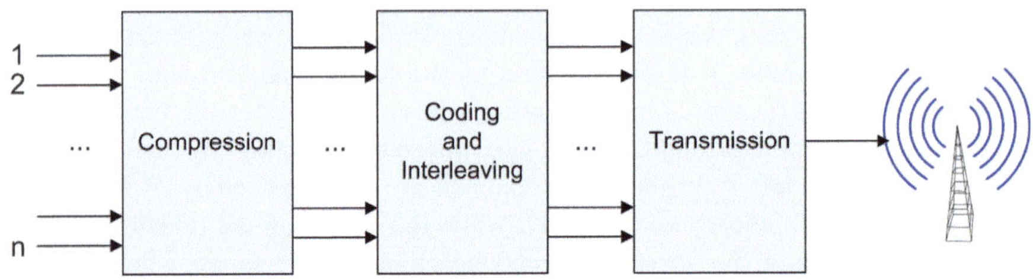


**Fig. 3.14** A block scheme of DAB system

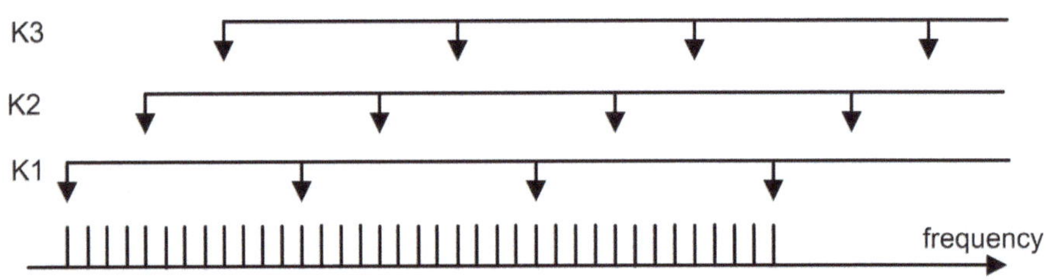


**Fig. 3.15** An illustration of channel interleaving

## 3.5.1 Orthogonal Frequency-Division Multiplexing (OFDM)

OFDM is used to achieve more efficient bandwidth utilization for DAB systems. Binary data sequence is first divided into pairs of bits, which are then forwarded to the QPSK modulator (some systems use QAM or other modulation schemes). This means that two bits are mapped into one of the four phase values, as illustrated in the diagram in Fig. 3.16.

This produces the complex QPSK symbols. If the changes in the phase of the received signal are used instead of the phase itself, the scheme is called the

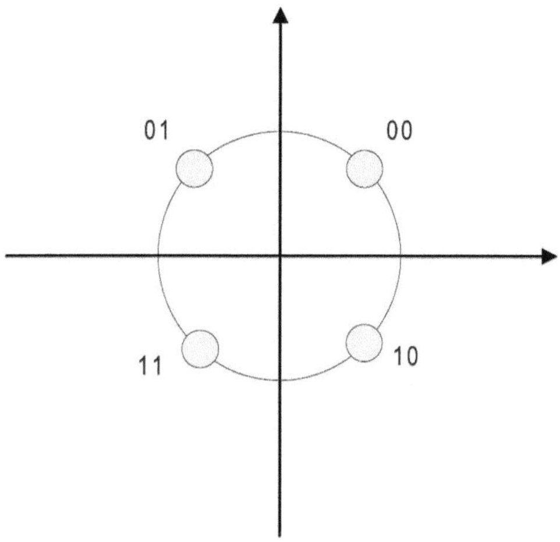


| Bits | Coordinates (I+j·Q) | Phase |
|---|---|---|
| 00 | 0.707+j·0.707 | 45° |
| 01 | -0.707+j·0.707 | 135° |
| 11 | -0.707-j·0.707 | 225° |
| 10 | 0.707-j·0.707 | 315° |

**Fig. 3.16** Diagram and table for QPSK modulation

differential QPSK (DQPSK). It depends on the difference between successive phases. In DQPSK the phase-shifts are 0°, 90°, 180°, 270°, corresponding to the data '00', '01', '11', '10'.

Each of the symbols obtained after QPSK modulation is multiplied by a sub-carrier frequency:

$$s_k(t) = A_k e^{j\phi_k} e^{j2\pi f_k t}, \tag{3.8}$$

where $A_k$ and $\phi_k$ are the amplitude and the phase of a QPSK symbol. For example, symbols obtained by using the QPSK modulation have the constant amplitude and their phases can have one of four possible values. If we assume that we have $N$ sub-carriers, then one OFDM symbol will be in the form:

$$s(t) = \frac{1}{\sqrt{N}} \sum_{k=0}^{N-1} A_k e^{j(2\pi f_k t + \phi_k)}, \quad 0 < t < T, \tag{3.9}$$

where $f_k = f_0 + k\Delta f = f_0 + k\frac{1}{NT_s}$, $T_s$ is the length of the symbols (e.g., the QPSK symbols), while $T = N{\cdot}T_s$ is the OFDM symbol duration. The carrier frequency is $f_0$, while the sub-carriers are separated by $1/T$. The sub-carriers are transmitted in mutually orthogonal frequencies, so that the subcarriers are peak centered at the positions where other sub-carries pass through zero (Fig. 3.17). Note that the OFDM symbol corresponds to the definition of the inverse Fourier transform. Comparing to the previously used form of the Fourier transform, $\omega_k$ is replaced by $2\pi f_k$, and consequently $1/N$ is replaced by $1/\sqrt{N}$.

The spectrum of an individual sub-carrier is of the form $\sin(x)/x$ and it is centered at the sub-carrier frequency.

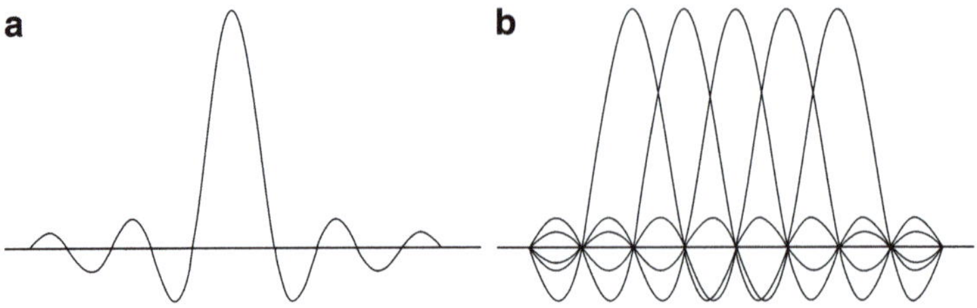


**Fig. 3.17** An OFDM spectrum: (**a**) one sub-carrier, (**b**) five sub-carriers

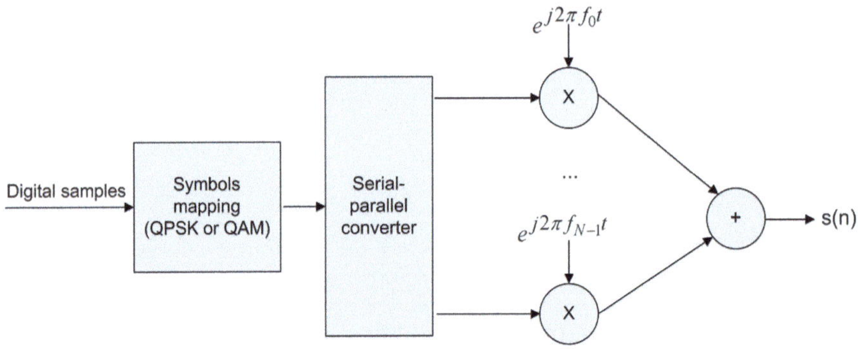


**Fig. 3.18** A simplified block diagram of OFDM system

A simplified scheme including QPSK and OFDM modulator is given in Fig. 3.18. Note that an OFDM system should include additional elements, such as pilot symbols, guard intervals, etc., but here we only deal with the basic OFDM concepts.

We saw that the OFDM modulation can be performed by calculating the inverse Fourier transform. Demodulation is achieved by dividing the signal into the parts that are equal in duration to OFDM symbols. Then, the Fourier transform is performed and we can identify the sub-carrier frequencies. The resulting signal is obtained by calculating the phases of the components on the sub-carrier frequencies.

## 3.6 Examples

3.1. Starting from the sequence:
{1,2,3,4,5,6,7,8,9,10,11,12,13,14,15,16},
perform a simple interleaving procedure defined as follows: the sequence is divided into 4-samples segments, and then the first interleaved block is formed by taking the first elements from each segment, the second block is

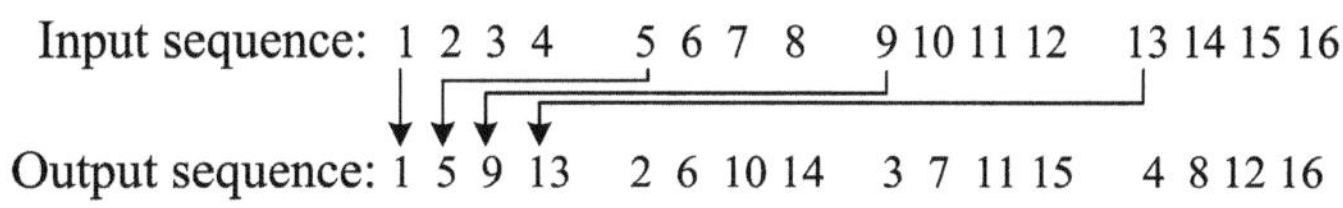


**Fig. 3.19** Example of interleaving

formed from the elements on the second position, and so on. Determine the output sequence.

Solution is given in Fig. 3.19.

3.2. Consider the following input sequence:

{1,2,3,4,5,6,7,8,9,10,11,12,13,14,15,16}.

The interleaving procedure is defined as follows:

- The input samples are placed to the $4 \times 4$ matrix, by filling the matrix rows.
- The matrix rows are reordered according to the principle 4-2-3-1 (the new order of rows).
- The columns are reordered by using the same rule.

Determine the output sequence obtained by reading the columns of the resulting matrix.

Solution:

$$\begin{array}{cccc} 1 & 2 & 3 & 4 \\ 5 & 6 & 7 & 8 \\ 9 & 10 & 11 & 12 \\ 13 & 14 & 15 & 16 \end{array} \Rightarrow \begin{array}{cccc} 13 & 14 & 15 & 16 \\ 5 & 6 & 7 & 8 \\ 9 & 10 & 11 & 12 \\ 1 & 2 & 3 & 4 \end{array} \Rightarrow \begin{array}{cccc} 16 & 14 & 15 & 13 \\ 8 & 6 & 7 & 5 \\ 12 & 10 & 11 & 9 \\ 4 & 2 & 3 & 1 \end{array}$$

Output sequence is: {16,8,12,4,14,6,10,2,15,7,11,3,13,5,9,1}.

3.3. Consider a 16-bit audio stereo signal and calculate how much does the bit rate change when passing through the first three CD encoding stages (CIRC coding, Generating control word, EFM), if the starting bit rate at the input of the coder is $1.4112 \cdot 10^6$ b/s.

Solution:

*CIRC coding*: At the input of the CIRC coder we have 24 words (8 bits each). The coder C2 generates 4 $Q$ words (8 bits each), while the coder C1 generates 4 $P$ words (8 bits each).

At the output of the CIRC coder we have 32 words.

24 words produce the bit rate equal to $1.4112 \cdot 10^6$ b/s $\Rightarrow$

32 words produce the following bit rate:

$$(32/24) \cdot 1.4112 \cdot 10^6 \text{ b/s} = 1.8816 \cdot 10^6 \text{ b/s}.$$

*Generating control word*: The 8-bit control word $(P, Q, R, S, T, U, V, W)$ is assigned to each block.

Before generating the control word, the bit rate was $1.8816 \cdot 10^6$ b/s. At the end of this stage the bit rate becomes:

$$(33/32) \cdot 1.8816 \cdot 10^6 \text{ b/s} = 1.9404 \cdot 10^6 \text{ b/s}.$$

*EFM*: In this stage 8-bit symbols are firstly extended into 14-bit symbols, and then 3 additional bits are embedded between 14-bit combinations. Hence, instead of 8-bit words, we have 17-bit words at the output of the EFM coder, which results in the bit rate:

$$(17/8) \cdot 1.9404 \cdot 10^6 \text{ b/s} = 4.1233 \cdot 10^6 \text{ b/s}.$$

3.4. Consider the Super audio CD with sampling frequency 2.8224 MHz and 1-bit DSD coding. It is recommended that the 74 min of an audio is stored within 4.7 GB. Is it enough memory to store the considered 6-channel audio format?

Solution:

$f_s = 2.8224$ MHz $\Rightarrow 2.8224 \cdot 10^6$ samples/s.

74 min $= 74 \cdot 60$ s $= 4440$ s

*Memory requirements*: $6 \cdot 4440 \cdot 2.8224 \cdot 10^6 \cdot 1$ b $= 75188 \cdot 10^6$ b,

$$\frac{75188 \cdot 10^6}{8} = 9.4 \cdot 10^9 \text{ B} = 8.75 \text{ GB}.$$

Hence, 4.7 GB is not enough to store 74 min of the considered audio format.

3.5. In the case of DVD, the samples of direct left, right and central channel are coded by using 24 bits, while the sampling frequency is 192 KHz. The samples of the three environmental channels are coded by using 16 bits (the sampling frequency is 96 KHz). Calculate the memory requirements for storing 10 min of audio on DVD.

Solution:

In the considered case, we have:

- 3 channels with $192 \cdot 10^3$ samples/s, each coded by using 24 bits
- 3 channels with $96 \cdot 10^3$ samples/s, each coded by using 16 bits

*Memory requirements*: $3 \cdot 24 \cdot 192 \cdot 10^3 \cdot 10 \cdot 60 + 3 \cdot 16 \cdot 96 \cdot 10^3 \cdot 10 \cdot 60$
$= 8294400 \cdot 10^3 + 2764800 \cdot 10^3 = 11059200 \cdot 10^3$ b
$(11059200 \cdot 10^3/8)/(1024^3) = 1.29$ GB

3.6. Having in mind that the sector of a CD contains 98 frames, each frame contains 588 bits, and each sample is coded by using 16 bits (the sampling frequency is 44.1 KHz, stereo channel), calculate the number of sectors that are processed/read within 2 s.

Solution:

One sector of a CD contains 98 frames and the following number of bits:

$$98 \cdot 588 \text{ b} = 57624 \text{ b}.$$

The bit rate for a considered stereo signal is:

$$2 \cdot 44100 \cdot 16 \text{ b/s} = 1411200 \text{ b/s}.$$

Hence, the total number of sectors that are read in 2 s is: $2 \cdot 1411200/57624 = 49$ sectors.

3.7. By using the CD bit rate equal to $4.3218 \cdot 10^6$ b/s, calculate the number of bits which are used for $(P,Q,R,S,T,U,V,W)$ words within 1 s of the audio signal stored on CD?

Solution:

The total number of frames: $\dfrac{4.3218 \cdot 10^6 \text{ b/s}}{588 \text{ b/frames}} = 7350 \text{ frames/s}$

The total number of sectors is: $\dfrac{7350}{98} = 75.$

Each sector contains one of each word type: $P,Q,R,S,T,U,V,W$. Hence the total number of bits used to represent these words is:

$$75 \text{ sectors} \cdot 98 \text{ b/word} \cdot 8 \text{ words} = 58800 \text{ b}$$

3.8. Consider a 16-bit sequence: 1001101010101010, which is fed to the QPSK modulator. The resulting QPSK sequence duration is $T = 4.984$ ms and it is the input of OFDM block. Assuming that the carrier frequency is $f_0 = 1$ KHz, while the inverse Fourier transform is calculated in 768 point, determine the frequencies of sub-carriers $f_1$ and $f_2$.

Solution:

At the output of QPSK modulator, we obtain the sequence of 8 symbols. The QPSK symbol duration is:

$$T_S = \frac{T}{8} = \frac{4.984}{8} \text{ms} = 0.623 \text{ ms}.$$

The frequency of the $k$-th sub-carrier is given by:

$$f_k = f_0 + k\Delta f = f_0 + k\frac{1}{NT_S},$$

Hence, the frequencies of the first two sub-carriers are obtained as:

$$f_1 = f_0 + \Delta f = f_0 + \frac{1}{NT_S} = 1\ \text{KHz} + \frac{1}{768 \cdot 0.623\ \text{ms}} = 1002.09\ \text{Hz},$$

$$f_2 = f_0 + 2\Delta f = f_0 + 2\frac{1}{NT_S} = 1\ \text{KHz} + \frac{2}{768 \cdot 0.623\ \text{ms}} = 1004.18\ \text{Hz}.$$

3.9. Determine the frequency of the sub-carrier $k = 5$ within a certain OFDM system, if the carrier frequency is $f_0 = 2400$ MHz, while the symbols rate is $f_S = 2$ MHz and the total number of sub-carriers is $N = 200$.

Solution:

The frequency of the $k$-th sub-carrier can be calculated as:

$$f_k = f_0 + k\Delta f = f_0 + k\frac{1}{NT_S},$$

where: $f_S = \frac{1}{T_S}$. The frequency of the 5-th sub-carrier is then:

$$f_5 = f_0 + 5\Delta f = f_0 + 5\frac{f_S}{N} = 2400\ \text{MHz} + 5\frac{2}{200}\text{MHz} = 2400.05\ \text{MHz}.$$

3.10. Determine the number of sub-carriers in the OFDM system if the OFDM symbol duration is 3.2 μs, while the total transmission bandwidth is $B = 20$ MHz.

Solution:

For a given OFDM symbol duration $NT_S = 3.2$ μs, we can calculate sub-carrier spacing:

$$\Delta f = \frac{1}{NT_S} = \frac{1}{3.2\,\mu\text{s}} = 312.5\ \text{KHz}.$$

The number of sub-carriers in the OFDM system can be obtained as:

$$N_{sc} = \frac{B}{\Delta f} = 64.$$

## References

1. Bahai A, Saltzberg BR, Ergen M (2004) Multi-carrier digital communications, 2nd edn. Springer, New York
2. Bosi M, Goldberg RE (2003) Introduction to digital audio coding and standards. Springer, New York
3. Frederiksen FB, Prasad R (2002) An overview of OFDM and related techniques towards development of future wireless multimedia communications. IEEE Radio and Wireless Conference, RAWCON 2002: 19–22
4. Furht B (2008) Encyclopedia of Multimedia, 2nd edn. Springer, New York

5. Immink KAS (2002) A survey of codes for optical disk recording. IEEE J Selected Areas Commun 19(4):756–764
6. Hoeg W, Lauterbach T (2003) Digital audio broadcasting: principles and applications of Digital Radio. John Wiley and Sons, Hoboken, NJ
7. Li YG, Stuber GL (2006) Orthogonal Frequency Division Multiplexing for Wireless Communications. Springer, New York
8. Lin TC, Truong TK, Chang HC, Lee HP (2011) A future simplification of procedure for decoding nonsystematic Reed-Solomon codes using the Berlekamp-Massey algorithm. IEEE Trans Commun 59(6):1555–1562
9. Maes J, Vercammen M, Baert L (2002) Digital audio technology, 4th edn. Focal Press, Oxford, In association with Sony
10. Mandal M (2003) Multimedia signals and systems. Springer, New York
11. Orović I, Zarić N, Stanković S, Radusinović I, Veljović Z (2011) Analysis of power consumption in OFDM systems. J Green Eng 1(1):477–489
12. Painter T (2000) Perceptual coding of digital audio. Proc IEEE 88(4):451–513
13. Roth R (2006) Introduction to coding theory. Cambridge University Press, Cambridge
14. Shieh W, Djordjević I (2009) Orthogonal frequency division multiplexing for optical communications. Academic, San Diego, CA
15. Spanias A, Painter T, Atti V (2007) Audio signal processing and coding. Wiley-Interscience, Hoboken
16. Watkinson J (2001) The art of digital audio, 3rd edn. Focal Press, Oxford
17. Wicker SB, Bhargava VK (1999) Reed-Solomon codes and its applications. John Wiley and Sons, Hoboken, NJ

# Chapter 4
# Digital Image

## 4.1 Fundamentals of Digital Image Processing

An image can be represented as a two-dimensional analog function ***f*(*x*,*y*)**. After digitalization, a digital image is obtained and it is represented by a two-dimensional set of samples called pixels. Depending on the number of bits used for pixel representation, a digital image can be characterized as:

- Binary image—each pixel is represented by using one bit.
- Computer graphics—four bits per pixel are used.
- Grayscale image—eight bits per pixel are used.
- Color image—each pixel is represented by using 24 or 32 bits.

Increasing the number of bits reduces the quantization error, i.e., increases the SNR by 6 dB per bit.

Grayscale image with $N_1$ rows and $N_2$ columns contains $N_1 \times N_2$ spatially distributed pixels, and it requires $8 \times N_1 \times N_2$ bits for representation. Color images are represented by using three matrices (for three color channels). Hence, if 8 bits per pixel are used, we need $3 \times 8 \times N_1 \times N_2$ bits of memory to store a color image.

In addition to the spatial distribution of pixels which provides the information about the positions of grayscale values, a pixel value distribution in different image regions can be analyzed as well. Such a distribution can be described by the joint density function:

$$p(x_i) = \sum_{k=1}^{N} \pi_k p_k(x_i), \qquad i = 1, 2, \ldots, M, \tag{4.1}$$

where $x_i$ represents the gray level of the $i$-th pixel, $p_k(x_i)$ is the probability density function (pdf) for a region $k$, and $\pi_k$ is a weighting factor. The pdf for a region $k$ can be described by the generalized Gaussian function:



S. Stanković et al., *Multimedia Signals and Systems*,
DOI 10.1007/978-3-319-23950-7_4

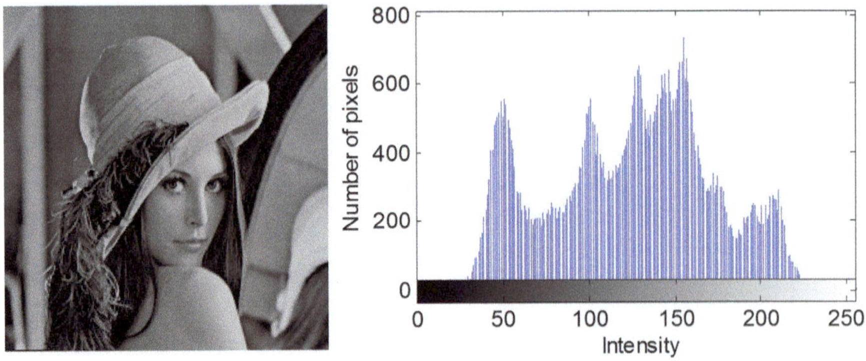


**Fig. 4.1** Histogram of "Lena" image

$$p_k(x_i) = \frac{\alpha\beta_k}{2\Gamma(1/\alpha)} e^{[-|\beta_k(x_i-\mu_k)|^{\alpha}]}, \quad \alpha > 0, \quad \beta_k = \frac{1}{\sigma_k}\left[\frac{\Gamma(3/\alpha)}{\Gamma(1/\alpha)}\right]^{\frac{1}{2}}, \tag{4.2}$$

where $\Gamma(\cdot)$ is the gamma function and $\mu_k$ represents the mean value. The variance $\sigma_k$ is used to calculate $\beta_k$. For $\alpha >> 1$, the pdf becomes uniform. For $\alpha = 2$, the Gaussian distribution is obtained, while for $\alpha = 1$ the Laplace distribution follows. The generalized Gaussian pdf is suitable, because it can be used to describe the image histogram. The image histogram provides important information about the occurrence of certain pixel values, and as such, plays an important role in image analysis. The histogram of a grayscale image "Lena" is given in Fig. 4.1.

## 4.2 Elementary Algebraic Operations with Images

Consider two images of the same dimensions, whose pixels at an arbitrary position $(i,j)$ are denoted as $\boldsymbol{a(i,j)}$ for the first and $\boldsymbol{b(i,j)}$ for the second image. Addition or subtraction of two images is done by adding or subtracting the corresponding pixels of an image, so that the resulting pixel is given in the form: $\boldsymbol{c(i,j) = a(i,j) \pm b(i,j)}$. Multiplying the image by a constant term $\boldsymbol{k}$ can be written as $\boldsymbol{c(i,j) = ka(i,j)}$. However, if we want to represent the result of these and other operations as a new image, we must perform quantization (i.e., rounding to integer values) and limit the results in the range of 0 to 255 (grayscale image is assumed).

Consider now the grayscale images "Baboon" and "Lena" (Fig. 4.2).

Let us perform the following operation: $\boldsymbol{c(i,j) = a(i,j) + 0.3b(i,j)}$, where $\boldsymbol{a(i,j)}$ denotes the pixel belonging to the "Lena" image, while $\boldsymbol{b(i,j)}$ belongs to the "Baboon" image. The result is the image shown in Fig. 4.3.

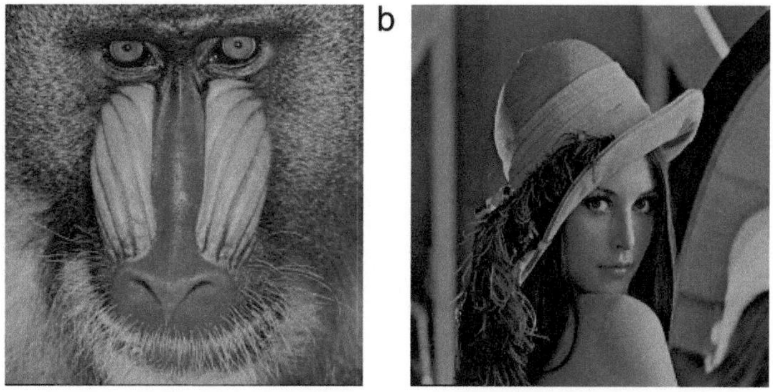

**Fig. 4.2** (**a**) Grayscale image "Baboon," (**b**) Grayscale image "Lena"

**Fig. 4.3** The resulting image obtained by adding 30 % of "Baboon" to "Lena"

**Fig. 4.4** Negative of "Lena" image

To obtain a negative of a grayscale image, we use the following relation:

$$\boldsymbol{n}(\boldsymbol{i},\boldsymbol{j}) = 255 - \boldsymbol{a}(\boldsymbol{i},\boldsymbol{j}).$$

The negative image of "Lena" is shown in Fig. 4.4.

Clipping (cutting the pixels values over a certain level $c_{\max}$ and below a certain level $c_{\min}$) is another mathematical operation used in image processing, and it is defined as:

**Fig. 4.5** Clipped "Lena" image

$$a(i,j) = \begin{cases} c_{\max}, & a(i,j) > c_{\max}, \\ a(i,j), & c_{\max} \geq a(i,j) \geq c_{\min}, \\ c_{\min}, & a(i,j) < c_{\min}. \end{cases} \tag{4.3}$$

For example, consider clipping of image "Lena" with $c_{min} = 100$, $c_{max} = 156$. The result of clipping is shown in Fig. 4.5.

## 4.3 Basic Geometric Operations

Translation of an image ***a*(*i*,*j*)** with dimensions $N_1 \times N_2$ can be represented as moving the pixels in one or both directions for a certain number of positions. In the example shown in Fig. 4.6, we translated the image by embedding 31 rows and 31 columns of black color (zero value), while omitting the last 31 rows and columns. In the case we would like a white surface to appear after translation, the zero values should be replaced by the maximum values (e.g., value 255).

For an image $a(x,y)$ the coordinates can be written by the vector $\begin{bmatrix} x \\ y \end{bmatrix}$. Then the image rotation can be defined by:

$$\begin{bmatrix} X \\ Y \end{bmatrix} = \begin{bmatrix} \cos\theta & -\sin\theta \\ \sin\theta & \cos\theta \end{bmatrix} \begin{bmatrix} x \\ y \end{bmatrix}, \tag{4.4}$$

where $\begin{bmatrix} X \\ Y \end{bmatrix}$ are the new coordinates after rotation (Fig. 4.7).

After image rotation, we need to transform points from the polar coordinate system to the rectangular coordinate system. In general, this transform is performed with certain approximations.

**Fig. 4.6** "Lena" image translated for 31 columns and 31 rows

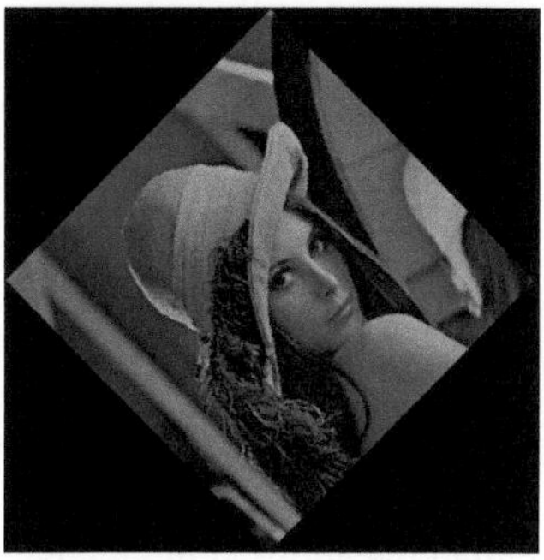

**Fig. 4.7** "Lena" image rotated by 45°

## 4.4 The Characteristics of the Human Eye

By considering the characteristics of human visual system, we can define different image processing algorithms that will meet important perceptual criteria.

One of the specific features of the human eye is sensitivity to the change of light intensity. Specifically, the eye does not perceive the changes in light intensity linearly, but logarithmically. It means that at lower intensity human eye can notice very small changes in brightness, while at high intensity even a much bigger change can hardly be registered.

There are two types of cells in the eye: elongated (rod cells or rods) and cone-like (cone cells or cones). There are about 125 million rods and about 5 million cones. The rods just detect the amount of light, while the cones detect colors. An eye is not equally sensitive to three primary colors: red, green and blue. The relative ratio of these sensitivities is:

$$Red : Green : Blue = 30\% : 59\% \ : \ 11\%$$

An eye is able to identify approximately between 40 and 80 shades of gray, while for color images it can recognize between 15 and 80 million colors. The light entering the eye is detected by the cones and rods. The image in the brain is actually obtained as the sum of images in primary colors. TV sets (CRT, LCD and plasma), monitors, video projectors follow the human three-color model.

It is interesting to note that various models are used to measure the image quality in different applications. Namely, the image quality can be represented by three dimensions: *fidelity, usefulness* and *naturalness*. For example, the usefulness is a major metric for medical imaging, the fidelity is the major metric for paintings, while the naturalness is used in virtual reality applications.

## 4.5 Color Models

Color is one of the most important image characteristics. It is generally invariant to translation, rotation and scaling. The color image can be modeled using various color systems. RGB is one of the commonly used color systems. It can be represented by the color cube as shown in Fig. 4.8. The gray level is defined by the line $R = G = B$. Although the RGB model is based on the human perception of colors, and thus, it has been used for displaying images (monitors, TV, etc.), other color systems have been also defined in order to meet various constraints that exist in the applications.

The RGB model is based on the fact that color can be viewed as a vector function of three coordinates for each position within the image. Sometimes this model is called the additive model, because the image is obtained by adding the components in primary colors. Each point in the image can be represented by the sum of values

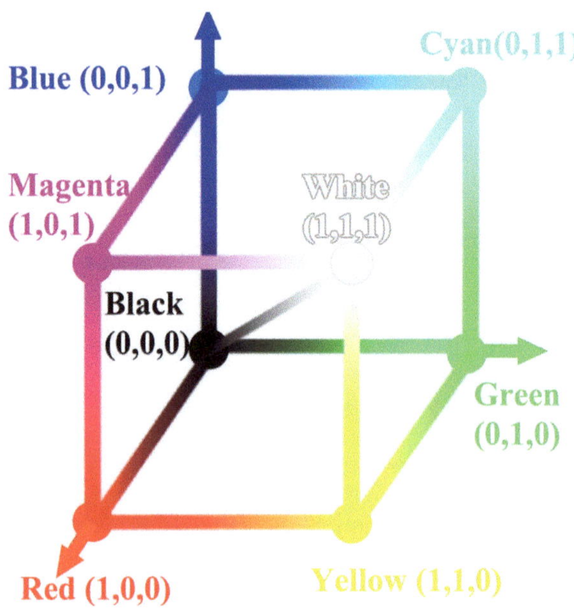


**Fig. 4.8** The color cube

of the three primary colors ($R$, $G$, $B$). A size of an RGB digital image depends on how many bits we use for quantization. For example, for $n = 8$ bits, the values range from 0 to 255. In the RGB model, the value 0 (coordinate = 0) means the absence of color, while the value 255 (coordinate = 1) denotes the color with maximum intensity. Thus, we conclude that (0,0,0) represents black and (1,1,1) represents white. When converting a color image to a grayscale one, the luminescence is calculated as the mean value of the RGB components. By combining two of the three primary colors ($R, G, B$), we get the colors used in the CMY color model, and white color as a sum of all three colors:

$$G + B = C(cyan);\; R + B = M(magenta);$$

$$R + G = Y(yellow);\; R + G + B = W(white).$$

In Fig. 4.8, the color cube is shown in rectangular coordinates. It illustrates the relative position of the RGB and CMY color model.

## 4.5.1 CMY, CMYK, YUV, and HSV Color

The coordinate system in the color space can be formed by using three noncollinear color vectors. Thus, if we choose the basis vectors as follows: $C$—Cyan, $M$—Magenta and $Y$—Yellow, the CMY color model is obtained. This model is basically the most commonly used in printers, because the white is obtained by the absence of colors. Even though, black is obtained by combining all three colors together, the printers usually have a separate cartridge for the black color. The CMY model including the black color is called the CMYK color model. $K$ is used to refer to the black color. The connection between the CMY and RGB models is evident from the color cube:

$$C = 1 - R,\;\; M = 1 - G,\;\; Y = 1 - B, \tag{4.5}$$

while the CMYK model can be obtained as:

$$K = \min(C, M, Y),\;\; C = C - K,\;\; M = M - K,\;\; Y = Y - K. \tag{4.6}$$

Another commonly used system is YUV. Here, the color is represented by three components: luminance ($Y$) and two chrominance components ($U$ and $V$). The YUV is obtained from the RGB by using the following equations:

$$\begin{aligned} Y &= 0.299R + 0.587G + 0.114B, \\ U &= 0.564(B - Y), \\ V &= 0.713(R - Y). \end{aligned} \tag{4.7}$$

It is interesting to note that in the case of $R = G = B$, we have $Y = R = G = B$ which is actually the luminance component, while $U = 0$, $V = 0$.

Special efforts have been made to define the color systems that are more uniform from the standpoint of perceptual sensitivity, such as L*u*v* and L*a*b* systems. Perceptually uniform means that two colors that are equally distant in the color space are equally distant perceptually, which is not the case with the RGB or CMY models (the calculated distance between two colors does not correspond with the perceived difference between the colors). In the L*a*b* model the perceptual color difference is represented by the Euclidean distance:

$$\begin{aligned} \Delta E^*_{ab} &= \left(\Delta L^{*2} + \Delta a^{*2} + \Delta b^{*2}\right)^{\frac{1}{2}} \text{ where} \\ \Delta L^* &= L_1^* - L_2^* \\ \Delta a^* &= a_1^* - a_2^* \\ \Delta b^* &= b_1^* - b_2^* \end{aligned} \tag{4.8}$$

The L*a*b* model can be obtained from the RGB by the following transformations:

$$\begin{bmatrix} X \\ Y \\ Z \end{bmatrix} = \begin{bmatrix} 0.490 & 0.310 & 0.200 \\ 0.177 & 0.813 & 0.011 \\ 0.000 & 0.010 & 0.990 \end{bmatrix} \begin{bmatrix} R \\ G \\ B \end{bmatrix}, \tag{4.9}$$

$$\begin{aligned} L* &= 25\left(\frac{100Y}{Y_0}\right)^{\frac{1}{3}} - 16, \\ a* &= 500\left[\left(\frac{X}{X_0}\right)^{\frac{1}{3}} - \left(\frac{Y}{Y_0}\right)^{\frac{1}{3}}\right], \\ b* &= 200\left[\left(\frac{Y}{Y_0}\right)^{\frac{1}{3}} - \left(\frac{Z}{Z_0}\right)^{\frac{1}{3}}\right]. \end{aligned} \tag{4.10}$$

The condition $1 \leq 100Y \leq 100$ should be satisfied in Eq. (4.10). The value $(X_0, Y_0, Z_0)$ represents reference white. On the basis of this system, we can introduce the HSV color system that is more oriented towards the perceptual model. The HSV color system is represented by a cylindrical coordinate system as shown in Fig. 4.9.

This system is based on the three coordinates: $H$, $S$ and $V$. H is a measure of the spectral composition of color, while $S$ provides information about the purity of color, or more accurately, it indicates how far is the color from the gray level, under the same amount of luminescence. $V$ is a measure of the relative luminescence. The component $H$ is measured by the angle around the $V$ axis, ranging from 0° (red) to

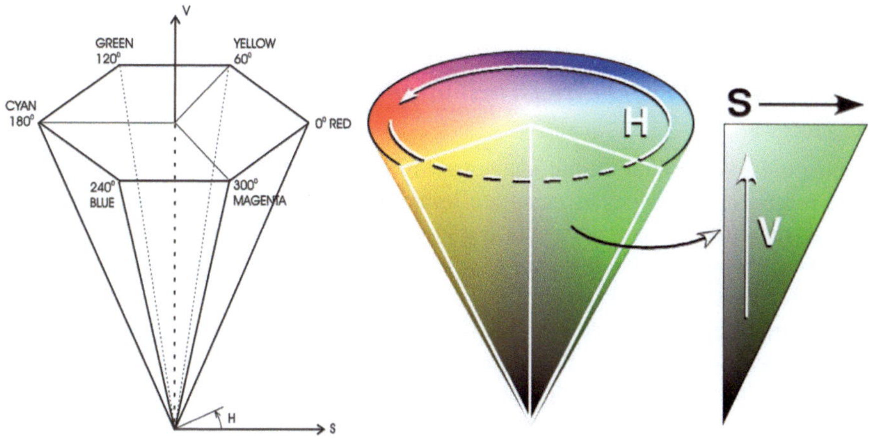


**Fig. 4.9** The HSV color model

360°. Along the $V$ axis, the luminance is changed from black to white. The value of the $H$, $S$ and $V$ can be defined by using the RGB model as follows:

$$H_1 = \cos^{-1}\left(\frac{\frac{1}{2}[(R-G)+(R-B)]}{\sqrt{(R-G)^2+(R-B)(G-B)}}\right), \quad \begin{array}{ll} H = H_1 & for\ B \le G, \\ H = 360^\circ - H_1 & for\ B > G, \end{array} \tag{4.11}$$

$$S = \frac{\max(R,G,B) - \min(R,G,B)}{\max(R,G,B)}, \tag{4.12}$$

$$V = \frac{\max(R,G,B)}{255}. \tag{4.13}$$

The HSV model is suitable for face detection and tracking algorithms. The thresholds that define the human face color are specified as:

$$\begin{aligned} 340^\circ \le H \le 360^\circ \ \text{ and } \ 0^\circ \le H \le 50^\circ, \\ S \ge 20\%, \\ V \ge 35\%. \end{aligned} \tag{4.14}$$

Having in mind the coordinate system of this color model, we may observe that the previously given intervals are wide, which may lead to the false detection of the object that actually does not represent the face, but have a similar color information. In order to avoid this possibility, additional analyses are required.

## 4.6 Filtering

### *4.6.1 Noise Probability Distributions*

Image noise may occur during image transmission over a communication channel. The most common types of noise are impulse noise and Gaussian noise. Impulse noise is manifested as a set of black and white pulses in the image (Fig. 4.10). It occurs as a result of atmospheric discharges, or due to electromagnetic field generated by various appliances.

If impulse noise takes two fixed values: $a$ (negative impulse) and $b$ (positive impulse), with equal probabilities $p/2$, we will have the two-sided impulse noise model. In this case, an image with impulse noise can be defined as:

$$f_I(i,j) = \begin{cases} a, & \text{with a probability } p/2; \\ b, & \text{with a probability } p/2; \\ f(i,j), & \text{with a probability } (1-p). \end{cases} \tag{4.15}$$

Thermal noise is usually modeled as the white Gaussian one and its distribution is given by:

$$P_g(x) = \frac{1}{\sqrt{2\pi}\sigma} e^{-\frac{(x-\mu)^2}{2\sigma^2}}, \tag{4.16}$$

where $\mu$ is the mean value, while $\sigma^2$ denotes the variance of noise ($\sigma$ is the standard deviation of noise). Figure 4.11 demonstrates "Lena" image affected by white Gaussian noise.

Beside the impulse and Gaussian noise, the uniformly distributed noise can appear. The gray level values of the noise are evenly distributed across a specific range. The quantization noise can be approximated by using uniform distribution. The corresponding pdf is defined as:

**Fig. 4.10** "Lena" affected by an impulse noise with density 0.05

**Fig. 4.11** "Lena" affected by zero-mean white Gaussian noise whose variance is equal to 0.02

$$P_u(x) = \begin{cases} \dfrac{1}{b-a}, & \text{for } a \le x \le b \\ 0, & \text{otherwise.} \end{cases} \tag{4.17}$$

The mean value and variance of the uniform density function are:
$\mu = (a+b)/2$ and $\sigma^2 = (b-a)^2/12$, respectively.
Radar images may contain noise characterized by the Rayleigh distribution:

$$P_R(x) = \begin{cases} \dfrac{2}{\beta}(x-\alpha)e^{-(x-\alpha)^2/\beta}, & \text{for } x \ge \alpha \\ 0, & \text{otherwise,} \end{cases} \tag{4.18}$$

with the mean equal to $\mu = \alpha + \sqrt{\pi\beta/4}$ and the variance $\sigma^2 = \beta(4-\pi)/4$.

## 4.6.2 *Filtering in the Spatial Domain*

Filtering of noisy images intends to reduce noise and to highlight image details. For this purpose, the commonly used filters in the spatial domain are the mean and median filters. Use of these filters depends on the nature of the noise that is present within the image. Spatial domain filters are especially suitable in the cases when additive noise is present.

### 4.6.2.1 Mean Filter

Mean filters are used to filter the images affected by the Gaussian white noise, since it is based on calculating the average pixel intensity within an image part captured by a specified window. The filter should use a small number of points within the window to avoid blurring of image details. Note that a larger window would provide better noise filtering.

Consider a window of the size $(2N_1+1)\times(2N_2+1)$. The signal $f(i,j)$ is affected by the noise $n(i,j)$ and the noisy signal is:

$$x(i,j) = f(i,j) + n(i,j). \tag{4.19}$$

The output of the arithmetic mean filter is defined by the relation:

$$x_f(i,j) = \frac{1}{(2N_1+1)(2N_2+1)} \sum_{n=i-N_1}^{i+N_1} \sum_{m=j-N_2}^{j+N_2} x(n,m), \tag{4.20}$$

where $x(n,m)$ is the pixel value within the window, while the impulse response of the filter is $h(i,j) = 1/((2N_1+1)(2N_2+1))$. The output of this filter is actually the mean value of pixels captured by the window. For a window size $3\times 3$, we deal with 9 points, while the $5\times 5$ window includes 25 points. From the aspect of noise reduction, the $5\times 5$ window will be more effective. However, it will introduce more smoothed edges and blurred image (Fig. 4.12).

As an example, "Lena" image affected by a zero-mean Gaussian noise with variance 0.02, and its filtered versions are shown in Fig. 4.13.

Instead of the arithmetic mean filter, the geometric mean can be used as well, where the filter output is given by:

$$x_f(i,j) = \left( \prod_{n=i-N_1}^{i+N_1} \prod_{m=j-N_2}^{j+N_2} x(n,m) \right)^{\frac{1}{(2N_1+1)(2N_2+1)}}. \tag{4.21}$$

The geometric mean filter introduces less blurring and preserves more image details (Figs. 4.14 and 4.15).

#### 4.6.2.2 Median Filter

Median filters are used to filter out the impulse noise. Consider a sequence with an odd number of elements. After sorting the elements, the median value is obtained as the central element. In a sequence with an even number of elements, the median is calculated as the mean of two central elements of the sorted sequence.

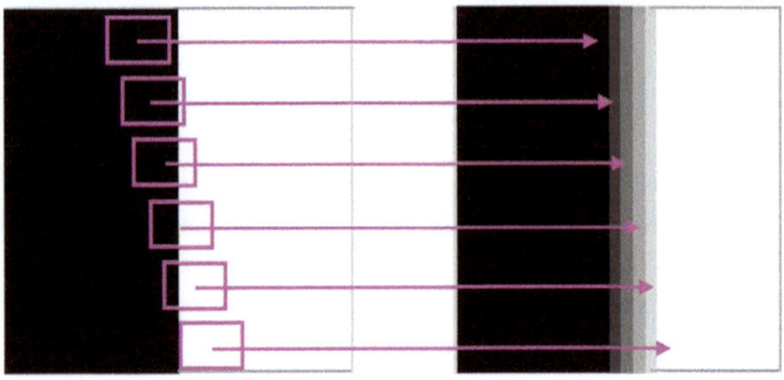

**Fig. 4.12** An illustration of blurring after applying mean filter on the image edge (the mask size used is $5\times 5$)

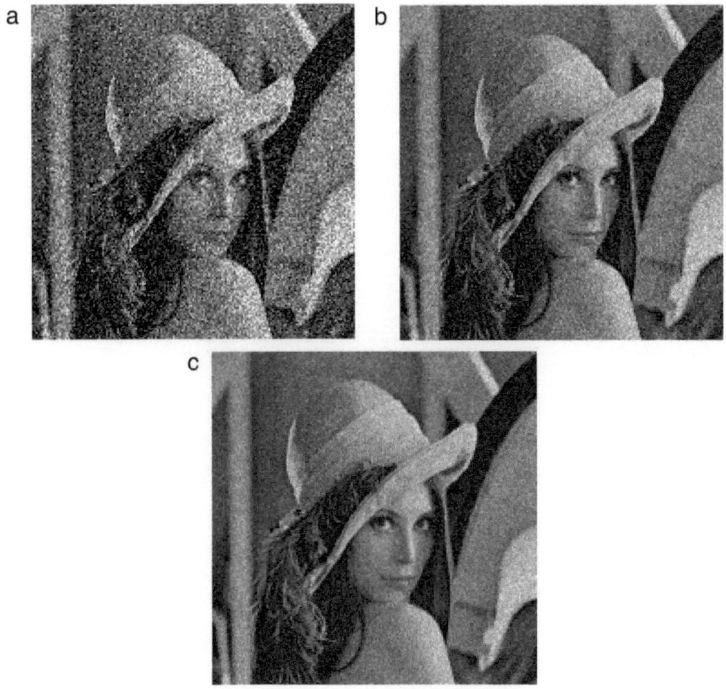

**Fig. 4.13** (**a**) "Lena" affected by Gaussian noise with zero mean and variance 0.02, (**b**) filtered image obtained by using mean filter of size $3 \times 3$, (**c**) filtered image obtained by using mean filter of size $5 \times 5$

## Example

The sequence is given as:

| 3 | 14 | 7 | 1 | 5 |
|---|---|---|---|---|

Sort the numbers in the ascending order:

| 1 | 3 | 5 | 7 | 14 |
|---|---|---|---|---|

and then the median is the central element 5.

Consider now a sequence with an even number of elements:

| 1 | 12 | 7 | 9 | 4 | 2 |
|---|---|---|---|---|---|

Sort the elements in ascending order:

| 1 | 2 | 4 | 7 | 9 | 12 |
|---|---|---|---|---|---|

4 and 7 are the two central elements, and the median is equal to their mean value, or 5.5.

**Fig. 4.14** (**a**) Original image, (**b**) noisy image (Gaussian noise with 0.05 mean and variance 0.025), (**c**) image filtered by using arithmetic mean of size 3 × 3, (**d**) image filtered by using geometric mean of size 3 × 3

The median filter is applied in image denoising by using a rectangular window that slides over the entire image. The elements captured by the window are reordered as a vector $\mathbf{x}$:$\{x(k), k\in[1,N]\}$ and then the median value $x_m$ for the vector is calculated as follows:

$$x_m = median(x(1), \ldots, x(k), \ldots, x(N)) = \begin{cases} x_s(\lfloor N/2 \rfloor + 1), & N \text{ is odd}, \\ \dfrac{x_s(N/2) + x_s(N/2+1)}{2}, & N \text{ is even}, \end{cases} \tag{4.22}$$

where $\mathbf{x_s}$ is the sorted version of $\mathbf{x}$, while $\lfloor \cdot \rfloor$ denotes the integer part of a positive number. Another way to calculate the median of a matrix is to calculate the median value for columns and then for rows (or vice versa). Generally, these two approaches usually do not produce exactly the same result.

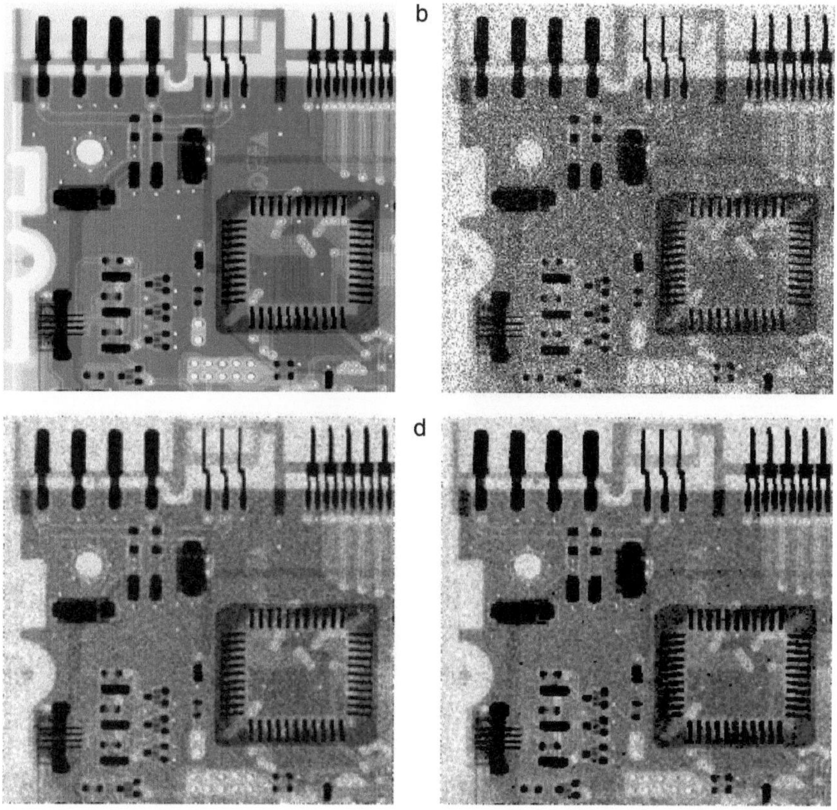


**Fig. 4.15** (**a**) Original image, (**b**) noisy image (Gaussian noise with 0.05 mean and variance 0.025), (**c**) image filtered by using arithmetic mean of size $3 \times 3$, (**d**) image filtered by using geometric mean of size $3 \times 3$

Suppose that the filter window covers $(2N_1 + 1)\ (2N_2 + 1)$ pixels. The pixel $x(i, j)$ is the central one in the filter window. From all the pixels within the window, we form a matrix:

$$\begin{bmatrix} x(i-N_1, j-N_2) & \dots & x(i-N_1, j) & \dots & x(i-N_1, j+N_2) \\ \vdots & & \vdots & & \vdots \\ x(i, j-N_2) & \dots & x(i, j) & \dots & x(i, j+N_2) \\ \vdots & & \vdots & & \vdots \\ x(i+N_1, j-N_2) & \dots & x(i+N_1, j) & \dots & x(i+N_1, j+N_2) \end{bmatrix}$$

The first step is to sort the elements within columns and to determine the median for each column. The second step uses the median values of columns and calculates the median again. Mathematically, it is described as:

**Fig. 4.16** (**a**) "Lena" affected by impulse noise with density 0.05, (**b**) denoised image (median filter of size 3 × 3 is used), (**c**) denoised image (5 × 5 median filter)

$$\begin{aligned} q_n &= median\{x(i-N_1,j), x(i-N_1+1,j), \ldots, x(i+N_1,j)\}, \quad \text{for } n=j, \\ q(i,j) &= median\{q_n;\ n \in (j-N_2, \ldots, j+N_2)\}. \end{aligned} \tag{4.23}$$

Therefore, $q(i,j)$ represents the output of separable median filter. An application of the median filter to image "Lena" affected by the impulse noise is illustrated in Fig. 4.16.

The α-trimmed mean filter has been introduced as a good compromise between the median and arithmetic mean filter. Namely, after sorting the windowed pixels, we discard a few lowest and highest samples, while the remaining pixels are averaged. The $\alpha$-trimmed mean filter can be defined as:

$$x_\alpha(i,j) = \frac{1}{(N-2[\alpha N])} \sum_{n=[\alpha N]+1}^{N-[\alpha N]} x_s(n), \tag{4.24}$$

where $x_s(n)$ is the vector of sorted pixels from the window $N_1 \times N_2, N = N_1 N_2$, $[\alpha N]$ denotes rounding to the greatest integer not greater than $\alpha N$. The parameter $\alpha$ takes the value from the range: $0 \leq \alpha < 0.5$. Note that this filter form corresponds to the median for $\alpha = 0.5$ (for odd $N$), while for $\alpha = 0$ it performs as a moving average filter. Alternatively, we can apply the same operation separately on the rows and columns as follows:

$$x_\alpha(i,j) = \frac{1}{(N_1 - 2[\alpha N_1])(N_2 - 2[\alpha N_2])} \sum_{n=[\alpha N_1]+1}^{N_1-[\alpha N]} \sum_{m=[\alpha N_2]+1}^{N_2-[\alpha N]} x(m,n). \tag{4.25}$$

### 4.6.3 Filtering in the Frequency Domain

Filters in the frequency domain are designed on the basis of a priori knowledge about signal frequency characteristics. The most significant frequency content of images is mostly concentrated at low frequencies. Therefore, in many applications, the images are usually filtered with low-pass filters. The ideal rectangular separable low-pass filter has the following transfer function:

$$H(\omega_1,\omega_2) = \begin{cases} 1, & |\omega_1| \le W_1 \text{ and } |\omega_2| \le W_2 \\ 0, & \text{otherwise.} \end{cases} \tag{4.26}$$

A band-pass filter can be defined as:

$$H(\omega_1,\omega_2) = \begin{cases} 1, & W_{11} \le |\omega_1| \le W_{12}\,,\ W_{21} \le |\omega_2| \le W_{22} \\ 0, & \text{otherwise.} \end{cases} \tag{4.27}$$

In addition to rectangular, a circular low-pass filter can be used:

$$H(\omega_1,\omega_2) = \begin{cases} 1, & \omega_1^2 + \omega_2^2 \le W, \\ 0, & \text{otherwise.} \end{cases} \tag{4.28}$$

Filtering images with a high-pass filter provides high-frequency components that contain the image details:

$$H(\omega_1,\omega_2) = \begin{cases} 1, & (|\omega_1| > W_1 \text{ and } |\omega_2| > W_2) \text{ or } \left(\omega_1^2 + \omega_2^2 > W\right), \\ 0, & \text{otherwise.} \end{cases} \tag{4.29}$$

***L-estimate space-varying filtering***

Certain types of images such as interferograms, textures and surface images, are characterized by a specific frequency content that can be efficiently processed using the concept of space-varying filtering based on the space/spatial-frequency (SSF) representation. Particularly, in the presence of mixed Gaussian and heavy-tailed noise, the filter formulation based on the robust L-estimate forms of SSF representations can be used. The L-estimate representation is exploited to identify important components and to filter them out from the scattered noise. The filtering approach comprises two filter forms as follows:

1. ***α-trimmed approach in SSF domain.*** The calculation of the 2D L-estimate STFT of image $f(x, y)$ using $N \times N$ rectangular unit window is performed in this phase:

$$STFT_L(x,y,k_x,k_y) = \sum_{p=0}^{N-1}\sum_{q=0}^{N-1} a_p a_q \left(r_{p,q}(x,y,k_x,k_y) + j \cdot i_{p,q}(x,y,k_x,k_y)\right), \tag{4.30}$$

   where $r_{p,q}$ and $i_{p,q}$ are sequences sorted in non-decreasing order:

$$\begin{aligned} r_{p,q}(x,y,k_x,k_y) &\in \left\{\mathrm{Re}\left(f_{x,y,u,v}\right),\ u,v \in [-N/2, N/2-1)\right\}, \\ i_{p,q}(x,y,k_x,k_y) &\in \left\{\mathrm{Im}\left(f_{x,y,u,v}\right),\ u,v \in [-N/2, N/2-1)\right\}, \end{aligned} \tag{4.31}$$

   while

$$f_{x,y,u,v} = f(x+u, y+v)e^{-j2\pi\left(uk_x + vk_y\right)/N}. \tag{4.32}$$

   Due to the sorting operation, the coefficients corrupted by noisy pulses will be located at the ends of the sorted sequence. The weights $a_p$ and $a_q$ are designed in analogy with coefficients of α-trimmed filter and they will have zero values at the position corresponding to the ends of sorted sequence. It further means that the coefficients at the ends of sorted sequence will be set to zero and the mean value is calculated using the remaining ones.
2. ***2D Space-varying filtering***. The output of the 2D space-varying filtering (pseudo form) can be calculated according to:

$$y(x,y) = \sum_{k_x}\sum_{k_y} L_H(x,y,k_x,k_y) STFT_L(x,y,k_x,k_y). \tag{4.33}$$

   Recall that the corresponding 1D form is known as time-varying filtering which is defined in Chap. 1. The support function $L_H$ of the nonstationary space-varying filter can be defined in the form:

$$L_H(x,y,k_x,k_y) = \begin{cases} 1 & \text{for} \quad (x,y,k_x,k_y) \in D \\ 0 & \text{for} \quad (x,y,k_x,k_y) \notin D \end{cases}. \tag{4.34}$$

The support region $D$, in practical realizations, can be simply calculated as:

$$D = \{(x, y, k_x, k_y) : |SSFR_L(x, y, k_x, k_y)| > \xi\}, \tag{4.35}$$

where $\xi$ represents the energy floor, *SSFR* denotes a space/spatial-frequency representation, while $L$ indicates the L-estimate form (for instance, we can employ the 2D L-estimate spectrogram or the 2D L-estimate S-method).

### 4.6.4 Image Sharpening

A blurred noisy image in the Fourier domain can be written as:

$$X(u, v) = H(u, v)F(u, v) + N(u, v), \tag{4.36}$$

where $X(u,v)$ is the Fourier transform of blurred image, $H(u,v)$ is the impulse response of the system that induces blurring (degradation), $F(u,v)$ is the Fourier transform of the original image, and $N(u,v)$ is the Fourier transform of noise. For example, if the blurring is produced by the uniform linear motion between the image and the sensor (during image acquisition) along the $x$ axis, then the degradation function can be defined by:

$$H(u, v) = T\frac{\sin(\pi nu)}{\pi nu}e^{-j\pi nu}, \tag{4.37}$$

where $n$ is the distance of pixels displacement, while $T$ is the duration of the exposure. Sharpening of the image is achieved based on the following relation:

$$F(u, v) = \frac{X(u, v)}{H(u, v)}. \tag{4.38}$$

### 4.6.5 Wiener Filtering

The Wiener filter is defined in the theory of optimal signal estimation. It is based on the equation:

$$f_e(i, j) = L[f(i, j)], \tag{4.39}$$

where $L$ is a linear operator meaning that the estimated values are a linear function of the original (degraded) values. The estimated values are obtained such that the mean square error:

$$E\left\{(f(i, j) - f_e(i, j))^2\right\}, \tag{4.40}$$

is minimized. The Wiener filter in the frequency domain is obtained in the form:

$$H_w(u,v) = \frac{H^*(u,v)}{|H(u,v)|^2 + \frac{S_n(u,v)}{S_f(u,v)}}, \tag{4.41}$$

where $S_n(u,v)$ and $S_f(u,v)$ represent the power spectrum of the noise and the signal, respectively: $S_n(u,v) = |N(u,v)|^2$, $S_f(u,v) = |F(u,v)|^2$. $H^*(u,v)$ is the complex conjugate of the degradation function and for $H^*(u,v) = 1$, Eq. (4.41) becomes:

$$H_w(u,v) = \frac{S_f(u,v)}{S_f(u,v) + S_n(u,v)} = 1 - \frac{S_n(u,v)}{S_x(u,v)}. \tag{4.42}$$

It is assumed that the signal and noise are uncorrelated: $S_x(u,v) = S_f(u,v) + S_n(u,v)$. Note that when $S_n$ tends to zero, the filter function is approximately equal to 1 (no modification of the signal), while in the case when $S_f$ tends to zero, the filter function is zero. The filtered spectrum can be written as:

$$F_e(u,v) = H_w(u,v)X(u,v), \tag{4.43}$$

where $X(u,v)$ is the spectrum of noisy image. The noise should be measured in the time intervals when the signal is not present (e.g., consider a communication channel without signal during an interval of time), and then the estimated noise spectrum is available for the calculation of $H_w(u,v)$.

## 4.7 Enhancing Image Details

An image can be represented in terms of its illumination and reflectance components:

$$a(i,j) = a_i(i,j)a_r(i,j), \tag{4.44}$$

where $a_i$ is the illumination describing the amount of incident light on the observed scene, while $a_r$ is the reflectance component describing the amount of light reflected by the objects. It is usually assumed that the scene illumination varies slowly over space, while the reflectance varies rapidly especially on the transitions between different objects. Hence, the illumination and the reflectance components are associated with low and high frequencies, respectively. Usually, the goal is to extract the reflectance component and to minimize the illumination effect, which can be done by using the logarithm to transform multiplicative into additive procedure:

$$\log(a(i,j)) = \log(a_i(i,j)) + \log(a_r(i,j)). \tag{4.45}$$

Having in mind their frequency characteristics, we can separate these two components of images and emphasize the details.

## 4.8 Analysis of Image Content

The distribution of colors and textures are considered as two important features for the analysis of image content.

### 4.8.1 The Distribution of Colors

The distribution of colors can be described by using histogram. If we want to search an image database, we can achieve it by comparing the histogram of the sample image $Q$ and the histogram of each image $I$ from the database. Suppose that both histograms have $N$ elements. Comparison is done by calculating the total number of pixels that are common to both histograms:

$$S = \sum_{i=1}^{N} \min(I_i, Q_i). \tag{4.46}$$

This amount is often normalized by the total number of pixels in one of the two histograms. Having in mind that this method is computationally demanding, the modified forms have been considered. Namely, by using a suitable color model, an image can retain its relevant properties even with a coarser representation. Hence, a significant computational savings can be achieved.

A computationally efficient method for comparison of color images can be obtained if colors are represented by fewer bits. For example, if each color is reduced to 2 bits, then we have 64 possible combinations in the case of three colors.

The colorfulness of images can be described by using the color coherence vectors. Assume that the total number of colors is $N$, the color coherence vectors for images $Q$ and $I$ are given by:

$$\left[\left(\alpha_1^Q, \beta_1^Q\right), \ldots, \left(\alpha_N^Q, \beta_N^Q\right)\right] \text{ and } \left[\left(\alpha_1^I, \beta_1^I\right), \ldots, \left(\alpha_N^I, \beta_N^I\right)\right],$$

where $\alpha_i$ and $\beta_i$ represent the number of coherent and incoherent pixels for color $i$, respectively. Coherent pixels are those that belong to a region characterized by the same color. A difference between two images can be calculated by using the following formula:

$$\text{dist}(Q, I) = \sum_{i=1}^{N} \left( \left| \frac{\alpha^Q{}_i - \alpha^I{}_i}{\alpha^Q{}_i + \alpha^I{}_i + 1} \right| + \left| \frac{\beta^Q{}_i - \beta^I{}_i}{\beta^Q{}_i + \beta^I{}_i + 1} \right| \right). \tag{4.47}$$

### 4.8.2 Textures

A texture is an important characteristic of the image surface. There are different methods and metrics for texture analysis. For instance, the textures can be described by using the following properties: contrast, directionality and coarseness.

The ***contrast*** can be quantified by the statistical distribution of pixel intensities. It is expressed as:

$$Con = \frac{\sigma}{K^{1/4}}, \tag{4.48}$$

where $\sigma$ is the standard deviation, $K$ is the kurtosis, defined by:

$$K = \frac{\mu_4}{\sigma^4}, \tag{4.49}$$

while $\mu_4$ is the fourth moment about the mean. The presented definition is a global measure of the contrast obtained for the entire image.

***Coarseness*** represents a measure of texture granularity. It is obtained as the mean value calculated over windows of different sizes $2^k \times 2^k$, where $k$ is usually between 1 and 5. The windowing and averaging is done for each image pixel. Consider a pixel at the position $(x,y)$. The mean value within the window of size $2^k \times 2^k$ is defined as:

$$A_k(x, y) = \sum_{i=x-2^{k-1}}^{x+2^{k-1}-1} \sum_{j=y-2^{k-1}}^{y+2^{k-1}-1} \frac{a(i, j)}{2^{2k}}, \tag{4.50}$$

where $a(i,j)$ is the grayscale pixel value at the position $(i,j)$. Then the differences between mean values in the horizontal and vertical directions are calculated as follows:

$$\begin{aligned} D_{kh} &= \left|A_k\left(x + 2^{k-1}, y\right) - A_k\left(x - 2^{k-1}, y\right)\right|, \\ D_{kv} &= \left|A_k\left(x, y + 2^{k-1}\right) - A_k\left(x, y - 2^{k-1}\right)\right|. \end{aligned} \tag{4.51}$$

Using the above equations, we choose the value of $k$, which yields the maximum values for $D_{kh}$ and $D_{kv}$. The selected $k$ is used to calculate the ***optimization parameter***:

$$g(x, y) = 2^k. \tag{4.52}$$

Finally, the measure of granulation can be expressed in the form:

$$Gran = \frac{1}{mn} \sum_{i=1}^{m} \sum_{j=1}^{n} g(i, j). \tag{4.53}$$

In order to reduce the number of calculations, the measure of granularity can be calculated for lower image resolution.

***Directionality*** is the third important texture feature. As a measure of directionality, at each pixel we calculate a gradient vector, whose amplitude and angle are given as:

$$|\Delta G| = (|\Delta_H| + |\Delta_V|)/2,$$
$$\varphi = \arctan\left(\frac{\Delta_V}{\Delta_H}\right) + \frac{\pi}{2}, \qquad (4.54)$$

where horizontal $\Delta_H$ and vertical $\Delta_V$ differences are calculated over $3 \times 3$ window around a pixel. After determining the above parameters for each pixel, we can draw a histogram of angle values $\varphi$, taking only those pixels where $|\Delta G|$ is larger than a given threshold. The resulting histogram will have dominant peaks for highly directional images, while for non-directional images it will be flatter.

### 4.8.3 Co-occurrence Matrix

A simplified method to measure the contrast of textures can be performed by using the co-occurrence matrices. First, we form the co-occurrence matrices that show how many times the values $y$ appear immediately after the $x$ values. For example, consider the following sample matrix:

| 1 | 1 | 1 | 2 | 3 |
|---|---|---|---|---|
| 1 | 1 | 1 | 2 | 3 |
| 1 | 1 | 1 | 2 | 3 |
| 1 | 1 | 1 | 3 | 4 |

The corresponding co-occurrence matrix is then obtained as:

| $x$ $y$ | 1 | 2 | 3 | 4 |
|---|---|---|---|---|
| 1 | 8 | 3 | 1 | 0 |
| 2 | 0 | 0 | 3 | 0 |
| 3 | 0 | 0 | 0 | 1 |
| 4 | 0 | 0 | 0 | 0 |

Let us analyze the numbers in the matrix. The number 8 means that 1 occurs 8 times after 1, while 3 denotes that value 2 occurs after 1 three times. The expression for the measure of the texture contrast is given by:

$$Con = \sum_{x=0}^{N-1}\sum_{y=0}^{N-1}(x-y)^2 c(x,y), \qquad (4.55)$$

where $c(x,y)$ represents the elements of the co-occurrence matrix of size $N \times N$. If there are significant variations in the image, $c(x,y)$ will be concentrated outside the main diagonal and contrast measures will have greater values. The co-occurrence matrix with values concentrated on its diagonal corresponds to a homogeneous region.

There are other useful features that can be computed from the co-occurrence matrix, as listed below:

$$Energy : \sum_{x=0}^{N-1}\sum_{y=0}^{N-1} c^2(x,y),$$

$$Entropy : -\sum_{x=0}^{N-1}\sum_{y=0}^{N-1} c(x,y)\log_2 c(x,y),$$

$$Homogeneity : \sum_{x=0}^{N-1}\sum_{y=0}^{N-1} \frac{c(x,y)}{1+|x-y|},$$

$$Correlation : \sum_{x=0}^{N-1}\sum_{y=0}^{N-1} \frac{(x-\mu_x)(y-\mu_y)c(x,y)}{\sigma_x\sigma_y}.$$

### 4.8.4 Edge Detection

Edge detection plays an important role in a number of applications. Consider an image with pixels $a(i,j)$. Edges of the image should be obtained by simple differentiation. However, bearing in mind that the image is always more or less affected by noise, the direct application of differentiation is not effective. For this purpose, several algorithms have been defined, and among them the most commonly used one is based on the Sobel matrices (for vertical and horizontal edge). Specifically, the image is analyzed pixel by pixel using the Sobel matrix as a mask. The matrix elements are the weights that multiply the pixels within the mask. Then the sum is calculated by adding all the obtained values. The resulting value is compared with a threshold. If it is greater than the threshold, the central pixel belongs to the edge, and vice versa.

The Sobel matrices for vertical and horizontal edges are given by:

$$S_v = \begin{bmatrix} 1 & 0 & -1 \\ 2 & 0 & -2 \\ 1 & 0 & -1 \end{bmatrix} \quad S_h = \begin{bmatrix} 1 & 2 & 1 \\ 0 & 0 & 0 \\ -1 & -2 & -1 \end{bmatrix}$$

The edges are obtained by:

$$L(i,j) = \sum_{m=-1}^{1}\sum_{n=-1}^{1} a(i+m, j+n)S(m+2, n+2), \tag{4.56}$$

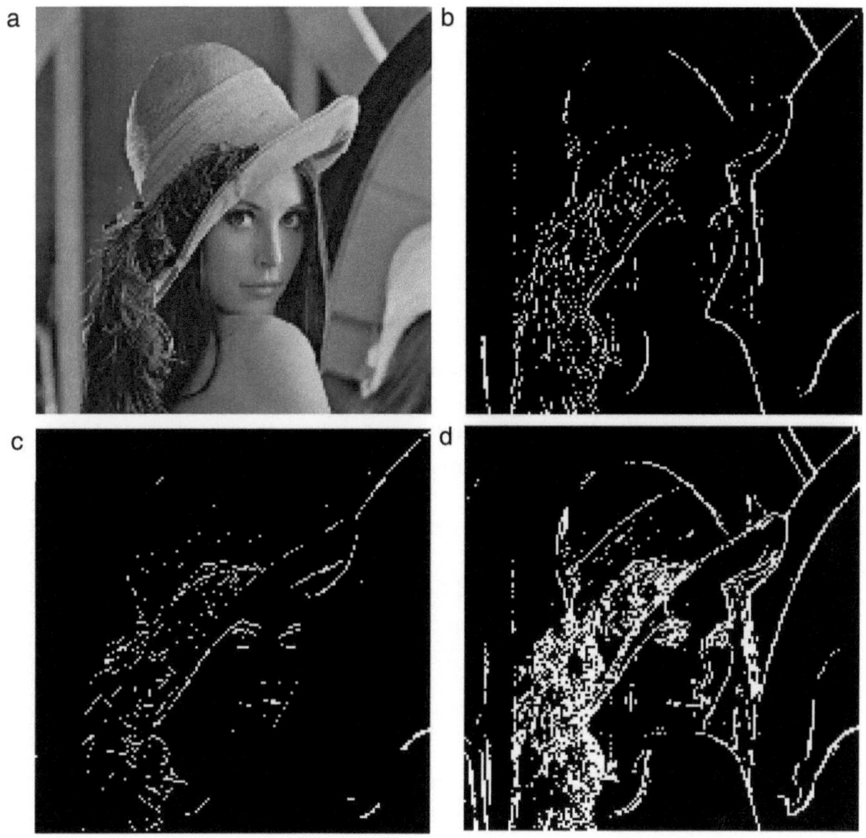


**Fig. 4.17** An illustration of edge detection: (**a**) original image, (**b**) $L_v$, (**c**) $L_h$, (**d**) $L$

where $S(m,n)$ is a filtering function (e.g., the Sobel matrix $S_h$ or $S_v$). After calculating $L_h$ and $L_v$ (using the horizontal and vertical matrix), the overall $L$ is calculated as follows:

$$L = \sqrt{{L_h}^2 + {L_v}^2}. \tag{4.57}$$

The obtained values (for all pixels) are compared with a threshold and the results are represented in a binary form. An example of edge detection is illustrated in Fig. 4.17. For simplicity, the threshold was set to 100 for the entire image.

However, the local threshold values are frequently used in practical applications. They are calculated based on the mean response of the edge detector around the current pixel. For example, a threshold value can be calculated as:

$$T(i,j) = \overline{L}(i,j)(1+p) = \frac{1+p}{2N+1}\sum_{k=i-N}^{i+N}\sum_{l=j-N}^{j+N} L(k,l), \tag{4.58}$$

where $p$ has a value between 0 and 1.

### 4.8.5 The Condition of the Global Edge (Edge Based Representation: A Contour Image)

An algorithm for edge-based image representation is described in the sequel. First, the image is normalized by applying the affine transformation which results in the square image of the size 64 × 64. Then the gradient is calculated for each pixel:

$$\partial_{i,j} = \frac{|\Delta p_{i,j}|}{|I_{i,j}|}, \tag{4.59}$$

where the numerator represents the maximum of the differences between the intensity of a given pixel and the intensity of its neighbors. The denominator is the local power of pixel intensities. The calculated gradient is compared with the sum of the mean and the variance of original image:

$$\partial_{i,j} \geq \mu + \sigma. \tag{4.60}$$

Pixels that fulfill the condition Eq. (4.60) are called the global edge candidates. Now, from pixels selected as the global edge candidates, we reselect the pixels for which:

$$\partial_{i,j} \geq \mu_{i,j} + \sigma_{i,j}, \tag{4.61}$$

holds, where $\mu$ and $\sigma$ are mean and variance of the local gradient to its neighbors. They are called the local edge candidates.

### 4.8.6 Dithering

One of the properties of the human eye is that when observing a small area from a long distance, it perceives just the overall intensity as a result of averaging granular details. This feature is used in dithering, where a group of points represents a color. Consider a simple example by using four points:

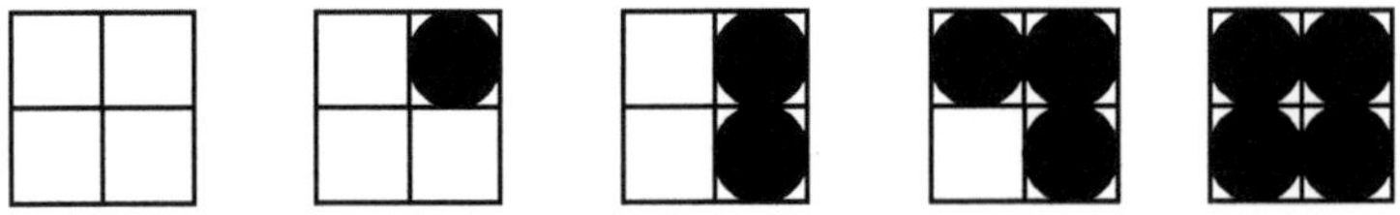

We see that with only two values, we can create 5 different colors (from pure white to pure black). If this 2 × 2 structure is used with the three primary colors, we can get 125 color combinations.

## 4.9 Image Compression

Multimedia information is very demanding on the memory space and usually needs much processing power. Additionally, it may require higher bit rates compared to the available bit rates of communication channels. All of these aspects lead to the inevitable use of compression algorithms. As already mentioned, data compression can be performed as lossless compression and lossy compression. This section considers compression algorithms for digital images. Special attention will be devoted to JPEG and JPEG2000 compression.

### *4.9.1 JPEG Image Compression Algorithm*

Note that the JPEG algorithm can achieve significant compression ratio while maintaining high image quality. Therefore, in this chapter we discuss in detail the elements of JPEG encoder. JPEG algorithm can be analyzed across several blocks used for image compression. These blocks can be summarized as follows: a block performing DCT on the $8 \times 8$ image blocks, quantization block, zig-zag matrix scanning and an entropy coding block (Fig. 4.18).

The DCT of an $8 \times 8$ image block is defined by:

$$DCT(k_1, k_2) = \frac{C(k_1)}{2}\frac{C(k_2)}{2}\sum_{i=0}^{7}\sum_{j=0}^{7} a(i,j)\cos\left(\frac{(2i+1)k_1\pi}{16}\right)\cos\left(\frac{(2j+1)k_2\pi}{16}\right) \tag{4.62}$$

where:

$$C(k_1) = \begin{cases} \frac{1}{\sqrt{2}}, & \text{for } k_1 = 0 \\ 1, & \text{for } k_1 > 0 \end{cases}, \quad C(k_2) = \begin{cases} \frac{1}{\sqrt{2}}, & \text{for } k_2 = 0 \\ 1, & \text{for } k_2 > 0 \end{cases}.$$

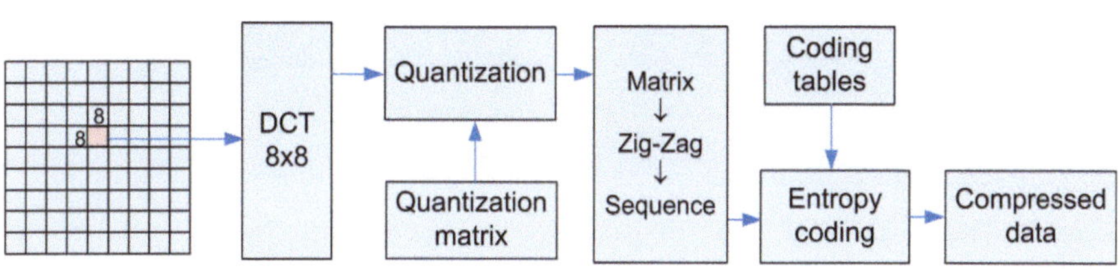


**Fig. 4.18** JPEG encoder block diagram

The DCT coefficient (0,0) is called the DC component, and it carries an information about the mean value of 64 coefficients. The remaining 63 coefficients are the AC coefficients.

The samples of the grayscale image whose values are in the range $[0, 2^n - 1]$ ($n$ is number of bits used to represent samples), are shifted to the range $[-2^{n-1}, 2^{n-1} - 1]$, and then the DCT is applied. Hence, in the case of 8-bit samples, the shifted range is $[-128, 127]$. The corresponding DCT coefficients will be in the range $[-1024, 1023]$ and they require additional 3 bits. To encode the DC coefficient of a current block, we subtract its value from the DC coefficient in the previous block, and then encode their difference.

Before introducing the quantization matrix, let us show that the most important transform coefficients of images are concentrated at low frequencies. For this purpose we will analyze the image "Lena" (of the size $256 \times 256$ pixels). After applying the DCT, we take the first $128 \times 128$ coefficients, then the first $64 \times 64$ coefficients, and finally the first $25 \times 25$ coefficients.

Applying the inverse DCT, we reconstruct the images shown in Fig. 4.19. Note that, although the number of coefficients is significantly decreased, the image retains much of the information.

**Fig. 4.19** (**a**) Original image "Lena" (**b**) image based on the first $128 \times 128$ DCT coefficients, (**c**) image based on the first $64 \times 64$ DCT coefficients, (**d**) image based on the first $25 \times 25$ DCT coefficients

The previous fact was extensively investigated in order to determine the optimal block size and to provide the efficient energy compaction within the smallest number of coefficients. Even though the $32 \times 32$ and $16 \times 16$ blocks slightly improve the coding gain compared to the $8 \times 8$ blocks, the JPEG compression still uses $8 \times 8$ blocks due to the easier calculation. Namely, the $8 \times 8$ blocks provide an optimal trade-off between the computational complexity, prediction gain and energy compaction, with as small artifacts as possible. Therefore, the algorithm for JPEG image compression firstly decomposes an image into $8 \times 8$ blocks. Next, the DCT is calculated for each $8 \times 8$ block. The DCT coefficients are divided by weighting coefficients, representing the elements of quantization matrix. Therefore, we have:

$$DCT_q(k_1,k_2) = \text{round}\left\{\frac{DCT(k_1,k_2)}{Q(k_1,k_2)}\right\}, \tag{4.63}$$

where $Q$ is a quantization matrix, while $DCT_q$ are the quantized coefficients. A simplified example for calculating coefficients of a matrix that can be used for quantization is given by the following code:

```
for i = 0 : N − 1
   for j = 0 : N − 1
Q(i + 1, j + 1) = 1 + [(1 + i + j)*quality];
   end
end
```

The *quality* parameter ranges from 1 to 25. Higher values denote better compression, but worse image quality. The compression matrix for *quality* = 2 is given in Fig. 4.20.

In practical applications, the quantization matrices are derived from the experimental quantization matrix given in Fig. 4.21. The experimental quantization matrix is defined for the 50 % compression ratio (quality factor $QF = 50$).

Using the matrix $Q_{50}$, we can obtain matrices for other compression degrees as follows:

$$Q_{QF} = \text{round}(Q_{50} \cdot q), \tag{4.64}$$

**Fig. 4.20** Quantization matrix obtained for *quality* = 2

| | | | | | | | |
|---|---|---|---|---|---|---|---|
| 3 | 5 | 7 | 9 | 11 | 13 | 15 | 17 |
| 5 | 7 | 9 | 11 | 13 | 15 | 17 | 19 |
| 7 | 9 | 11 | 13 | 15 | 17 | 19 | 21 |
| 9 | 11 | 13 | 15 | 17 | 19 | 21 | 23 |
| 11 | 13 | 15 | 17 | 19 | 21 | 23 | 25 |
| 13 | 15 | 17 | 19 | 21 | 23 | 25 | 27 |
| 15 | 17 | 19 | 21 | 23 | 25 | 27 | 29 |
| 17 | 19 | 21 | 23 | 25 | 27 | 29 | 31 |

**Fig. 4.21** Coefficients of the quantization matrix $Q_{50}$

| | | | | | | | |
|---|---|---|---|---|---|---|---|
| 16 | 11 | 10 | 16 | 24 | 40 | 51 | 61 |
| 12 | 12 | 14 | 19 | 26 | 58 | 60 | 55 |
| 14 | 13 | 16 | 24 | 40 | 57 | 69 | 56 |
| 14 | 17 | 22 | 29 | 51 | 87 | 80 | 62 |
| 18 | 22 | 37 | 56 | 68 | 109 | 103 | 77 |
| 24 | 35 | 55 | 64 | 81 | 104 | 113 | 92 |
| 49 | 64 | 78 | 87 | 103 | 121 | 120 | 101 |
| 72 | 92 | 95 | 98 | 112 | 100 | 103 | 99 |

**Fig. 4.22** Zig-zag reordering

| | | | | | | | |
|---|---|---|---|---|---|---|---|
| 0 | 1 | 5 | 6 | 14 | 15 | 27 | 28 |
| 2 | 4 | 7 | 13 | 16 | 26 | 29 | 42 |
| 3 | 8 | 12 | 17 | 25 | 30 | 41 | 43 |
| 9 | 11 | 18 | 24 | 31 | 40 | 44 | 53 |
| 10 | 19 | 23 | 32 | 39 | 45 | 52 | 54 |
| 20 | 22 | 33 | 38 | 46 | 51 | 55 | 60 |
| 21 | 34 | 37 | 47 | 50 | 56 | 59 | 61 |
| 35 | 36 | 48 | 49 | 57 | 58 | 62 | 63 |

where $q$ is defined as:

$$q = \begin{cases} 2 - 0.02QF & \text{for} \quad QF \geq 50\,, \\ \dfrac{50}{QF} & \text{for} \quad QF < 50\,. \end{cases}$$

The DCT coefficients of $8 \times 8$ blocks are divided by the corresponding coefficients of quantization matrices and rounded to the nearest integer values.

After quantization, the zig-zag reordering is applied to the $8 \times 8$ matrix to form a vector of 64 elements. This reordering allows the values to be sorted from the low-frequency coefficients towards the high-frequency ones. A schematic of zig-zag reordering is shown in Fig. 4.22.

Next, the entropy coding is applied based on the Huffman coding. Each AC coefficient is encoded with two symbols. The first symbol is defined as: $(a,b) =$ (*runlength,size*). The *runlength* provides the information about the number of consecutive zero coefficients preceding the non-zero AC coefficient. Since it is encoded with 4 bits, it can be used to represent no more than 15 consecutive zero coefficients. Hence, the symbol (15,0) represents 16 consecutive zero AC coefficients and it can be up to three (15,0) extensions. This symbol also contains information on the number of bits required to represent the coefficient value (*size*). The second symbol is the amplitude of the coefficient (which is in the range $[-1023,1024]$) that can be represented with up to 10 bits.

For example, if we have the following sequence of coefficients: 0,0,0,0,0,0,0,239, we code it as: (7,8) (239).

The symbol (0,0) denotes the end of the block (EOB).

Since there is a strong correlation between the DC coefficients from adjacent blocks, the differences between DC coefficients are coded instead of their values. The DC coefficients are in the range $[-2048, \; 2047]$ and are coded by two symbols: the first symbol is the number of bits (*size*) used to represent the amplitude, while the second symbol is the amplitude itself.

The amplitude for both DC and AC coefficients are encoded by using the variable-length integer code, as shown in the Table 4.1.

Consider an example of JPEG compression applied to the $8 \times 8$ block. The pixel values (Fig. 4.23a) from range [0,255] are initially shifted to range $[-128,127]$

**Table 4.1** Encoding of the coefficients amplitudes

| Amplitude range | Size |
|---|---|
| -1,1 | 1 |
| -3,-2,2,3 | 2 |
| -7,-6,-5,-4,4,5,6,7 | 3 |
| -15 ,. . . ,-8,8 ,. . . ,15 | 4 |
| -31 ,. . . ,-16,16 ,. . . ,31 | 5 |
| -63 ,. . . ,-32,32 ,. . . ,63 | 6 |
| -127 ,. . . ,-64,64 ,. . . ,127 | 7 |
| -255 ,. . . ,-128,128 ,. . . ,255 | 8 |
| -511 ,. . . ,-256,256 ,. . . ,511 | 9 |
| -1023 ,. . . ,-512,512 ,. . . ,1023 | 10 |

a

| 177 | 153 | 151 | 143 | 142 | 145 | 150 | 158 |
|---|---|---|---|---|---|---|---|
| 159 | 148 | 148 | 145 | 144 | 153 | 147 | 149 |
| 154 | 155 | 151 | 154 | 157 | 146 | 156 | 146 |
| 152 | 159 | 157 | 155 | 151 | 147 | 159 | 151 |
| 142 | 154 | 144 | 143 | 144 | 144 | 168 | 156 |
| 160 | 150 | 136 | 136 | 148 | 153 | 161 | 153 |
| 150 | 142 | 142 | 142 | 156 | 152 | 161 | 161 |
| 135 | 131 | 150 | 154 | 151 | 144 | 146 | 160 |

b

| 49 | 25 | 23 | 15 | 14 | 17 | 22 | 30 |
|---|---|---|---|---|---|---|---|
| 31 | 20 | 20 | 17 | 16 | 25 | 19 | 21 |
| 26 | 27 | 23 | 26 | 29 | 18 | 28 | 18 |
| 24 | 31 | 29 | 27 | 23 | 19 | 31 | 23 |
| 14 | 26 | 16 | 15 | 16 | 16 | 40 | 28 |
| 32 | 22 | 8 | 8 | 20 | 25 | 33 | 25 |
| 22 | 14 | 14 | 14 | 28 | 24 | 33 | 33 |
| 7 | 3 | 22 | 26 | 23 | 16 | 18 | 32 |

c

| 180 | -10 | 21 | 4 | 3 | 6 | -6 | 8 |
|---|---|---|---|---|---|---|---|
| 10 | 28 | 9 | 4 | -2 | 3 | 3 | -3 |
| -7 | 0 | 2 | 0 | 15 | 1 | 19 | 0 |
| 0 | 0 | 26 | 14 | -3 | 0 | -1 | 0 |
| 0 | 1 | 0 | -17 | 0 | 3 | 0 | 0 |
| 13 | 5 | 3 | 3 | 3 | 0 | -2 | 4 |
| 0 | 6 | 1 | 3 | 4 | 1 | -2 | 1 |
| 0 | -6 | 0 | -3 | 0 | 0 | -7 | 4 |

d

| 10 | 7 | 6 | 10 | 14 | 24 | 31 | 37 |
|---|---|---|---|---|---|---|---|
| 7 | 7 | 8 | 11 | 16 | 35 | 36 | 33 |
| 8 | 8 | 10 | 14 | 24 | 34 | 41 | 34 |
| 8 | 10 | 13 | 17 | 31 | 52 | 48 | 37 |
| 11 | 13 | 22 | 34 | 41 | 65 | 62 | 46 |
| 14 | 21 | 33 | 38 | 49 | 62 | 68 | 55 |
| 29 | 38 | 47 | 52 | 62 | 73 | 72 | 61 |
| 43 | 55 | 57 | 59 | 67 | 60 | 62 | 59 |

e

| 18 | -1 | 4 | 0 | 0 | 0 | 0 | 0 |
|---|---|---|---|---|---|---|---|
| 1 | 4 | 1 | 0 | 0 | 0 | 0 | 0 |
| -1 | 0 | 0 | 0 | 1 | 0 | 0 | 0 |
| 0 | 0 | 2 | 1 | 0 | 0 | 0 | 0 |
| 0 | 0 | 0 | -1 | 0 | 0 | 0 | 0 |
| 1 | 0 | 0 | 0 | 0 | 0 | 0 | 0 |
| 0 | 0 | 0 | 0 | 0 | 0 | 0 | 0 |
| 0 | 0 | 0 | 0 | 0 | 0 | 0 | 0 |

**Fig. 4.23** (**a**) $8 \times 8$ image block, (**b**) values of pixels after shifting to the range $[-128, 127]$, (**c**) DCT coefficients (rounded to integers) for the given block, (**d**) quantization matrix $QF = 70$, (**e**) DCT coefficients after quantization

(Fig. 4.23b). The values of DCT coefficients are shown in Fig. 4.23c. Note that, for the sake of simplicity, the DCT coefficients are rounded to the nearest integers. The quantization matrix is used with a *quality factor* $QF = 70\ \%$ (Fig. 4.23d) and the quantized DCT coefficients are shown in Fig. 4.23e.

The zig-zag sequence is obtained in the form:

18, −1, 1, −1, 4, 4, 0, 1, 0, 0, 0, 0, 0, 0, 0, 0, 0, 2, 0, 1, 0, 0, 0, 1, 1, 0, 0, 0, 0, 0, 0, −1, 0, 0, 0, 0, 0, 0, 0, 0, 0, 0, 0, 0, 0, 0, 0, 0, 0, 0, 0, 0, 0, 0, 0, 0, 0, 0, 0, 0, 0, 0

and by using symbols for DC and AC coefficients, we obtain the intermediate symbol sequence:

(5)(18), (0,1)(−1), (0,1)(1), (0,1)(−1) (0,3)(4), (0,3)(4), (1,1)(1), (10,2)(2), (1,1)(1), (3,1)(1), (0,1)(1), (6,1)(−1), (0,0)

The symbols for AC components ($(a,b) = (runlength,size)$) are coded by using the Huffman tables, specified by the JPEG standard and given at the end of this chapter (for luminance component). The symbols used in this example are provided in the Table 4.2.

The entire 8 × 8 block in encoded form is given by:

(101)(10010) (00)(0) (00)(1) (00)(0) (100)(100) (100)(100) (1100)(1) (1111111111000111)(10) (1100)(1) (111010)(1) (00)(1) (1111011)(0) (1010)

Note that in this example we have coded the DC coefficient value, not the DC coefficients difference, since we have examined a single block.

Decoding is performed using the blocks in Fig. 4.24.

We first return the sequence of samples into the matrix form. Next, we perform de-quantization followed by the inverse DCT. In other words, after we get:

**Table 4.2** Code words for the symbols obtained in the example

| Symbol $(a,b)$ | Code word |
|---|---|
| (0,1) | 00 |
| (0,3) | 100 |
| (1,1) | 1100 |
| (3,1) | 111010 |
| (6,1) | 1111011 |
| (10,2) | 1111111111000111 |
| (0,0) EOB | 1010 |

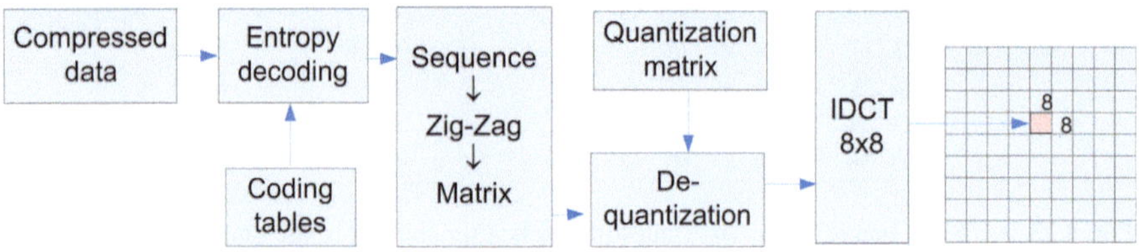


**Fig. 4.24** JPEG decoder

**Fig. 4.25** (**a**) Original "Lena" image, (**b**) "Lena" image after applying JPEG compression with $QF = 70\ \%$, (**c**) "Lena" image after JPEG compression with $QF = 25\ \%$, (**d**) "Lena" image after JPEG compression with $QF = 5\ \%$

$$DCT_{dq} = DCT_q(k_1, k_2) \cdot Q(k_1, k_2), \tag{4.65}$$

we apply the inverse DCT transformation:

$$a(i,j) = \sum_{k_1=0}^{7} \sum_{k_2=0}^{7} \frac{C(k_1)}{2} \frac{C(k_2)}{2} DCT_{dq}(k_1, k_2) \cos\left(\frac{(2i+1)k_1\pi}{16}\right) \cos\left(\frac{(2j+1)k_2\pi}{16}\right) \tag{4.66}$$

It is obvious that quantization/de-quantization procedures and rounding procedures introduce an error proportional to the quantization step.

In order to illustrate the efficiency of JPEG compression in terms of the compromise between the compression factor and image quality, the examples of compressed images with different qualities are shown in Fig. 4.25.

### 4.9.2 JPEG Lossless Compression

The JPEG lossless compression provides a compression ratio approximately equal to 2:1. It uses a prediction approach to encode the difference between the current pixel X and the one predicted from three neighboring pixels A, B, C, as illustrated below:

| | C | B | |
|---|---|---|---|
| | A | X | |

The prediction sample $X_p$ can be obtained using one of the formulas:

| Case | Prediction formula |
|---|---|
| 1 | $X_p = A$ |
| 2 | $X_p = B$ |
| 3 | $X_p = C$ |
| 4 | $X_p = A + B - C$ |
| 5 | $X_p = A + (B - C)/2$ |
| 6 | $X_p = B + (A - C)/2$ |
| 7 | $X_p = (A + B)/2$ |

Then the difference $\Delta X = X - X_p$ is encoded by using the Huffman code.

### 4.9.3 Progressive JPEG Compression

*Spectral Compression*

During the image transmission, it is often demanded that a receiver gradually improves the image resolution. Namely, a rough version of the image is firstly transmitted (which can be done with a high compression factor), and then we transmit the image details. This is achieved with progressive compression methods.

In the algorithm for progressive compression, coding is implemented by using several spectral bands. The bands are divided according to their importance. For example, the first band can be dedicated only to DC coefficients; the second band may be dedicated to the first two AC coefficients (AC1 and AC2); the third band contains the next four AC coefficients, while the fourth band may contain remaining coefficients.

*Successive approximations*

In this algorithm the coefficients are not initially sent with the original values, i.e., they are sent with fewer bits (lower resolution) and then refined. For example, in the first all DCT coefficients are scanned with 2 bits left out (divided by 4), and in the successive scans the least significant bits are added until all the bits are sent.

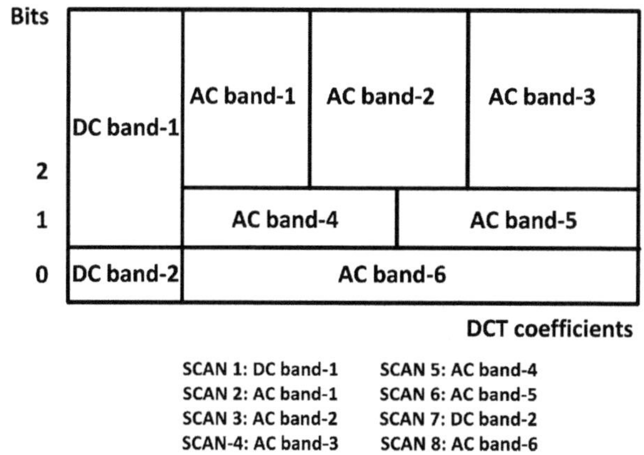


**Fig. 4.26** An example of using the combined progressive JPEG algorithm

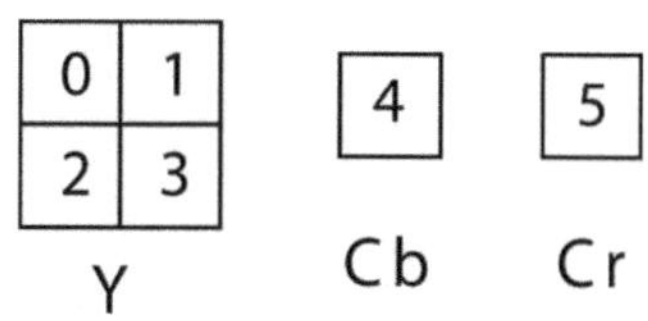


**Fig. 4.27** Y, Cb, and Cr component

*Combined progressive algorithms*

This algorithm combines the two previously described algorithms. Specifically, all the coefficients are grouped into the spectral bands (as in the spectral algorithm), and then the information from all bands is sent with different resolutions (in terms of the number of bits), as in the second algorithm. An example of this algorithm is shown in Fig. 4.26. Namely, an image divided into eight combined scans. In the first scan only the DC coefficients from **DC band-1** will be sent (with lower resolution, i.e., divided by two), then for example **AC band-1** coefficients, etc.

### 4.9.4 JPEG Compression of Color Images

JPEG compression of color images can be performed by compressing each color channel as described for the grayscale images. In JPEG compression, the RGB model can be transformed into the YCbCr space (Fig. 4.27). Y channel contains information about luminance, and Cb and Cr channels are related to the color along the axes red-green and yellow-blue, respectively. Then, each channel is treated separately, because it is not necessary to encode them with the same precision.

During decompression, the process is reversed: each channel is decoded individually and then the information is merged together. Lastly, we convert from the YCbCr to the RGB color space.

However, a drawback of this approach is that the color components appear sequentially until the complete image is displayed: the red color is displayed first,

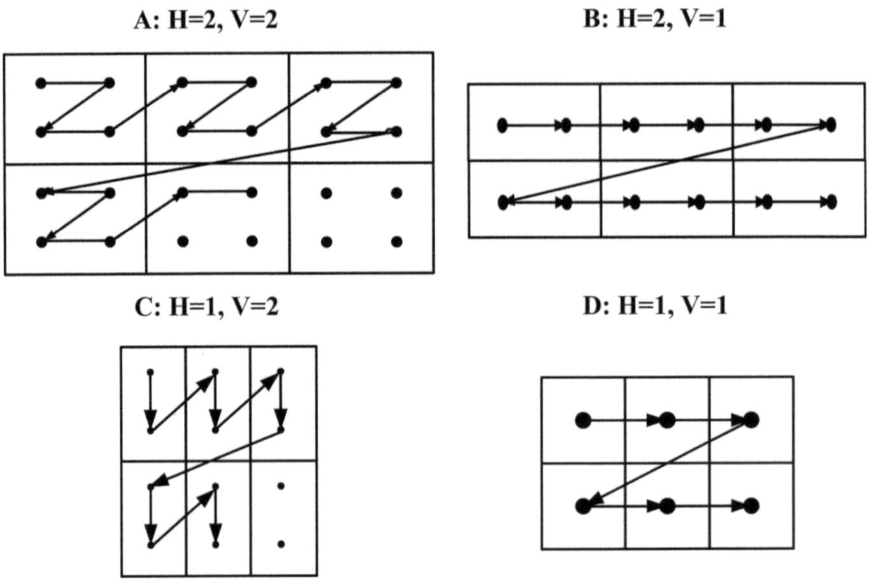


**Fig. 4.28** An interleaving procedure for JPEG color image compression

then the green color is joined, and finally the blue color. In the real applications where the image is decompressed and displayed simultaneously, the interleaved ordering approach is used to combine data from different color channels. Let us consider the case of an image with four components of different resolutions. In addition, each component is divided into rectangular regions with resolutions $\{H_i,V_i\}$. Specifically, factors $H_i$ and $V_i$ define the horizontal and vertical resolutions for each component (Fig. 4.28).

The coefficients are combined from the rectangular regions of different components. Each component has the same number of rectangular regions (e.g., 6 regions) as shown in Fig. 4.28. The basic coding units (Minimum Coded Units—MCU) are formed by using one region from each component. The coefficients in each region are sorted from left to right and from top to bottom. Each MCU can contain up to 10 coefficients.

In the example in Fig. 4.28, the basic MCUs for encoding are:

$$
\begin{aligned}
MCU1 &= a^1{}_{00}a^1{}_{01}a^1{}_{10}a^1{}_{11}a^2{}_{00}a^2{}_{01}a^3{}_{00}a^3{}_{10}a^4{}_{00}\\
MCU2 &= a^1{}_{02}a^1{}_{03}a^1{}_{12}a^1{}_{13}a^2{}_{02}a^2{}_{03}a^3{}_{01}a^3{}_{11}a^4{}_{01}\\
MCU3 &= a^1{}_{04}a^1{}_{05}a^1{}_{14}a^1{}_{15}a^2{}_{04}a^2{}_{05}a^3{}_{02}a^3{}_{12}a^4{}_{02}\\
MCU4 &= a^1{}_{20}a^1{}_{21}a^1{}_{30}a^1{}_{31}a^2{}_{10}a^2{}_{11}a^3{}_{20}a^3{}_{30}a^4{}_{10}
\end{aligned}
$$

The described procedure ensures that an image is always displayed with all color components.

### 4.9.5 JPEG2000 Compression

The standard JPEG algorithm based on the $8 \times 8$ DCT blocks often leads to visible block distortions (block effects). To avoid these effects, a JPEG2000 compression algorithm based on the wavelet transform is introduced.

In this algorithm, the entire system can be divided into three parts: (1) image preprocessing, (2) compression, and (3) encoding.

1. Image preprocessing contains some of the optional functions including:
   - Dividing large images into regions (this process is called tiling).
   - DC level shifting.
   - Color components transformation.

Dividing large images (usually larger than $512 \times 512$) into smaller rectangular areas that are separately analyzed is required in order to avoid large buffers in the implementation of the algorithm.

Similarly to the standard JPEG compression, DC level shifting ($\mathrm{I}(x,y) = \mathrm{I}(x,y) - 2^{n-1}$) is also used in order to obtain an image with dynamic range that is centered around zero. Values in the range $[0, 2^n - 1]$ are shifted to the values in the range $[-2^{n-1}, 2^{n-1} - 1]$ ($n$ is the number of bits used to represent pixel values).

This algorithm is defined for the color images consisting of three components. The JPEG2000 standard supports two color transforms: the reversible color transform (RCT) and the irreversible color transform (ICT). ICT is called "irreversible" since it needs to be implemented in floating or fix-point, which will cause rounding errors. RCT can be applied for both lossy and lossless compression, while ICT can be used only for lossy compression. ICT uses the YCbCr color space, while the RCT uses a modified YUV color space that does not introduce quantization errors.

The RCT transform (integer-to-integer) is defined as:

$$Y_r = \left\lfloor \frac{R + 2G + B}{4} \right\rfloor, \quad U_r = B - G, \quad V_r = R - G. \tag{4.67}$$

In the case of RCT, the pixels can be exactly reconstructed by using the inverse RCT defined as follows:

$$R = V_r + G, \; G = Y_r - \left\lfloor \frac{U_r + V_r}{4} \right\rfloor, \; B = U_r + G. \tag{4.68}$$

The ICT is actually based on YCbCr color model (real-to-real transform):

$$\begin{bmatrix} Y \\ C_b \\ C_r \end{bmatrix} = \begin{bmatrix} 0.299 & 0.587 & 0.114 \\ -0.168 & -0.331 & 0.5 \\ 0.5 & -0.41 & -0.08 \end{bmatrix} \begin{bmatrix} R \\ G \\ B \end{bmatrix}, \tag{4.69}$$

$$\overset{\xleftarrow{\;l\;}}{\cdots I_m I_{m-1} I_{m-2} \cdots I_{k+2} I_{k+1}} \,|\, \underset{\substack{\uparrow\\ k}}{I_k} I_{k+1} I_{k+2} \cdots I_{m-2} I_{m-1} \underset{\substack{\uparrow\\ m}}{I_m} \,|\, \overset{\xrightarrow{\;r\;}}{I_{m-1} I_{m-2} \cdots I_{k+2} I_{k+1} I_k \cdots}$$

**Fig. 4.29** Expansion of a sequence of pixels

while the inverse transform is given as:

$$\begin{bmatrix} R \\ G \\ B \end{bmatrix} = \begin{bmatrix} 1.0 & 0.0 & 1.402 \\ 1.0 & -0.344136 & -1.714136 \\ 1.0 & 1.772 & 0.0 \end{bmatrix} \begin{bmatrix} Y \\ C_b \\ C_r \end{bmatrix}. \tag{4.70}$$

2. Compression

JPEG 2000 algorithm uses two types of the wavelet transforms. These are (9,7) floating-point wavelets (irreversible) and (5,3) integer wavelet transform (reversible). Only the (5,3) integer transform, which is fully reversible, can be used for lossless compression.

Consider a sequence of pixels denoted as $I_k$, $I_{k+1}$, $I_{k+2}$, . . ., $I_m$ that belong to an image row. To calculate the wavelet transform, it is necessary to use a few pixels with indices less than $k$ and greater than $m$. Before applying the wavelet transform, the considered area has to be expanded. An easy way to extend a sequence of pixels ($I_k$, $I_{k+1}$, $I_{k+2}$, . . ., $I_{m-2}$, $I_{m-1}$, $I_m$) is illustrated in Fig. 4.29.

After expanding the sequence of pixels, the wavelet coefficients are calculated. For the (5,3) integer wavelet transform, the coefficients are calculated according to:

$$\begin{aligned} d_{j-1,i} &= I_{j,2i+1} - \left\lfloor \frac{I_{j,2i+2} + I_{j,2i}}{2} \right\rfloor, \\ s_{j-1,i} &= I_{j,2i} + \left\lfloor \frac{d_{j-1,i} + d_{j-1,i-1} + 2}{4} \right\rfloor. \end{aligned} \tag{4.71}$$

Here, $d$ coefficients represent high frequency components, while $s$ coefficients represent low frequency components.

In the case of the (9,7) floating-point wavelet transform, wavelet coefficients are obtained as follows:

$$\begin{aligned} P_{j,2i+1} &= I_{j,2i+1} + \alpha\left(I_{j,2i} + I_{j,2i+2}\right), \\ P_{j,2i} &= I_{j,2i} + \beta\left(P_{j,2i-1} + P_{j,2i+1}\right), \\ d'_{j-1,i} &= P_{j,2i-1} + \gamma\left(P_{j,2i-2} + P_{j,2i}\right), \\ s'_{j-1,i} &= P_{j,2i} + \delta\left(d'_{j-1,i} + d'_{j-1,i+1}\right), \\ d_{j-1,i} &= -Kd'_{j-1,i}, \\ s_{j-1,i} &= (1/K)s'_{j-1,i}, \end{aligned} \tag{4.72}$$

where the constants (wavelet filter coefficients) for the JPEG2000 algorithm are: $\alpha = -1.586134342$, $\beta = -0.052980118$, $\gamma = 0.882911075$, $\delta = 0.443506852$, $K = 1.230174105$. As in (4.69), $d$ are details, and $s$ are low-frequency coefficients.

Consider a simple example with five consecutive pixels $I_{j,2i-2}, I_{j,2i-1}, I_{j,2i}, I_{j,2i+1}, I_{j,2i+2}$. The (9,7) wavelet coefficients can be calculated in four steps:

The first step is: $$P_{j,2i-1} = I_{j,2i-1} + \alpha I_{j,2i-2} + \alpha I_{j,2i}, \quad P_{j,2i+1} = I_{j,2i+1} + \alpha I_{j,2i} + \alpha I_{j,2i+2},$$

The second step is: $$P_{j,2i-2} = I_{j,2i-2} + \beta P_{j,2i-3} + \beta P_{j,2i-1}, \quad P_{j,2i} = I_{j,2i} + \beta P_{j,2i+1} + \beta P_{j,2i-1},$$

The third step is: $$d'_{j-1,i} = P_{j,2i-1} + \gamma\left(P_{j,2i-2} + P_{j,2i}\right), \quad d'_{j-1,i+1} = P_{j,2i+1} + \gamma\left(P_{j,2i} + P_{j,2i+2}\right),$$

The fourth step is: $$s'_{j-1,i} = P_{j,2i} + \delta\left(d'_{j-1,i} + d'_{j-1,i+1}\right), \quad s'_{j-1,i+1} = P_{j,2i+2} + \delta\left(d'_{j-1,i+1} + d'_{j-1,i+2}\right).$$

At the end, $d'$ coefficients are scaled by the parameter -$K$ and $s'$ coefficients are scaled by the parameter 1/$K$. This procedure is illustrated schematically in Fig. 4.30.

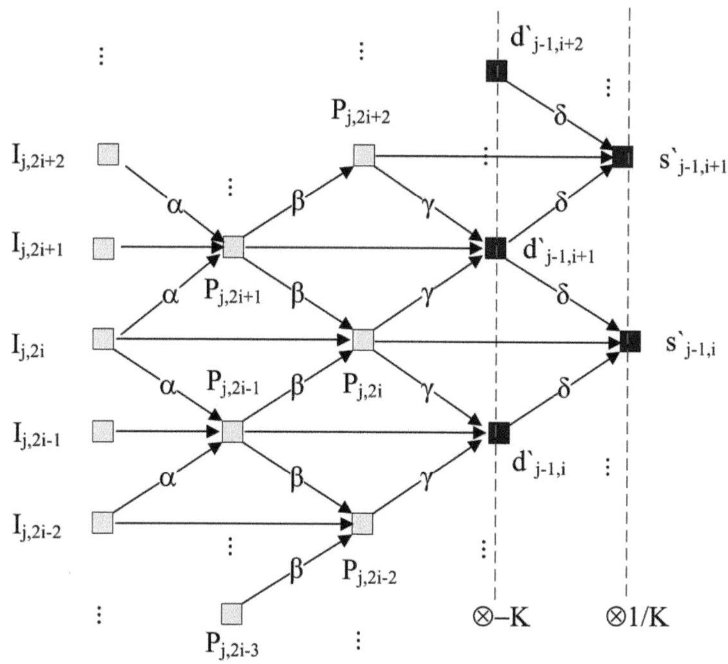


**Fig. 4.30** Calculating the wavelet coefficients

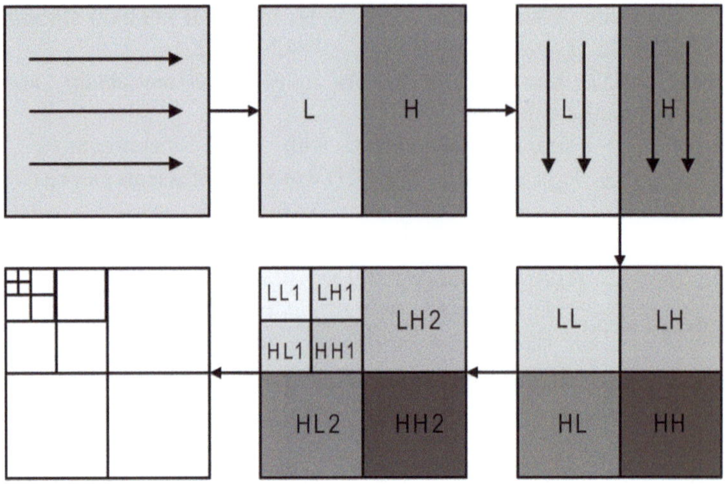


**Fig. 4.31** An illustration of applying the wavelet transform to two-dimensional signals

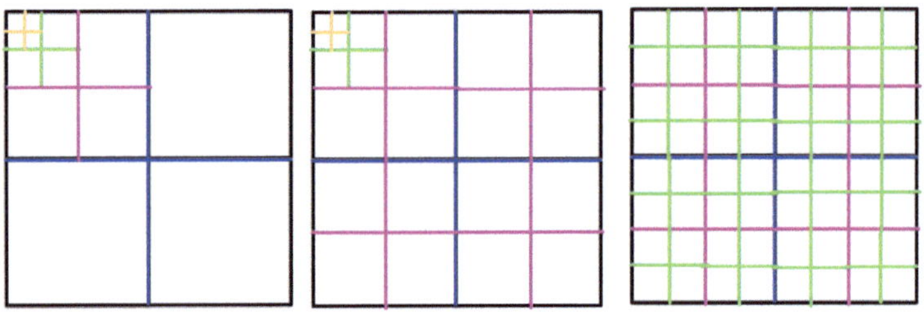

**Fig. 4.32** Organization of subbands for JPEG2000

The considered one-dimensional wavelet transform is usually applied to the image rows and then to the columns, as it is illustrated in Fig. 4.31.

Subbands can be organized in one of three ways as illustrated in Fig. 4.32.

#### 4.9.5.1 JPEG2000 Quantization

For each subband denoted by $b$, a quantization step $\Delta_b$ is used for the coefficients in the subband. Quantization is defined as follows:

$$Q_b(u,v) = sign(C_b(u,v))\left\lfloor \frac{|C_b(u,v)|}{\Delta_b} \right\rfloor, \tag{4.73}$$

where $C_b(u,v)$ is the original DWT coefficient from the subband $b$. The operator $\lfloor \cdot \rfloor$ represents rounding to integer number. Hence, the value of quantized coefficient is: $Q_b(u,v)\Delta_b$. The quantization step is defined as follows:

$$\Delta_b = 2^{R_b - \varepsilon_b}\left(1 + \frac{\mu_b}{2^{11}}\right), \tag{4.74}$$

where $R_b$ represents the nominal dynamic range of the subband $b$. The parameter $\mu_b$ is the 11-bit mantissa and $\varepsilon_b$ is the 5-bit exponent of the quantization step ($0 \le \mu_b < 2^{11},\ 0 \le \varepsilon_b < 2^5$). The exponent–mantissa pairs ($\mu_b$,$\varepsilon_b$) can be explicitly signaled in the bit stream syntax for every subband. The dynamic range depends on the number of bits used to represent the original image tile component and on the choice of the wavelet transform. For reversible compression, the quantization step-size is required to be one.

The inverse quantization is defined as:

$$R_Q(u,v) = \begin{cases} (Q(u,v) + \delta)\Delta_b, & \text{for } Q(u,v) > 0, \\ (Q(u,v) - \delta)\Delta_b, & \text{for } Q(u,v) < 0, \\ 0, & \text{for } Q(u,v) = 0. \end{cases} \tag{4.75}$$

The reconstruction parameter is usually given by $0 \le \delta < 1$ and the most commonly used value is $\delta = 0.5$.

#### 4.9.5.2 Coding the Regions of Interest

One of the important techniques in the JPEG2000 algorithm is coding the regions of interest (ROI). ROI is expected to be encoded with better quality than the other regions (e.g., image background). Coding of the ROI aims to scale the ROI coefficients and place them in the higher bit planes (comparing to the bit planes of other coefficients). Hence, the ROI coefficients will be progressively transmitted before the non-ROI coefficients. Consequently, ROI will be decoded before other image parts and with higher accuracy (Fig. 4.33).

The method based on scaling is implemented as follows:

1. First, the wavelet transform is calculated.
2. Then we form a mask indicating the set of coefficients that belong to ROI. Specifically, the ROI mask is mapped from the pixels domain into each subband in the wavelet domain (Fig. 4.34).

**Fig. 4.33** Face is coded as ROI

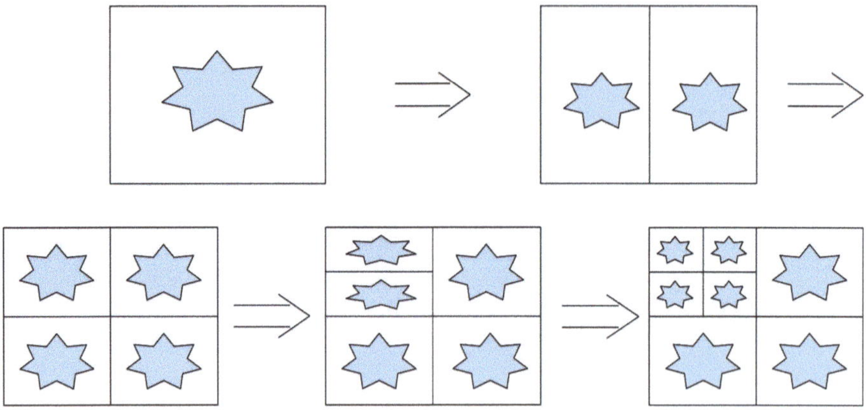

**Fig. 4.34** ROI mask mapped into subbands

3. Wavelet coefficients are quantized.
4. ROI coefficients are scaled and shifted to higher bit planes (*MAXSHIFT method*).
5. Finally, the coefficients are encoded.

The value of the scaling factor has to be sufficiently large to ensure that the lowest value in the ROI is greater than the largest value of the surrounding non-ROI coefficients. The scaling factor is transmitted with the coefficients in order to be able to reconstruct the original ROI values. An illustration of this process is shown in Fig. 4.35.

The most significant bit of each coefficient is indicated by "1" (the first non-zero bit in each column), while the following bits are denoted by "x" (could be either 0 or 1). The coefficients that belong to the ROI are shaded in gray. The bit planes that remain after scaling the ROI are filled by "0".

There are several ways to set the ROI masks. As shown in Fig. 4.36a, the low-frequency coefficients can be transmitted together with the ROI coefficients if the ROI masks are placed in all other subbands. Also, ROI regions can be only used in specific subbands as shown in Fig. 4.36b.

*Areas and code blocks*
To achieve more efficient coding, each wavelet decomposition subband can be further divided into precincts. The size of the precincts may vary at different levels of decomposition, but can be usually expressed as a power of 2. Areas on the same positions in different subbands are shaded in gray in Fig. 4.37. Each area is further divided into the code-blocks whose dimensions are also a power of 2. This division provides memory efficient implementation. Simple coders will use this division. On the other hand, sophisticated coders will use a large number of code-blocks to ensure that the decoder performs progressive decompression, as well as to provide higher bit rate, image zooming, and other operations when decoding only parts of the image.

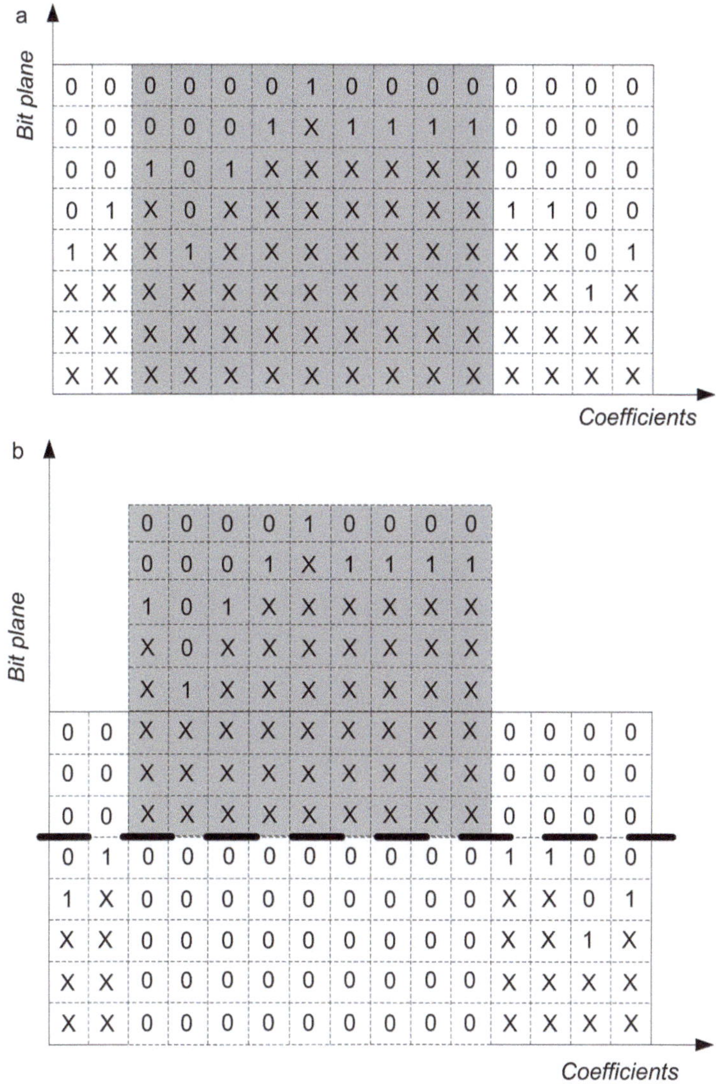

**Fig. 4.35** (**a**) Quantized wavelet coefficients before scaling, (**b**) quantized wavelet coefficients after scaling

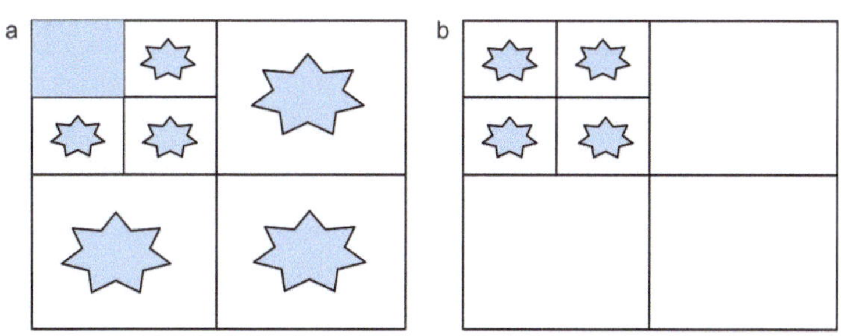

**Fig. 4.36** (**a**) The low-frequency subband is coded together with ROI regions in other subbands, (**b**) arbitrary ROI mask

**Fig. 4.37** An example of subbands, precincts and code-blocks partition

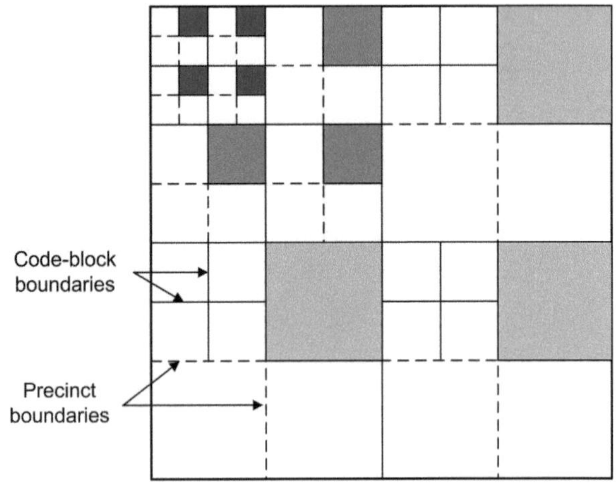


The highlighted precincts in Fig. 4.37 correspond roughly to the same $N/2 \times N/2$ region in the original image (of size $N \times N$). Note that the code-blocks in all subbands are of the same size, except when their size is constrained by the subband precinct size, as in the low-frequency subbands in Fig 4.37.

#### 4.9.5.3 Entropy Coding

The wavelet coefficients are coded by bit-planes using the arithmetic encoding scheme. The encoding is done from the most significant to the least significant bit-plane. We also have to determine the bit context, where the probability of occurrence for each bit is estimated. The bit value and its probability are forwarded to an arithmetic coder. Hence, unlike many other compression algorithms which encode the coefficients of images, the JPEG2000 encodes the sequences of bits.

Each wavelet coefficient should have an indicator of importance (1 if the coefficient is significant, and 0 if not). At the beginning of coding, all wavelet coefficients are set to be insignificant. If there are bit-planes that contain all zero values, the number of these bit-planes is stored in the bit stream.

A most significant non-zero bit plane is encoded in one pass which is called the *cleanup* pass. Each subsequent bit plane is encoded within three passes. Here, each bit-plane is divided into tracks containing four lines (rows), while the tracks are scanned by using the zig-zag order (from left to right), as shown in Fig. 4.38.

At the beginning, all bits from the most significant non-zero bit-plane are fed to the encoder (they are encoded in one pass). During this step, the coefficient is denoted as significant if its bit is equal to 1. The coefficient remains significant until the end of the encoding process.

The remaining bit-planes are encoded one after the other (from most to least significant bit-planes), and within the three passes:

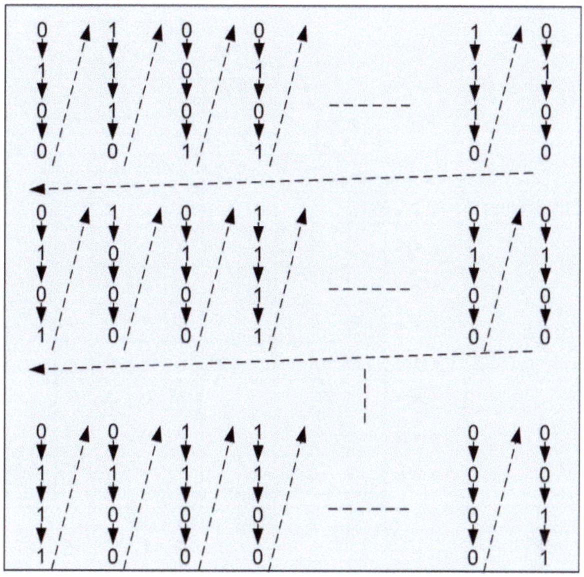

**Fig. 4.38** The bit-planes scan method

- The first pass involves the coefficients denoted as insignificant which have at least one significant neighbor (one of its eight nearest neighbors is denoted as significant). Their bits from the observed bit plane are forwarded to the coder. As previously described, the coefficients whose bit is 1 are declared significant (a significance indicator is set to 1).
- The second pass encodes the bits of the coefficients that became significant in earlier steps, when passing through a previous bit planes.
- The third step considers the bits omitted in the previous two steps. If the bit value in the third step is 1, the corresponding coefficient is declared significant.

**Example:** We illustrate the bit-planes encoding process by considering separately only four isolated wavelet coefficients whose values are 9, 1, 2, and −6, Fig. 4.39. The coefficients are encoded with 9 bits (1 sign bit and 8 bits for value).

Therefore, $9 = 0 \mid 00001001$, $1 = 0 \mid 00000001$, $2 = 0 \mid 00000010$, $-6 = 1 \mid 00000110$

Bit plane containing the sign bits is considered separately and it is ignored at the beginning. If we observe just the four given coefficients, the first four bit planes (planes from 7 to 4) are zero, so the encoding starts from the third plane.

*Plane 3:* The bit-plane 3 is the most significant non-zero bit-plane and it is encoded in a single pass. One bit from each coefficient is brought to the encoder. Note that in the plane 3, a bit for the coefficient with value 9 has the logical value of 1 and the coefficient is declared as significant. The sign bit for the coefficient 9 is encoded immediately after this bit.

*Plane 2:* The bit-plane 2 is encoded after plane 3. We first encode the coefficients that are insignificant, but they are the neighbors to the significant ones.

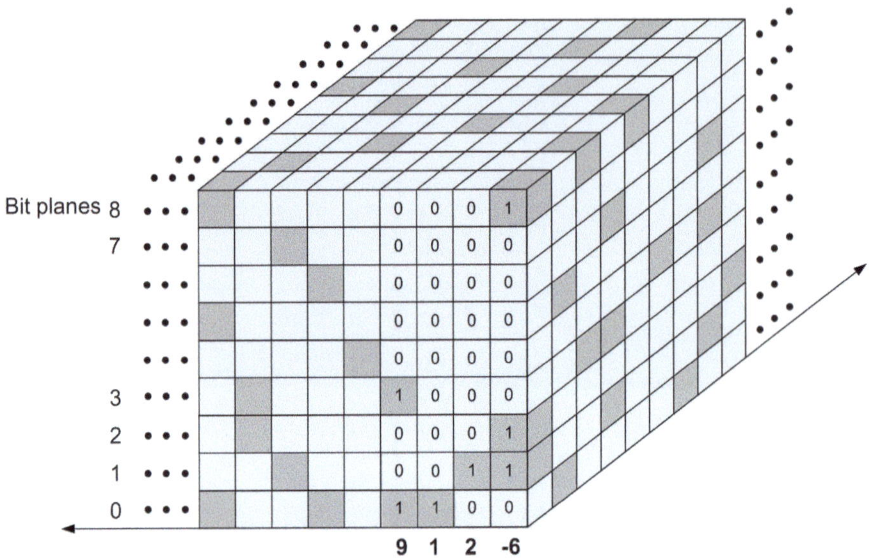


**Fig. 4.39** Part of bit planes that contain four considered coefficients

The coefficient 1 is insignificant, but it is the adjacent coefficient to 9, which is significant. Therefore the bit 0 of the coefficient 1 is passed to the encoder. Note that none of the coefficients have been declared significant at this stage. The second step includes bits belonging to the significant coefficients. The coefficient 9 is significant, and thus its bit 0 is passed to the encoder. Coefficients 2 and −6 are not significant, and they are not located next to the significant coefficients. Hence, they will be encoded in the third step. Since the bit of the coefficient −6 (plane 2) has value 1, coefficients −6 is declared significant, and its sign bit is encoded as well.

*Plane 1:* The bits of the coefficients 1 and 2 are coded in the first pass (they are neighbors of significant coefficients), while the bits of the significant coefficients 9 and −6 will be encoded in the second pass. The bit for coefficient 2 is 1, so this coefficient becomes significant (its sign bit is passed to the encoder as well).

*Plane 0:* This plane is encoded last. The bit of coefficient 1 is encoded in the first pass. The coefficient 1 becomes significant and its sign bit is encoded. The coefficients 9, 2, and −6 are significant, and their bits are encoded in the second pass.

*Arithmetic coding*
The sequence of bits from each plane is forwarded to an arithmetic coder. Here, we use the bit context to determine a probability of a binary symbol. One simple example of using arithmetic coding with known probabilities of symbols occurrences is illustrated in Fig. 4.40. Suppose that we want to encode the binary sequence 1011, where the probability of occurrence of symbol “1” is equal to 0.6, while for the symbol “0” it is 0.4.

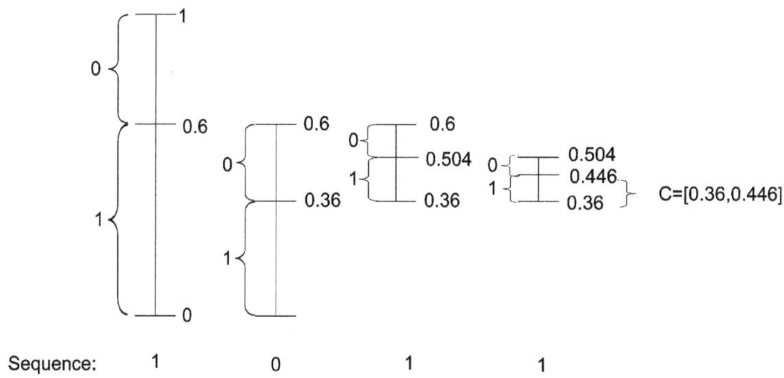


**Fig. 4.40** An example of arithmetic coding

**Fig. 4.41** Significance indicators for eight neighboring coefficients

| D0 | V0 | D1 |
|---|---|---|
| H0 | X | H1 |
| D2 | V1 | D3 |

At the beginning, we have the interval [0, 1], which is divided into two intervals according to the probability of symbol occurrence, as shown in Fig. 4.40. The lower interval [0, 0.6], used to denote the symbol "1", is then divided again into two intervals with the same proportions as in the previous case. The second symbol is "0", and therefore the upper interval [0.36, 0.6] is divided into two new intervals.

We continue this process until the whole sequence is encoded. At the end, we get C = [0.36, 0.446]. Finally, the sequence 1011 is coded by using one value from the obtained interval, e.g., value 0.4. The interval should be available during the decoding process. Note that the encoding is usually applied to much longer sequences.

*Determining the context*

The bit context is used in each step to estimate the probability for bit encoding. The following procedure describes how to determine the context of bits in the first pass. We consider eight neighboring coefficients around the wavelet coefficient X and their current indicators of importance (1 indicating a significant coefficient, and 0 indicating the insignificant one). Let us assume that the indicators of neighboring coefficients are denoted as in Fig. 4.41.

The context is selected based on the criteria listed in Table 4.3. Note that there are 9 contexts, and the criteria depend on the subband (LL, LH, HL, HH), which is encoded. For example, the context 0 represents the coefficient without any significant neighbor. JPEG2000 standard defines similar rules for determining the bit context for the other two passes.

Based on the estimated bit context, the probability estimation process (for encoding the bit) is done by using lookup tables.

**Table 4.3** Nine contexts for the first pass

| LL and LH subbands | | | |
|---|---|---|---|
| $\sum H_i$ | $\sum V_i$ | $\sum D_i$ | Context |
| 2 | | | 8 |
| 1 | ≥1 | | 7 |
| 1 | 0 | ≥1 | 6 |
| 1 | 0 | 0 | 5 |
| 0 | 2 | | 4 |
| 0 | 1 | | 3 |
| 0 | 0 | ≥2 | 2 |
| 0 | 0 | 1 | 1 |
| 0 | 0 | 0 | 0 |
| HL subband | | | |
| $\sum H_i$ | $\sum V_i$ | $\sum D_i$ | Context |
| | 2 | | 8 |
| ≥1 | 1 | | 7 |
| 0 | 1 | ≥1 | 6 |
| 0 | 1 | 0 | 5 |
| 2 | 0 | | 4 |
| 1 | 0 | | 3 |
| 0 | 0 | ≥2 | 2 |
| 0 | 0 | 1 | 1 |
| 0 | 0 | 0 | 0 |
| HH subband | | | |
| $\sum (H_i + V_i)$ | | $\sum D_i$ | Context |
| | | ≥3 | 8 |
| ≥1 | | 2 | 7 |
| 0 | | 2 | 6 |
| ≥2 | | 1 | 5 |
| 1 | | 1 | 4 |
| 0 | | 1 | 3 |
| ≥2 | | 0 | 2 |
| 1 | | 0 | 1 |
| 0 | | 0 | 0 |

### 4.9.6 Fractal Compression

Having in mind that nowadays the devices (e.g., printers) constantly increase the resolution, it is necessary to provide that once compressed image can be decompressed at any resolution. This can be accomplished by using the fractal compression. In fractal compression, the entire image is divided into pieces (fractals). Using an affine transformation, we are able to mathematically rotate, scale, skew, and translate a function, and thus certain fractals can be used to "cover" the whole image.

Affine transformations for two-dimensional case are given by the relation:

$$W(x, y) = (ax + by + e, cx + dy + f), \tag{4.76}$$

or in the matrix form:

$$W\begin{pmatrix} x \\ y \end{pmatrix} = \begin{pmatrix} a & b \\ c & d \end{pmatrix}\begin{pmatrix} x \\ y \end{pmatrix} + \begin{pmatrix} e \\ f \end{pmatrix}, \tag{4.77}$$

where the matrix with elements $a, b, c, d$ determines the rotation, skew, and scaling, while the matrix with elements $e$ and $f$ defines the translation.

In the algorithm for fractal compression the entire image is firstly divided into nonoverlapping regions. Then each region is divided into a number of predefined shapes (e.g., rectangles, squares, or triangles) as shown in Fig. 4.42.

The third step is to determine the affine transformations that closely match the domain regions. In the final step the image is recorded in the FIF format (Fractional Image Format) and it contains information about regions selection and the coefficients of affine transformation (for each region).

### 4.9.7 Image Reconstructions from Projections

Image reconstruction based on projections has important applications in various fields (e.g., in medicine when dealing with computer tomography, which is widely used in everyday diagnosis).

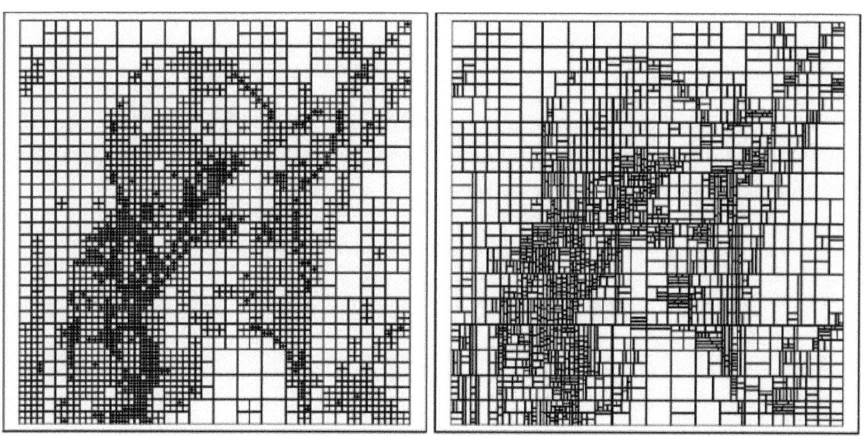

**Fig. 4.42** "Lena" image divided into different fractals

A theoretical approach to this problem is presented below. Consider an object in space, which can be described by the function *f*(*x*,*y*). The projection of function *f*(*x*,*y*) along an arbitrary direction $\overline{AB}$ (defined by an angle $\varphi$) can be defined as follows:

$$p_\varphi(u) = \int_{\overline{AB}} f(x,y)dl, \tag{4.78}$$

where $u = x\cos\varphi + y\sin\varphi$. Thus, the Eq. (4.78) can be written as:

$$p_\varphi(u) = \int_{-\infty}^{\infty}\int_{-\infty}^{\infty} f(x,y)\delta(x\cos\varphi + y\sin\varphi - u)dxdy. \tag{4.79}$$

The Fourier transform of the projection is given by:

$$P_\varphi(\omega) = \int_{-\infty}^{\infty} p_\varphi(u)e^{-j\omega u}du. \tag{4.80}$$

Furthermore, the two-dimensional Fourier transform of *f*(*x*,*y*) is defined as:

$$F(\omega_x,\omega_y) = \int_{-\infty}^{\infty}\int_{-\infty}^{\infty} f(x,y)e^{-j(\omega_x x+\omega_y y)}dxdy. \tag{4.81}$$

As a special case, we can observe $F(\omega_x, \omega_y)$ for $\omega_y = 0$:

$$F(\omega_x,0) = \int_{-\infty}^{\infty}\int_{-\infty}^{\infty} f(x,y)e^{-j\omega_x x}dxdy = \int_{-\infty}^{\infty} p_0(x)e^{-j\omega_x x}dx = P_0(\omega). \tag{4.82}$$

Hence, the Fourier transform of a two-dimensional object along the axis $\omega_y = 0$ is equal to the Fourier transform along the projection angle $\varphi = 0$. Consider now what happens in the rotated coordinate system:

$$\begin{pmatrix} u \\ l \end{pmatrix} = \begin{pmatrix} \cos\varphi & \sin\varphi \\ -\sin\varphi & \cos\varphi \end{pmatrix}\begin{pmatrix} x \\ y \end{pmatrix}. \tag{4.83}$$

In this case, Eq. (4.80) can be written as:

$$P_\varphi(\omega) = \int_{-\infty}^{\infty} p_\varphi(u)e^{-j\omega u}du = \int_{-\infty}^{\infty}\int_{-\infty}^{\infty} f(u,l)e^{-j\omega u}dudl$$
$$= \int_{-\infty}^{\infty}\int_{-\infty}^{\infty} f(x,y)e^{-j\omega(x\cos\varphi + y\sin\varphi)}dxdy = F(\omega,\varphi), \tag{4.84}$$

where:

$$F(\omega,\varphi) = F(\omega_x,\omega_y)\Big|_{\begin{array}{l}\omega_x = \omega\cos\varphi\\ \omega_y = \omega\sin\varphi\end{array}}.$$

Now, let us summarize the previous considerations. If we have the object projections, then we can determine their Fourier transforms. The Fourier transform of a projection represents the transform coefficients along the projection line of the object. By varying the projection angle from 0° to 180° we obtain the Fourier transform along all the lines (e.g., we get the Fourier transform of the entire object), but in the polar coordinate system. To use the well-known FFT algorithm, we have to switch from polar to rectangular coordinate system. Then, the image of the object is obtained by calculating the inverse Fourier transform.

The transformation from the polar to the rectangular coordinate system can be done by using the nearest neighbor principle, or by using some other more accurate algorithms that are based on the interpolations.

## 4.10 Examples

4.1. Calculate the memory requirements for an image of size 256 × 256 pixels, in the case of:

(a) Binary image,
(b) Grayscale image,
(c) Color image.

Solution:

(a) In the case of binary image each sample is represented by a single bit, and thus the required memory space is (in bits):
$256 \cdot 256 \cdot 1 = 65536$ b.

(b) Grayscale image is usually represented by 8 bits per pixel, thus having 256 grayscale levels. The memory requirements for such a kind of image are: $256 \cdot 256 \cdot 8 = 524288$ b.

(c) Color image usually contains three different matrices for each color channel and requires three time higher memory space than the grayscale image: 256·256·8·3 = 1572864 b.

4.2. If the values of R, G, and B components in the RGB systems are known and for a certain pixel they are given by: R = 0.5, G = 0.2, B = 0.8, determine the corresponding values of the components in the YUV color model.

Solution:

$$
\begin{aligned}
Y &= 0.299R + 0.587G + 0.114B \\
U &= 0.564(B - Y) \\
V &= 0.713(R - Y)
\end{aligned}
$$

$$
\begin{aligned}
\mathrm{Y} &= 0.299 \cdot 0.5 + 0.587 \cdot 0.2 + 0.114 \cdot 0.8 = 0.36 \\
\mathrm{U} &= 0.564 \cdot (0.8 - 0.358) = 0.25 \\
\mathrm{V} &= 0.713 \cdot (0.5 - 0.358) = 0.1
\end{aligned}
$$

4.3. Write a Matlab code which will load color image (e.g., *lena.jpg*), determine the image size, and then convert the color image into a grayscale version by using the Matlab built-in function *rgb2gray*, as well as by using the formula: $Grayscale = \frac{R_{value}+G_{value}+B_{value}}{3}$.

Solution:

```
I=imread('lena.jpg'); % load image
size(I)        % image size
ans =
512 512 3
I1=rgb2gray(I);
figure, imshow(I1)   % show image
I=double(I);
I2=(I(:,:,1)+I(:,:,2)+I(:,:,3))/3;
figure, imshow(uint8(I2));
```

*Note: The color channels are obtained as: I(:,:,1), I(:,:,2), I(:,:,3).*

4.4. Write a code in Matlab that will create a negative of image "*cameraman.tif*".

Solution:

```
I=imread('cameraman.tif');
I=double(I);
N=255-I;
figure, imshow(uint8(I))
figure, imshow(uint8(N)) (Fig. 4.43)
```

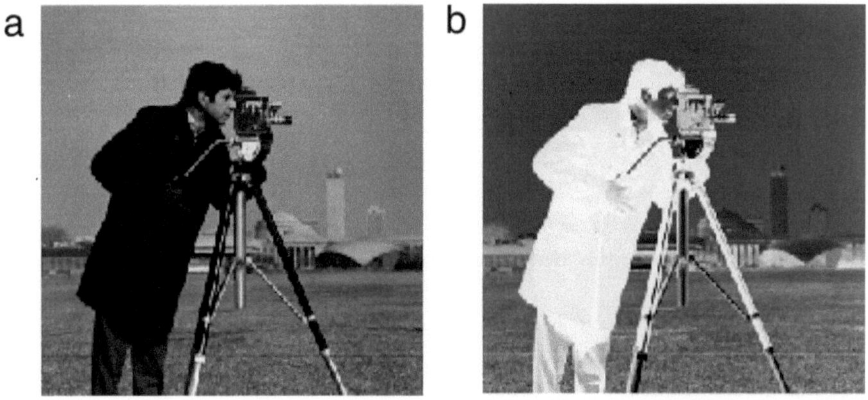

**Fig. 4.43** (**a**) Original image "Cameraman," (**b**) negative of image "Cameraman"

**Fig. 4.44** Image lightening and darkening

4.5. Write a code in Matlab that will provide a simple image darkening and brightening procedure by decreasing/increasing original pixels values for 40 %.

Solution:

```
I=imread('cameraman.tif');
I=double(I);
I_B=I+0.4*I;   % brightening
figure(1), imshow(uint8(I_B))
I_D=I-0.4*I;   % darkening
figure(2), imshow(uint8(I_D)) (Fig. 4.44)
```

4.6. Starting from the grayscale image "*cameraman.tif*", make a version of binary image by setting the threshold on value 128.

Solution:
A binary image will have values 255 at the positions $(i,j)$ where the original image has values above the threshold. On the remaining positions the pixels in the binary image will be 0:

$$B(i,j) = \begin{cases} 255, & I(i,j) > threshold \\ 0, & \text{otherwise} \end{cases}.$$

The Matlab code that transforms grayscale into binary image is given in the sequel.

```
I=imread('cameraman.tif');
[m,n]=size(I);
for i=1:m
  for j=1:n
    if I(i,j)>128
    I(i,j)=255;
    else
    I(i,j)=0;
    end
  end
end
  figure, imshow(I) (Fig. 4.45)
```

4.7. Consider a color image "*lena.jpg*". Transform the image into grayscale one and add a white Gaussian noise with variance 0.02.

Solution:

```
I=imread('lena.jpg');
I=rgb2gray(I);
I1=imnoise(I,'gaussian',0,0.02);
figure, imshow(uint8(I1))
```

**Fig. 4.45** Binary image "Cameraman"

4.8. Calculate the mean and median values for vectors:

(a) $v_1 = [\ 12\ 22\ 16\ 41\ -3]$; b) $v_2 = [12\ 9\ 22\ 16\ 41\ -3]$;

Solution:

(a) $v_1 = [\ 12\ 22\ 16\ 41\ -3]$;
mean = 17.6
sorted_$v_1 = [-3\ 12\ 16\ 22\ 41]$;
median = 16.

(b) $v_2 = [12\ 9\ 22\ 16\ 41\ -3]$
mean = 16.17
sorted_ $v_2 = [-3\ 9\ 12\ 16\ 22\ 41]$
median = (12 + 16)/2 = 14.

4.9. By using the Matlab function *imnoise*, add the impulse noise ('salt & pepper' with a density 0.1) to the image "*lena.jpg*". Then perform the image filtering by using the two-dimensional median filter realized by Matlab function *medfilt2*.

Solution:

```
I=imread('lena.jpg');
I=rgb2gray(I);
figure, imshow(I)
In=imnoise(I,'salt & pepper',0.1);
figure, imshow(In)
If=medfilt2(In);
figure, imshow(If) (Fig. 4.46)
```

4.10. Write your own code for median filtering in Matlab: the filtering should be applied to image "*cameraman.tif*" which is corrupted by the impulse noise with density 0.1. Use the window of size 3 × 3. It is necessary to include the image boundaries as well.

Solution:

```
I=imread('cameraman.tif');
In=imnoise(I,'salt & pepper',0.03);
```

**Fig. 4.46** (**a**) Original image, (**b**) noisy image, (**c**) filtered image

**Fig. 4.47** (**a**) Noisy image, (**b**) filtered image

```
[m,n]=size(In);
IM=zeros(m,n);
In=double(In);
for i=1:m
  for j=1:n
  b=In(max(i,i-1):min(m,i+1),max(j,j-1):min(n,j+1));
  b=b(:);
  IM(i,j)=median(b);
  end
end
figure(1),imshow(uint8(In))
figure(2),imshow(uint8(IM))(Fig. 4.47)
```

4.11. Write a Matlab code that filters an image corrupted by Gaussian noise with zero mean and variance equal to 0.01. The window of size $5 \times 5$ is used.

Solution:

```
I=imread('cameraman.tif');
In=imnoise(I,'gaussian',0,0.01);
M=In;
[m,n]=size(In);
a=double(In);
for i=1:m
    for j=1:n
    b=a(max(i,i-2):min(m,i+2),max(j,j-2):min(n,j+2));
    c=b(:);
    M(i,j)=mean(c);
    end
end
```

**Fig. 4.48** (**a**) Noisy image (Gaussian noise), (**b**) filtered image

```
figure(1),imshow(In)
figure(2),imshow(uint8(M))(Fig. 4.48)
```

4.12. For a given matrix of size $5 \times 5$, determine the corresponding co-occurrence matrix and the measure of contrast.

$$\begin{matrix} 12 & 11 & 11 & 11 & 14 \\ 12 & 11 & 11 & 11 & 14 \\ 12 & 11 & 11 & 14 & 14 \\ 13 & 11 & 11 & 14 & 14 \\ 13 & 13 & 12 & 12 & 12 \end{matrix}$$

Solution:
Co-occurrence matrix $c(x,y)$ is obtained in the following form:

| **x/y** | **11** | **12** | **13** | **14** |
|---|---|---|---|---|
| **11** | **6** | **0** | **0** | **4** |
| **12** | **3** | **2** | **0** | **0** |
| **13** | **1** | **1** | **1** | **0** |
| **14** | **0** | **0** | **0** | **2** |

The measure of contrast is given by:

$$Con = \sum_{x=0}^{3} \sum_{y=0}^{3} (x - y)^2 c(x, y) = 44.$$

4.13. For a given block of $8 \times 8$ DCT coefficients and the given JPEG quantization matrix $Q$, perform the quantization, zig-zag scanning and determine the coded sequence.

$$D = \begin{bmatrix} 80 & 50 & 26 & 10 & 33 & 11 & 0 & 0 \\ 22 & -28 & 34 & 10 & 0 & 0 & 0 & 0 \\ 14 & 10 & 17 & 11 & 5 & 0 & 5 & 0 \\ 56 & 17 & 20 & 12 & 0 & 12 & 8 & 0 \\ 10 & 12 & 8 & 3 & 2 & 0 & 7 & 0 \\ 10 & 13 & 17 & 3 & 0 & 2 & 2 & 0 \\ 6 & 0 & 5 & 10 & 14 & 0 & 0 & 0 \\ 0 & 0 & 0 & 0 & 0 & 0 & 0 & 0 \end{bmatrix} \quad Q = \begin{bmatrix} 3 & 5 & 7 & 9 & 11 & 13 & 15 & 17 \\ 5 & 7 & 9 & 11 & 13 & 15 & 17 & 19 \\ 7 & 9 & 11 & 13 & 15 & 17 & 19 & 21 \\ 9 & 11 & 13 & 15 & 17 & 19 & 21 & 23 \\ 11 & 13 & 15 & 17 & 19 & 21 & 23 & 25 \\ 13 & 15 & 17 & 19 & 21 & 23 & 25 & 27 \\ 15 & 17 & 19 & 21 & 23 & 25 & 27 & 29 \\ 17 & 19 & 21 & 23 & 25 & 27 & 29 & 31 \end{bmatrix}$$

Solution:
DCT coefficients from the $8 \times 8$ block are divided by the quantization matrix and rounded to the integer values, as follows:

$$D_q = \text{round}(D/Q) = \begin{bmatrix} 27 & 10 & 4 & 1 & 3 & 1 & 0 & 0 \\ 4 & -4 & 4 & 1 & 0 & 0 & 0 & 0 \\ 2 & 1 & 2 & 1 & 0 & 0 & 0 & 0 \\ 6 & 2 & 2 & 1 & 0 & 1 & 0 & 0 \\ 1 & 1 & 1 & 0 & 0 & 0 & 0 & 0 \\ 1 & 1 & 1 & 0 & 0 & 0 & 0 & 0 \\ 0 & 0 & 0 & 0 & 1 & 0 & 0 & 0 \\ 0 & 0 & 0 & 0 & 0 & 0 & 0 & 0 \end{bmatrix}$$

After performing the zig-zag scanning of the matrix $D_q$, the sequence is obtained in the form:

27, 10, 4, 2, −4, 4, 1, 4, 1, 6, 1, 2, 2, 1, 3, 1, 0, 1, 2, 1, 1, 0, 1, 1, 1, 0, 0, 0, 0, 0, 0, 0, 0, 1, 0, 0, 0, 0, 0, 0, 1, 0, 0, 0, 0, 0, 0, 0, 0, 0, 1, 0, 0, 0, ...

The intermediate symbol sequence is given by:

(5)(27), (0,4)(10), (0,3)(4), (0,2)(2), (0,3)(−4), (0,3)(4), (0,1)(1), (0,3)(4), (0,1)(1), (0,3)(6), (0,1)(1), (0,2)(2), (0,2)(2), (0,1)(1), (0,2)(3), (0,1)(1), (1,1)(1), (0,2)(2), (0,1)(1), (0,1)(1), (1,1)(1), (0,1)(1), (0,1)(1), (8,1)(1), (6,1)(1), (9,1)(1), (0,0).

The code words for the symbols $(a,b)$ are given in the table:

| Symbol $(a,b)$ | Code word |
|---|---|
| (0,1) | 00 |
| (0,2) | 01 |
| (0,3) | 100 |
| (0,4) | 1011 |
| (1,1) | 1100 |
| (6,1) | 1111011 |
| (8,1) | 111111000 |
| (9,1) | 111111001 |
| (0,0) EOB | 1010 |

Hence, the coded sequence is:

(101)(11011) (1011)(1010) (100)(100) (01)(10) (100)(011) (100)(100) (00) (1) (100)(100) (00)(1) (100)(110) (00)(1) (01)(10) (01)(10) (00)(1) (01) (11) (00)(1) (1100)(1) (01)(10) (00)(1) (00)(1) (1100)(1) (00)(1) (00) (1) (111111000)(1) (1111011)(1) (111111001)(1) (1010)

4.14. Consider four 8-bit coefficients A, B, C and D with values A = −1, B = −3, C = 11, D = 5. Explain the encoding procedure using the bit-planes concept as in JPEG2000 compression.

Solution:

| Bit plane sign | A | B | C | D |
|---|---|---|---|---|
| 8 | 1 | 1 | 0 | 0 |
| 7 | 0 | 0 | 0 | 0 |
| 6 | 0 | 0 | 0 | 0 |
| 5 | 0 | 0 | 0 | 0 |
| 4 | 0 | 0 | 0 | 0 |
| 3 | 0 | 0 | 1 | 0 |
| 2 | 0 | 0 | 0 | 1 |
| 1 | 0 | 1 | 1 | 0 |
| 0 | 1 | 1 | 1 | 1 |

The coding starts from the bit-plane 3 since it is the first bit-plane that contains the non-zero coefficients.

***Bit-plane 3****:* This bit-plane is encoded within a single encoding pass. C is declared significant, since its bit has value 1. The sign bit is encoded immediately after bit 1.

***Bit-plane 2***: The coefficients B and D are insignificant, but they are neighbors of the significant coefficient C. Hence, their bits are encoded in the first pass. The corresponding bit of coefficient D is 1, and thus D is declared significant (sign bit is encoded as well).

The bits of significant coefficients are encoded in the second pass, i.e., the bit of the coefficient C. Bit of the coefficient A is encoded in the third pass.

***Bit-plane 1***: The bit of the coefficient B is encoded in the first pass, since B is a neighbor of the significant coefficient C (the sign bit of B is encoded after its bit 1). The bits of significant coefficients C and D are encoded in the second pass. The bit of coefficient A is encoded in the third pass.

***Bit-Plane 0***: Bit of coefficient A is encoded in the first pass, and since the bit is 1, the sign bit is encoded as well. The bits of coefficients B, C and D are encoded in the second pass.

## Appendix: Matlab Codes for Some of the Considered Image Transforms

### *Image Clipping*

```
I=imread('lena512.bmp');
I=I(1:2:512,1:2:512);
I=double(I);
for i=1:256
  for j=1:256
    if I(i,j)<100
      I(i,j)=100;
    elseif I(i,j)>156
        I(i,j)=156;
    end
  end
end
I=uint8(I);
imshow(I)
```

### *Transforming Image Lena to Image Baboon*

```
Ia=imread('lena512.bmp');
Ia=Ia(1:2:512,1:2:512);
Ia=double(Ia);
Ib=imread('baboon.jpg');
Ib=rgb2gray(Ib);
Ib=double(Ib);
for i=1:10
  Ic=(1-i/10)*Ia+(i/10)*Ib;
imshow(uint8(Ic))
pause(0.5)
end
```

### *Geometric Mean Filter*

```
clear all
I=imread('board.tif');
I=imnoise(I,'gaussian',0,0.025);
I=double(I);
```

```
[m,n]=size(I);
Im=zeros(size(I));
for i=1:m
    for j=1:n
        a=I(max(i,i-1):min(m,i+1),max(j,j-1):min(n,j+1));
        Im(i,j)=geomean(a(:));
    end
end
figure(1), imshow(uint8(I))
figure(2), imshow(uint8(Im))
```

## *Consecutive Image Rotations (Image Is Rotated in Steps of 5° up to 90°)*

```
I=imread('lena512.bmp');
I=I(1:2:512,1:2:512);
for k=5:5:90
  I1=imrotate(I,k,'nearest');
  imshow(I1)
  pause(1)
end
```

## *Sobel Edge Detector Version1*

```
I=imread('cameraman.tif');
subplot(221),imshow(I)
edge_h=edge(I,'sobel','horizontal');
subplot(222),imshow(edge_h)
edge_v=edge(I,'sobel','vertical');
subplot(223),imshow(edge_v)
edge_b=edge(I,'sobel','both');
subplot(224),imshow(edge_b)
```

## *Sobel Edge Detector Version2: with an Arbitrary Global Threshold*

```
clear all
I=imread('lena512.bmp');
I=I(1:2:512,1:2:512);
[m,n]=size(I);
```

```
I=double(I);
H=[1 2 1; 0 0 0; -1 -2 -1];
V=[1 0 -1; 2 0 -2; 1 0 -1];
Edge_H=zeros(m,n);
Edge_V=zeros(m,n);
Edges=zeros(m,n);
thr=200;
for i=2:m-1
    for j=2:n-1
Lv=sum(sum(I(i-1:i+1,j-1:j+1).*V));
Lh=sum(sum(I(i-1:i+1,j-1:j+1).*H));
L=sqrt(Lv^2+Lh^2);
if Lv>thr
    Edge_V(i,j)=255;
end
if Lh>thr
    Edge_H(i,j)=255;
end
if L>thr
    Edges(i,j)=255;
end
    end
end
figure, imshow(uint8(Edge_H))
figure, imshow(uint8(Edge_V))
figure, imshow(uint8(Edges))
```

## *Wavelet Image Decomposition*

```
I=imread('lena512.bmp');
I=double(I);
n=max(max(I));
%First level decomposition
[S1,H1,V1,D1]=dwt2(I,'haar');
S1=wcodemat(S1,n);
H1=wcodemat(H1,n);
V1=wcodemat(V1,n);
D1=wcodemat(D1,n);
dec2d_1 = [S1 H1; V1 D1];
%Next level decomposition
I=S1;
[S2,H2,V2,D2]=dwt2(I,'haar');
```

```
S2=wcodemat(S2,n);
H2=wcodemat(H2,n);
V2=wcodemat(V2,n);
D2=wcodemat(D2,n);
dec2d_2 = [S2 H2; V2 D2];
dec2d_1 = [dec2d_2 H1; V1 D1];
imshow(uint8(dec2d_1))
```

## JPEG Image Quantization

```
I=imread('lena.jpg');
I=rgb2gray(I);
I=double(I(1:2:512,1:2:512));
Q50=[16 11 10 16 24 40 51 61;
      12 14 19 26 58 60 55;
      14 13 16 24 40 57 69 56;
      14 17 22 29 51 87 80 62;
      18 22 37 56 68 109 103 77;
      24 35 55 64 81 104 113 92;
      49 64 78 87 103 121 120 101;
      72 92 95 98 112 100 103 99];
QF=70;
q=2-0.02*QF; %q=50/QF;
Q=round(Q50.*q);
I1=zeros(256,256);
for i=1:8:256-7
  for j=1:8:256-7
      A=I(i:i+7,j:j+7);
      dct_block=dct2(A);
      dct_Q=round(dct_block./Q).*Q;
      I1(i:i+7,j:j+7)=idct2(dct_Q);
  end
end
figure(1), imshow(uint8(I))
figure (2), imshow (uint8(I1))
```

**Table 4.4** Symbols and corresponding code words for AC luminance components

| (*a*,*b*) | Code word | (*a*,*b*) | Code word |
|---|---|---|---|
| (0,0) | 1010 | (3,9) | 1111111110010100 |
| (0,1) | 00 | (3,A) | 1111111110010101 |
| (0,2) | 01 | (4,1) | 111011 |
| (0,3) | 100 | (4,2) | 1111111000 |
| (0,4) | 1011 | (4,3) | 1111111110010110 |
| (0,5) | 11010 | (4,4) | 1111111110010111 |
| (0,6) | 1111000 | (4,5) | 1111111110011000 |
| (0,7) | 11111000 | (4,6) | 1111111110011001 |
| (0,8) | 1111110110 | (4,7) | 1111111110011010 |
| (0,9) | 1111111110000010 | (4,8) | 1111111110011011 |
| (0,A) | 1111111110000011 | (4,9) | 1111111110011100 |
| (1,1) | 1100 | (4,A) | 1111111110011101 |
| (1,2) | 11011 | (5,1) | 1111010 |
| (1,3) | 1111001 | (5,2) | 11111110111 |
| (1,4) | 111110110 | (5,3) | 1111111110011110 |
| (1,5) | 11111110110 | (5,4) | 1111111110011111 |
| (1,6) | 1111111110000100 | (5,5) | 1111111110100000 |
| (1,7) | 1111111110000101 | (5,6) | 1111111110100001 |
| (1,8) | 1111111110000110 | (5,7) | 1111111110100010 |
| (1,9) | 1111111110000111 | (5,8) | 1111111110100011 |
| (1,A) | 1111111110001000 | (5,9) | 1111111110100100 |
| (2,1) | 11100 | (5,A) | 1111111110100101 |
| (2,2) | 11111001 | (6,1) | 1111011 |
| (2,3) | 1111110111 | (6,2) | 111111110110 |
| (2,4) | 111111110100 | (6,3) | 1111111110100110 |
| (2,5) | 1111111110001001 | (6,4) | 1111111110100111 |
| (2,6) | 1111111110001010 | (6,5) | 1111111110101000 |
| (2,7) | 1111111110001011 | (6,6) | 1111111110101001 |
| (2,8) | 1111111110001100 | (6,7) | 1111111110101010 |
| (2,9) | 1111111110001101 | (6,8) | 1111111110101011 |
| (2,A) | 1111111110001110 | (6,9) | 1111111110101100 |
| (3,1) | 111010 | (6,A) | 1111111110101101 |
| (3/2) | 111110111 | (7,1) | 11111010 |
| (3,3) | 111111110101 | (7,2) | 111111110111 |
| (3,4) | 1111111110001111 | (7,3) | 1111111110101110 |
| (3,5) | 1111111110010000 | (7,4) | 1111111110101111 |
| (3,6) | 1111111110010001 | (7,5) | 1111111110110000 |
| (3,7) | 1111111110010010 | (7,6) | 1111111110110001 |
| (3,8) | 1111111110010011 | (7,7) | 1111111110110010 |
| (7,8) | 1111111110110011 | (C,1) | 1111111010 |
| (7,9) | 1111111110110100 | (C,2) | 1111111111011001 |
| (7,A) | 1111111110110101 | (C,3) | 1111111111011010 |
| (8,1) | 111111000 | (C,4) | 1111111111011011 |
| (8,2) | 111111111000000 | (C,5) | 1111111111011100 |
| (8,3) | 1111111110110110 | (C,6) | 1111111111011101 |
| (8,4) | 1111111110110111 | (C,7) | 1111111111011110 |
| (8,5) | 1111111110111000 | (C,8) | 1111111111011111 |
| (8,6) | 1111111110111001 | (C,9) | 1111111111100000 |
| (8,7) | 1111111110111010 | (C,A) | 1111111111100001 |
| (8,8) | 1111111110111011 | (D,1) | 11111111000 |

(continued)

Table 4.4 (continued)

| (a,b) | Code word | (a,b) | Code word |
|---|---|---|---|
| (8,9) | 1111111110111100 | (D,2) | 1111111111100010 |
| (8,A) | 1111111110111101 | (D,3) | 1111111111100011 |
| (9,1) | 111111001 | (D,4) | 1111111111100100 |
| (9,2) | 1111111110111110 | (D,5) | 1111111111100101 |
| (9,3) | 1111111110111111 | (D,6) | 1111111111100110 |
| (9,4) | 1111111111000000 | (D,7) | 1111111111100111 |
| (9,5) | 1111111111000001 | (D,8) | 1111111111101000 |
| (9,6) | 1111111111000010 | (D,9) | 1111111111101001 |
| (9,7) | 1111111111000011 | (D,A) | 1111111111101010 |
| (9,8) | 1111111111000100 | (E,1) | 1111111111101011 |
| (9,9) | 1111111111000101 | (E,2) | 1111111111101100 |
| (9,A) | 1111111111000110 | (E,3) | 1111111111101101 |
| (A,1) | 111111010 | (E,4) | 1111111111101110 |
| (A,2) | 1111111111000111 | (E,5) | 1111111111101111 |
| (A,3) | 1111111111001000 | (E,6) | 1111111111110000 |
| (A,4) | 1111111111001001 | (E,7) | 1111111111110001 |
| (A,5) | 1111111111001010 | (E,8) | 1111111111110010 |
| (A,6) | 1111111111001011 | (E,9) | 1111111111110011 |
| (A,7) | 1111111111001100 | (E,A) | 1111111111110100 |
| (A,8) | 1111111111001101 | (F,0) | 1111111001 |
| (A,9) | 1111111111001110 | (F,1) | 1111111111110101 |
| (A,A) | 1111111111001111 | (F,2) | 1111111111110110 |
| (B,1) | 1111111001 | (F,3) | 1111111111110111 |
| (B,2) | 1111111111010000 | (F,4) | 1111111111111000 |
| (B,3) | 1111111111010001 | (F,5) | 1111111111111001 |
| (B,4) | 1111111111010010 | (F,6) | 1111111111111010 |
| (B,5) | 1111111111010011 | (F,7) | 1111111111111011 |
| (B,6) | 1111111111010100 | (F,8) | 1111111111111100 |
| (B,7) | 1111111111010101 | (F,9) | 1111111111111101 |
| (B,8) | 1111111111010110 | (F,A) | 1111111111111110 |
| (B,9) | 1111111111010111 | | |
| (B,A) | 1111111111011000 | | |

# References


1. Askelof J, Larsson Carlander M, Christopoulos C (2002) Region of interest coding in JPEG2000. Signal Process Image commun 17:105–111
2. Baret HH, Myers K (2004) Foundation of image science. John Willey and Sons, Hoboken, NJ
3. Bednar J, Watt T (1984) Alpha-trimmed means and their relationship to median filters. IEEE Trans Acoustics, Speech Signal Process 32(1):145–153
4. Daubechies I (1992) Ten Lectures on Wavelets. Society for industrial and applied mathematics
5. Djurovic I (2006) Digitalna obrada slike. Univerzitet Crne gore, Elektrotehnicki fakultet Podgorica
6. Dyer M, Taubman D, Nooshabadi S, Kumar Gupta A (2006) Concurrency techniques for arithmetic coding in JPEG2000. IEEE Trans Circuits Syst I 53(6):1203–1213
7. Fisher Y (ed) (1995) Fractal image compression: theory and application to digital images. Springer Verlag, New York
8. Furth B, Smoliar S, Zhang H (1996) Video and image processing in Multimedia systems. Kluwer Academic Publishers, Boston

9. González RC, Woods R (2008) Digital image processing. Prentice Hall, Upper Saddle River, NJ
10. Kato T, Kurita T, Otsu N, Hirata KA (1992) Sketch retrieval method for full color image database-query by visual example. Proceedings 11th IAPR International Conference on Pattern Recognition, vol I, Computer Vision and Applications: 530–533
11. Khan MI, Jeoti V, Khan MA (2010) Perceptual encryption of JPEG compressed images using DCT coefficients and splitting of DC coefficients into bitplanes. International Conference on Intelligent and Advanced Systems (ICIAS): 1–6
12. Man H, Docef A, Kossentini F (2005) Performance analysis of the JPEG2000 image coding standard. Multimedia Tools Appl J 26(1):27–57
13. Mandal M (2003) Multimedia signals and systems. Springer, New York
14. Orovic I, Lekic N, Stankovic S (2015) An Analogue-Digital Hardware for L-estimate Space-Varying Image Filtering. Circuits, Systems and Signal Processing
15. Orovic I, Stankovic S (2014) "L-statistics based Space/Spatial-Frequency Filtering of 2D signals in heavy tailed noise," Signal Processing, Volume 96, Part B, 190–202
16. Percival DB, Walden AT (2006) Wavelet methods for time series analysis. Cambridge University Press, Cambridge
17. Qi YL (2009) A Relevance Feedback Retrieval Method Based on Tamura Texture. Second International Symposium on Knowledge Acquisition and Modeling, KAM '09: 174-177
18. Rabbani M, Joshi B (2002) An overview of the JPEG2000 still image compression standard. Signal Process Image Commun 17(1):3–48
19. Salomon D, Motta G, Bryant D (2009) Handbook of data compression. Springer, New York
20. Schelkens P, Skodras A, Ebrahimi T (2009) The JPEG 2000 Suite. John Wiley & Sons, Hoboken, NJ
21. Stanković L (1990) Digitalna Obrada Signala. Naučna knjiga
22. Stankovic L, Stankovic S, Djurovic I (2000) Space/spatial-frequency analysis based filtering. IEEE Trans Signal Process 48(8):2343–2352
23. Steinmetz R (2000) Multimedia systems. McGrawHill, New York
24. Stollnitz EJ, DeRose TD, Salesin DH (1995) Wavelets for computer graphics: a primer, part 1. IEEE Computer Graphics Appl 15(3):76–84
25. Stollnitz EJ, DeRose TD, Salesin DH (1995) Wavelets for computer graphics: a primer, part 2. IEEE Computer Graphics Appl 15(4):75–85
26. Strutz T (2009) Lifting Parameterisation of the 9/7 Wavelet Filter Bank and its Application in Lossless Image Compression. ISPRA'09: 161–166
27. Strutz T (2009) Wavelet filter design based on the lifting scheme and its application in lossless image compression. WSEAS Trans Signal Process 5(2):53–62
28. Tamura H, Shunji M, Takashi Y (1978) Textural features corresponding to visual perception. IEEE Trans Systems Man Cybernetics 8(6):460–473
29. Taubman D, Marcellin M (2002) JPEG2000: standard for interactive imaging. Proc IEEE 90:1336–1357
30. Thyagarajan KS (2006) Digital image processing with application to digital cinema. Elsevier Focal Press, Oxford
31. T.81: Information technology—Digital compression and coding of continuous-tone still images—Requirements and guidelines
32. Veterli M, Kovačević J (1995) Wavelets and subband coding. Prentice Hall, Upper Saddle River, NJ
33. Wagh KH, Dakhole PK, Adhau VG (2008) Design and Implementation of JPEG2000 Encoder using VHDL. Proceedings of the World Congress on Engineering, vol. I WCE 2008
34. Young I, Gerbrands J, Vliet LV (2009) Fundamentals of Image Processing. Delft University of Technology

# Chapter 5
# Digital Video

Unlike digital audio signals that are sampled in time or digital images sampled in the spatial domain, a digital video signal is sampled in both space and time, as illustrated in Fig. 5.1.

Time sample is called frame. The sampling rate is 25 frames/s or 30 frames/s. Instead of a frame, two fields can be used, one containing even and the other odd lines. In this case, the sampling rate is 50 fields/s or 60 fields/s.

The color space widely used for video representation is YCbCr color space. The luminance or brightness component, denoted by Y, contains intensity information, whereas chrominance components Cb and Cr provide the information about hue and saturation. The YCbCr color model is related to the RGB model as follows:

$$\begin{aligned} Y &= 0.299R + 0.587G + 0.114B, \\ Cb &= 0.564(B - Y), \\ Cr &= 0.713(R - Y). \end{aligned}$$

The human eye is more sensitive to change in brightness than in color. Thus, Cb and Cr components can be transmitted at lower rates compared to Y component. Different sampling schemes are available depending on the resolution of the luminance Y and the chrominance components Cb and Cr. They have been known as: 4:4:4, 4:2:2, and 4:2:0.

The 4:4:4 scheme means that all components are used with the full resolution: each pixel contains Y, Cb, and Cr component, as shown in Fig. 5.2a. For 4:2:2 scheme, the Cb and Cr components are represented with a twice lower resolution compared to Y. Observing the four pixels, we see that for 4 Y samples, there are 2 Cb and 2 Cr samples, Fig. 5.2b. Lastly, the 4:2:0 sampling scheme has a four times lower resolution for Cb and Cr components comparing to the Y component. Thus, among four pixels only one contains Cb and Cr component, Fig. 5.2c.

Now, let us compute a required number of bits per pixel for each of the considered sampling schemes. Again we consider four neighboring pixels. In the


S. Stanković et al., *Multimedia Signals and Systems*,
DOI 10.1007/978-3-319-23950-7_5

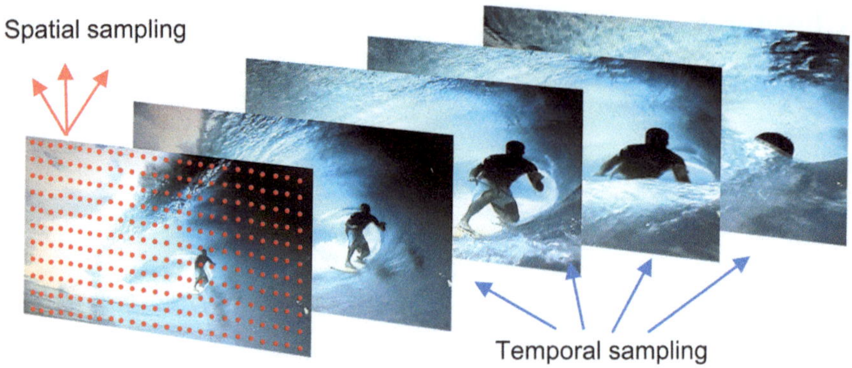


**Fig. 5.1** An illustration of video signal sampling

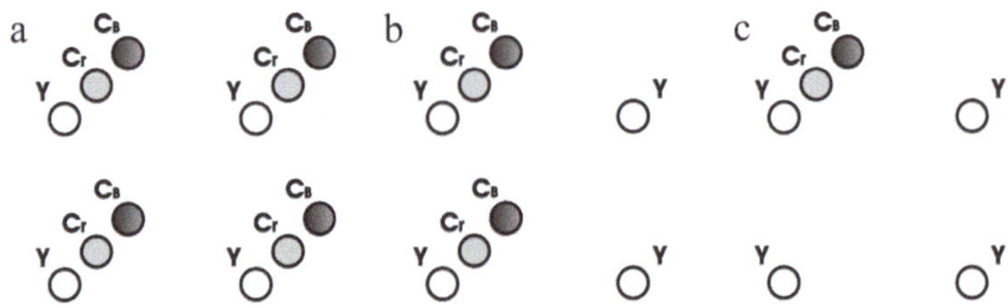


**Fig. 5.2** Sampling schemes: (**a**) 4:4:4, (**b**) 4:2:2, (**c**) 4:2:0

4:4:4 scheme, all pixels contain three components (YCbCr), and if each of them requires 8 bits, we have:

$4 \times 3 \times 8 = 96$ b in total, i.e., $96/4 = 24$ b/pixel.

In the 4:2:2 scheme, two pixels contain three components and the other two pixels contain only one component. Hence, the average number of bits per pixel is calculated as:

$2 \times 3 \times 8 + 2 \times 1 \times 8 = 64$ b in total, i.e., $64/4 = 16$ b/pixel.

In analogy to the previous case, for the 4:2:0 scheme we obtain:

$1 \times 3 \times 8 + 3 \times 1 \times 8 = 48$ b in total, i.e., $48/4 = 12$ b/pixel.

In the sequel, we consider one simple example to illustrate how the sampling schemes 4:4:4 and 4:2:0 influence the amount of data. The frame size is $352 \times 288$. In the first case (4:4:4) we have: $352 \times 288$ pixel $\times$ 24b/pixel = 2.433 Mb. In the second case (4:2:0): $352 \times 288$ pixel $\times$ 12 b/pixel = 1.216 Mb.

## 5.1 Digital Video Standards

The standard for digital video broadcasting is ITU-R BT.601-5 (International Telecommunication Union, Radiocommunications Sector—ITU-R). This standard specifies:

- 60 fields/s for NTSC or 50 fields/s for PAL system
- NTSC requires 525 lines per frame (858 luminance samples per line, $2 \times 429$ chrominance samples per line), while PAL system requires 625 lines (864 luminance samples per line, $2 \times 432$ chrominance samples per line), with 8 b/sample in both systems.

The bit rate of video data is 216 Mb/s in both cases.

In addition to the considered sampling schemes, an important resolution parameter is video signal format. The most frequently used formats are:

- High Definition (full HD) 1080p $1920 \times 1080$;
- HD 720p $1280 \times 720$;
- Standard Definition (SD) $640 \times 480$;
- 4CIF with the resolution $704 \times 576$ (it corresponds to broadcasting standard);
- CIF $352 \times 288$;
- QCIF $176 \times 144$;
- SubQCIF $128 \times 96$.

The video signal bit rate depends on the video frame format. The mentioned bit rate of 216 Mb/s corresponds to the quality used in standard television. Only 187 s of such a signal can be stored on a 4.7 GB DVD. For the CIF format with 25 frames/s and the 4:2:0 scheme, we achieve the bit rate of 30 Mb/s. Similarly, for the QCIF format with 25 frames/s, the bit rate is 7.6 Mb/s. Consider now these video bit rates in the context of the ADSL network capacity. For example, typical bit rates in the ADSL networks are 1–2 Mb/s. Hence, it is obvious that the signal must be compressed in order to be transmitted over the network.

Since we will deal with video compression later on, here we only mention that the compression algorithms belong to the ISO (International Standard Organization) and ITU (International Telecommunication Union) standards. The MPEG algorithms belong to the ISO standard, while the ITU standards cover VCEG algorithms. In order to improve the compression ratio and the quality of the compressed signal, compression algorithms have been improved over time, so today we have: MPEG-1, MPEG-2, MPEG-4, MPEG-7, and MPEG-21. The VCEG standards include: H.261, H.263, H.264 (widely used in many application), and the most recent standard H.265 (which is not yet finalized).

## 5.2 Motion Parameters Estimation in Video Sequences

Motion estimation is an important part of video compression algorithms. One of the simplest methods for motion estimation is a block matching technique. Namely, we consider a block of pixels from the current frame, and in order to estimate its position (motion vector), we compare it with the blocks within a predefined region in the reference frame. As a comparison parameter, it is possible to use the mean square error (MSE) or the sum of absolute errors (SAE):

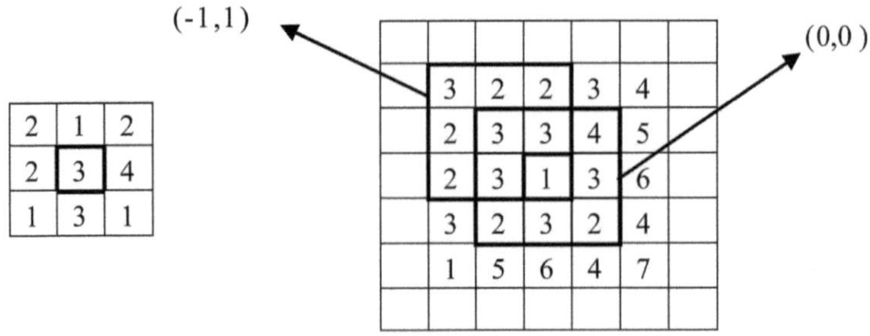


**Fig. 5.3** Motion vector estimation for a 3 × 3 block

$$MSE = \frac{1}{N^2}\sum_{i=1}^{N}\sum_{j=1}^{N}\left(C_{i,j} - R_{i,j}\right)^2, \tag{5.1}$$

$$SAE = \sum_{i=1}^{N}\sum_{j=1}^{N}\left|C_{i,j} - R_{i,j}\right|, \tag{5.2}$$

where $R_{i,j}$ and $C_{i,j}$ are the pixels in the reference and current frame, respectively, and the frame size is $N \times N$. Hence, MSE or SAE are calculated for a set of neighboring blocks in the reference frame. The minimal error obtained for the best-matching block is compared with a certain threshold. If the minimal error is below a certain predefined threshold, the best-matching block position determines the motion vector. This vector indicates the motion of the considered block within the two frames. If the difference between current and reference block is too large, then the current block should be encoded without exploiting temporal redundancy.

Let us illustrate the block matching procedure on a simplified example of a 3 × 3 block (larger blocks are used in practical applications), shown in Fig. 5.3.

Compute the MSE for the central position (0,0):

$$\begin{aligned} MSE_{00} = &\left\{(2-3)^2 + (1-3)^2 + (2-4)^2 + (2-3)^2 + (3-1)^2 + \right.\\ &\left.(4-3)^2 + (1-2)^2 + (3-3)^2 + (1-2)^2\right\}/9 = 1.89. \end{aligned} \tag{5.3}$$

In analogy with Eq. (5.3), the MSEs for other positions are obtained as:

$$\begin{array}{rl} (-1,-1) \to 4.11 & (1,0) \to 4.22 \\ (0,-1) \to 4.44 & (-1,1) \to 0.44 \\ (1,-1) \to 9.44 & (0,1) \to 1.67 \\ (-1,0) \to 2.56 & (1,1) \to 4 \end{array}$$

We see that $\min\{MSE_{nk}\} = MSE_{-1,1}$ and the vector (−1,1) is selected as a motion vector since this position gives the best match candidate for current block.

A procedure for motion vectors estimation in the case of larger blocks is analyzed in the sequel, and it is known as the full search algorithm. It compares

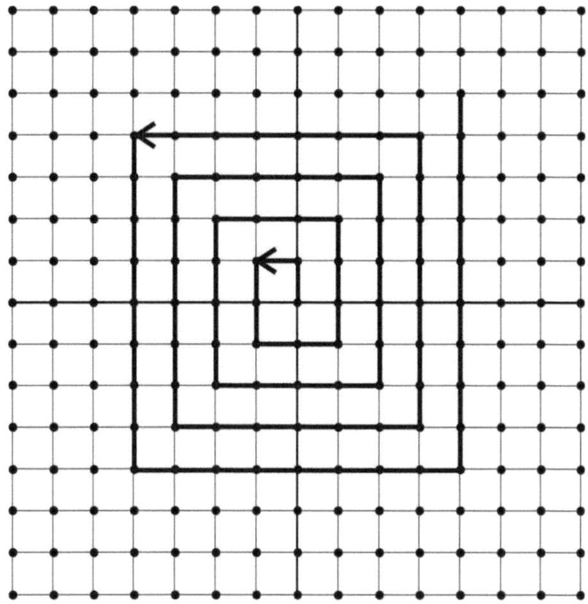

**Fig. 5.4** Illustration of full search

blocks of size 16 × 16, within the search area of 31 × 31 pixels. It means that the search is done over 15 pixels on each side from the central position (0,0), Fig. 5.4. This method is computationally demanding, since we need to calculate $31 \cdot 31 = 961$ MSEs for 16 × 16 blocks.

Therefore, fast search algorithms are defined to reduce the number of calculations, still providing sufficient estimation accuracy. The search procedure based on the three steps algorithm is shown in Fig. 5.5. In the first step, we observe the eight positions at the distance of $p$ pixels (e.g., $p = 4$) from the central point (0,0). The MSEs are calculated for all nine points (denoted by 1 in Fig. 5.5). The position that provides the lowest MSE becomes the central position for the next step.

In the second step, we consider locations on a distance $p/2$ from the new central position. Again, the MSEs are calculated for eight surrounding locations (denoted by 2). The position related to the lowest MSE is a new central position. In the third step, we consider another 8 points around the central position, with the step $p/4$. The position with minimal MSE in the third step determines the motion vector.

Another interesting search algorithm has been known as the logarithmic search (Fig. 5.6). In the first iteration, it considers the position that form a "+" shape (positions denoted by 1 in Fig. 5.6). The position with the smallest MSE is chosen for the central point. Then, in the second iteration, the same formation is done around the new central point and the MSE is calculated. The procedure is repeated until the same position is chosen twice in two consecutive iterations. Afterwards, the search continues by using the closest eight points (denoted by 5). Finally, the position with the lowest MSE defines the motion vector.

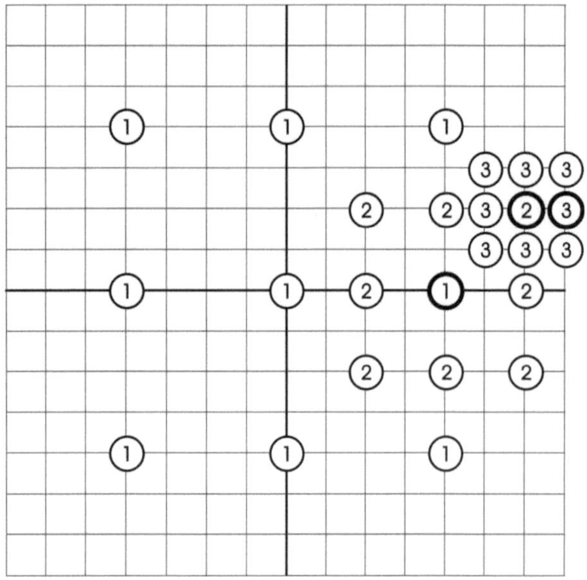

**Fig. 5.5** Illustration of three-step search

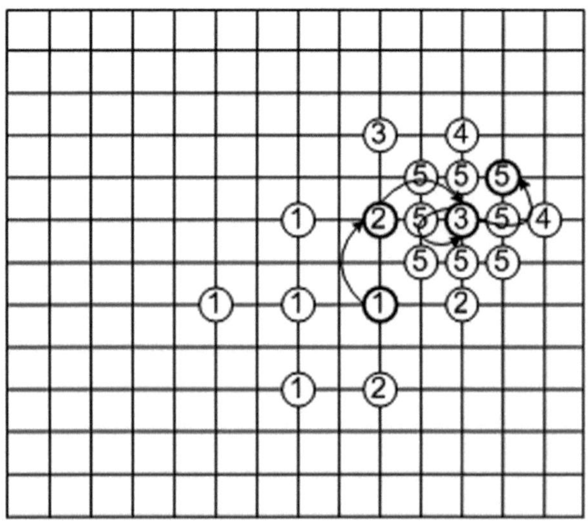

**Fig. 5.6** Illustration of the logarithmic search

The motion vectors search procedures can be combined with other motion parameters estimation (e.g., velocity) algorithms to speed up the algorithms.

Video encoding procedure performs the motion estimation and motion compensation for each block in the current video frame, based on the following steps:

Motion Estimation

- Calculate the MSE or SAE between the current block and a neighboring blocks in the reference frame
- Choose the region that provides the lowest error, i.e., the best matching between the current frame and blocks in the reference frame
- Determine the motion vector as a position of the matching region

Motion Compensation

- Calculate the difference block (prediction error) by subtracting the current block and the matching block (from the reference frame). The difference block should contain small values and low entropy, and thus it is suitable for coding.

Finally, to complete the process for the current block, it is necessary to encode and transmit the difference block and the motion vector.

## 5.3 Digital Video Compression

Compression algorithms are of great importance for digital video signals. In fact, as previously demonstrated, uncompressed video contains large amount of data that requires significant transmission and storage capacities. Hence, the powerful MPEG algorithms are developed and used.

### *5.3.1 MPEG-1 Video Compression Algorithm*

The primary purpose of the MPEG-1 algorithm was to store 74 min of digital video recording on a CD, with a bit rate 1.4 Mb/s. This bit rate is achieved by using the MPEG-1 algorithm, but with a VHS (Video Home System) video quality. A low video quality obtained by the MPEG-1 algorithm was one of the main drawbacks that prevented a wide use of this algorithm. However, the MPEG-1 algorithm served as a basis for the development of the MPEG-2 and was used in some Internet applications, as well.

The main characteristics of the MPEG-1 algorithm are the CIF format ($352 \times 288$) and the YCbCr 4:2:0 sampling scheme. The basic coding units are $16 \times 16$ macroblocks. Therefore, the $16 \times 16$ macroblocks are used for the Y component while, given the 4:2:0 scheme, the $8 \times 8$ macroblocks are used for the Cb and Cr components.

#### 5.3.1.1 Structure of Frames

The MPEG-1 algorithm consists of I, B and P frames. The I frame is firstly displayed, followed by B and then by P frames. The scheme continuously repeats as shown in Fig. 5.7.

The I frames are not encoded by using motion estimation. Thus, the I frames use only *intra coding*, where the blocks are compared within the same frame. P is *inter coded* frame and it is based on the forward prediction. It means that this frame is coded by using motion prediction from the reference I frame.

B frame is the intercoded frame as well, but unlike the P frame, the forward and backward motion predictions (bidirectional prediction) are used. In other words, bidirectional temporal prediction uses two reference frames: the past and the future reference frame.

Let us consider the following example. Assume that we have a video scene in which there is a sudden change in the background at the position of the second B frame. In this case, it is much more efficient to code the first B frame with respect to I frame, while the second B frame should be coded with respect to the P frame. Having in mind the role of individual frames in video decoding, the sequence of frames used for transmission is depicted in Fig. 5.8. Note that this is just an example, but other similar structures of frames sequences can be used in practice.

So an I frame is transmitted first, followed by P and then by B frames. For the considered case, the frame transfer order is:

I1 P4 B2 B3 P7 B5 ...

To reconstruct the video sequence, we use the following order:

I1 B2 B3 P4 B5 ...

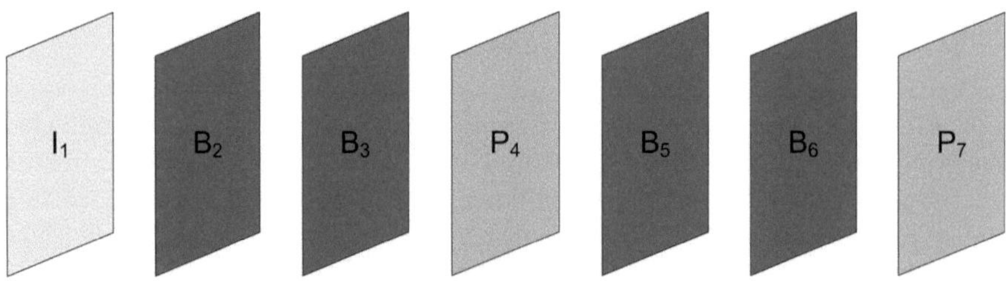


**Fig. 5.7** Structure of frames in MPEG-1 algorithm

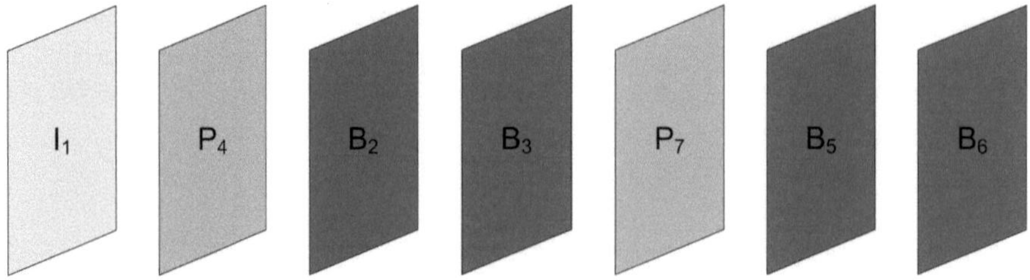


**Fig. 5.8** The order of I, B and P frames during transmission

#### 5.3.1.2 Inter Coding for Exploiting the Temporal Redundancy

The temporal redundancy in MPEG-1 algorithm is exploited by computing the prediction error. In order to perform the motion estimation, the frame is divided into $16 \times 16$ macroblocks ($16 \times 16$ luminance component and associated chrominance components). For the P frames, the forward prediction is performed for each macroblock in the current frame, such that the current macroblock is matched with a certain macroblock in the referent frame (prediction macroblock). The reference frames for a certain P frame should be either previous P or I frame. The position of the best-matching prediction block defines a motion vector. Note that the MPEG-1 algorithm allows the motion estimation with half pixel precision to provide an improved prediction. This concept is discussed in detail later in this chapter. The difference between the current macroblock and the prediction block in the reference frame represents the prediction error (usually called residual).

In the case of B frames, a $16 \times 16$ macroblock (for luminance component) in the current frame can be predicted using $16 \times 16$ macroblock from the past reference frame (forward prediction), or future reference frame (backward prediction). Also, the average value of $16 \times 16$ macroblocks (one from each reference picture) can be used. In the case of bidirectional prediction, each macroblock may have up to two motion vectors. Reference frames for B frames are either P or I frames.

***Transform and coding***: The prediction error is firstly transformed using 2D DCT applied to the $8 \times 8$ residual blocks. The 2D DCT transform is used to de-correlate spatial redundancy. Furthermore, the DCT coefficients are quantized using the quantization matrices and quantization scale to obtain a demanded compression ratio for specific applications. MPEG 1 algorithm uses different quantization matrices for intra and inter coding (Fig. 5.9). The quantized DCT coefficients from the $8 \times 8$ blocks for intra mode are given by:

$$DCT_Q(i,j) = \text{round}\left(\frac{8 \times DCT(i,j)}{Q_1(i,j) \times scale}\right), \tag{5.4}$$

$$Q_1 = \begin{bmatrix} 8 & 16 & 19 & 22 & 26 & 27 & 29 & 34 \\ 16 & 16 & 22 & 24 & 27 & 29 & 34 & 37 \\ 19 & 22 & 26 & 27 & 29 & 34 & 34 & 38 \\ 22 & 22 & 26 & 27 & 29 & 34 & 37 & 40 \\ 22 & 26 & 27 & 29 & 32 & 35 & 40 & 48 \\ 26 & 27 & 29 & 32 & 35 & 40 & 48 & 58 \\ 26 & 27 & 29 & 34 & 38 & 46 & 56 & 69 \\ 27 & 29 & 35 & 38 & 46 & 56 & 69 & 83 \end{bmatrix} \quad Q_2 = \begin{bmatrix} 16 & 16 & 16 & 16 & 16 & 16 & 16 & 16 \\ 16 & 16 & 16 & 16 & 16 & 16 & 16 & 16 \\ 16 & 16 & 16 & 16 & 16 & 16 & 16 & 16 \\ 16 & 16 & 16 & 16 & 16 & 16 & 16 & 16 \\ 16 & 16 & 16 & 16 & 16 & 16 & 16 & 16 \\ 16 & 16 & 16 & 16 & 16 & 16 & 16 & 16 \\ 16 & 16 & 16 & 16 & 16 & 16 & 16 & 16 \\ 16 & 16 & 16 & 16 & 16 & 16 & 16 & 16 \end{bmatrix}$$

**Fig. 5.9** Default quantization matrix for intra coding ($Q_1$) and inter coding ($Q_2$)

while in the case of inter coding we have:

$$DCT_Q(i,j) = \mathrm{round}\left(\frac{8 \times DCT(i,j)}{Q_2(i,j) \times scale}\right). \tag{5.5}$$

The quantization "*scale*" parameter ranges from 1 to 31. Larger numbers give better compression, but worse quality. It is derived from the statistics of each block to control the loss of information depending on the block content or the bit rate requirements.

After the quantization, the entropy coding is applied based on a Variable Length Coder (VLC). The motion vectors are coded and transmitted along with the prediction error.

*The data structure of MPEG-1 algorithms*

The data in MPEG-1 are structured in several levels.

1. **Sequence layer**. The level of sequence contains information about the image resolution and a number of frames per second.
2. **Group of pictures layer**. This level contains information about I, P, and B frames. For example, a scheme consisted of 12 frames can be: 1 I frame, 3 P frames, and 8 B frames.
3. **Picture layer**. It carries information on the type of pictures (e.g., I, P, or B frame), and defines when the picture should be displayed in relation to other pictures.
4. **Slice layer**. Pictures consist of slices, which are further composed of macroblocks. The slice layer provides information about slice position within the picture.
5. **Macroblock layer**. The macroblock level consists of six $8 \times 8$ blocks (four $8 \times 8$ blocks represent the information about luminance and two $8 \times 8$ blocks are used to represent colors).
6. **Block layer**. This level contains the quantized DCT transform coefficients from $8 \times 8$ blocks.

### *5.3.2 MPEG-2 Compression Algorithm*

The MPEG-2 is a part of the ITU-R 601 standard and it is still present in digital TV broadcasting. MPEG-2 is optimized for data transfer at bit rates 3–5 Mb/s. The standard consists of the MPEG-1 audio algorithm, MPEG-2 video algorithm, and a system for multiplexing and transmission of digital audio/video signals. The picture resolutions in MPEG-2 can vary from CIF ($352 \times 288$ with 25 or 30 frames per second) to HDTV with $1920 \times 1080$ with 50 and 60 frames per second. Unlike the MPEG-1 video which mainly works with so-called progressive full frames, most of the frames in MPEG-2 are interlaced: a single frame is split into two half-frames (with odd and even numbered lines) called fields. MPEG-2 can encode either frame pictures or field pictures (two fields are encoded separately). The term picture is used in a general sense, because a picture can be either a frame or a field. Consequently, we have two types of predictions: frame-based and field-based prediction, depending whether the prediction is done using the reference frame or reference field. Frame-based prediction uses one $16 \times 16$ block, while field-based prediction uses two $16 \times 8$ blocks. In field pictures, the DCT coefficients are organized as independent fields. In frame pictures, the DCT can be performed on the macroblocks of frames (frame DCT coding) or fields (field DCT coding). In other words, a frame picture can have frame-coded macroblocks and/or field-coded macroblocks. Frame DCT coding is similar to MPEG-1: the luminance $16 \times 16$ macroblock is divided into four $8 \times 8$ DCT blocks, where each block contains the lines from both fields. In Field DCT, the luminance macroblock is also divided into four $8 \times 8$ DCT blocks, but the top two luminance blocks contain only the samples from the upper field (odd lines), while the two bottom luminance components contain the samples from bottom field (even lines). This representation provides a separate motion estimation/compensation for the two fields, which means that each of the fields will have its own motion vector. Therefore, in order to improve prediction efficiency, MPEG-2 supports $16 \times 8$ motion compensation mode. The first motion vector is determined for $16 \times 8$ block in the first field and the second motion vector is determined for $16 \times 8$ block in the second field. Another mode is dual-prime adaptive motion prediction (for P pictures only). When prediction of a current field is based on two adjacent reference fields (top and bottom), the motion vectors usually look very similar. If we code them independently, it would cause redundant information. The Dual prime motion prediction tends to send only the minimal differential information about motion vectors, by exploiting the similarity between motion vectors of adjacent fields. Hence, the first motion vector is formed by prediction of the same parity reference field. Instead of the second motion vector predicted from the opposite parity reference field, only the differential information (difference motion vector) needs to be sent as a correction of the former motion vector. In order to reduce the noise, the motion-compensated prediction is obtained as the average value of two preliminary predictions from two adjacent fields of opposite parity (odd and even).

Therefore, we can define 5 different prediction modes in MPEG-2 algorithm.

1. **Frame prediction for frame pictures**
   This mode is the same as in the case of MPEG-1. The prediction is done on the macroblocks $16 \times 16$. In the case of P frames, the prediction is made using blocks from the reference frame, resulting in one motion vector for macroblock. For the prediction of B frames, one may use the previous, future or averaged past and future reference frames, to determine the two motion vectors (forward and backward).
2. **Field prediction for field pictures**
   It is similar to the previous mode except that the macroblocks (both current and reference) are made of pixels that belong to the same field (top or bottom). In the case of P pictures, the reference macroblocks may belong to any of the two most

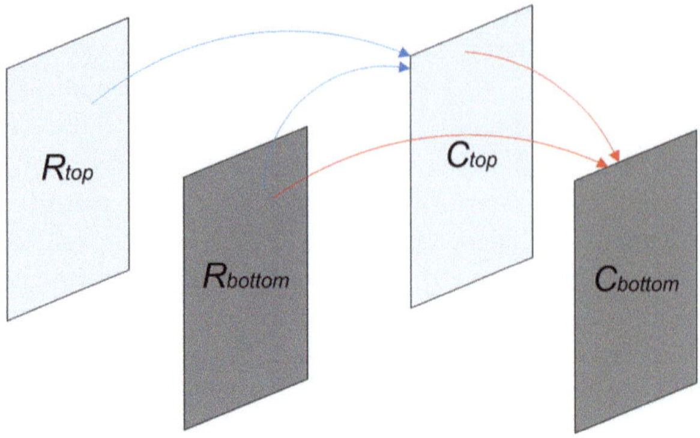


**Fig. 5.10** Field prediction for field pictures (P picture macroblocks)

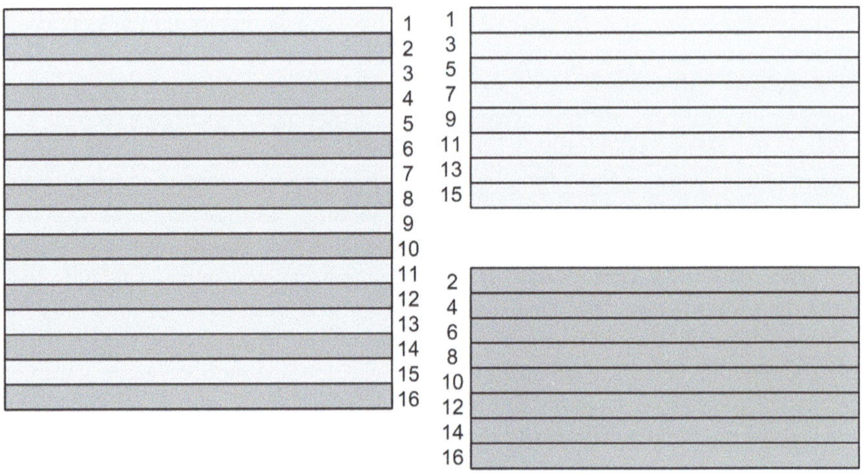


**Fig. 5.11** 16 × 16 macroblock divided into two 16 × 8 fields

recent fields. Let us denote the top and bottom fields in the reference frame as: $R_{top}$ and $R_{bottom}$, while in the current frame we have $C_{top}$ and $C_{bottom}$, appearing in the order: $[R_{top}, R_{bottom}, C_{top}, C_{bottom}]$, Fig. 5.10. Then, the prediction for the current macroblock $C_{top}$ of a P frame, may come either from $R_{top}$ or $R_{bottom}$. The prediction of $C_{bottom}$ can be done from its two recent fields: $C_{top}$ or $R_{bottom}$.

Similarly, for B pictures, the prediction macroblocks are taken from the two most recent reference pictures (I or P). Due to the forward and backward prediction, there will be two motion vectors for each B field.

3. **Field prediction for frame pictures**
   In this prediction mode, a 16 × 16 macroblock of a frame picture is divided into two fields (top and bottom), and field prediction is done independently for each 16 × 8 block from the field (Fig. 5.11). In the case of P pictures, this will results in two motion vectors for each 16 × 16 macroblock. The 16 × 8 predictions can be taken from the two most recently decoded reference pictures. Unlike the

previous case, $16 \times 8$ field prediction cannot belong to the same frame. In the case of B pictures, again we have forward and backward motion. It means that for $16 \times 16$ macroblock made of two $16 \times 8$ fields, we can have two or four motion vectors (one or two per field). The $16 \times 8$ predictions may be taken from either field of the two most recently decoded reference pictures.

4. **Dual-prime prediction for P pictures**
   Dual-prime is used only for P picture and when there are no B pictures between these predicted P pictures and reference pictures. Two independently coded predictions are made: one for the top 8 lines (top filed), another for the 8 bottom field lines. In this prediction mode, there is only one motion vector coded in full format and a small differential motion vector correction. The components of differential motion vector can take the values: −0.5, 0, or 0.5. The first motion vector and its correction are used to obtain the second motion vector. Then, the two motion vectors are used to obtain two predictions from the two adjacent reference fields (top and bottom). The average value of the two predictions represents the final prediction.
5. **$16 \times 8$ motion compensation for field pictures**
   Unlike the case 2 (*Field prediction for field pictures*), where a frame macroblock is split into two field blocks (top and bottom), in this mode, a $16 \times 16$ field macroblock is split into upper half and lower half (two $16 \times 8$ blocks). Then the separate field prediction is carried out for each. Two motion vectors are transmitted for each P picture macroblock, and two or four motion vectors for the B picture macroblock.

***Quantization in MPEG-2***

Another novelty in MPEG-2 algorithm is that the algorithm supports both the linear and nonlinear quantization of the DCT coefficients, which is the advantage over MPEG-1 algorithm. The nonlinear quantization increases the precision at high bit rates by using lower step sizes, while for the lower bit rates it employs larger step sizes.

### *5.3.3 MPEG-4 Compression Algorithm*

The MPEG-4 compression algorithm is designed for low bit rates. The main difference in comparison to the MPEG-1 and MPEG-2 is reflected in the object-based coding and content-based coding. Hence, the algorithm uses an object as the basic unit instead of a frame (the entire scene is split into the objects and background). Equivalently to the frame, in the MPEG-4 we have a video object plane. This concept provides higher compression ratio, because the interaction between objects is much higher than among frames.

MPEG-4 video algorithm with very low bit rate (MPEG-4 VLBV) is basically identical to the H.263 protocol for video communications. The sampling scheme is 4:2:0 and it supports formats 16CIF, 4CIF, CIF, QCIF, SubQCIF, with 30 frames/s.

The motion parameters estimation is performed for $16 \times 16$ or $8 \times 8$ blocks. The DCT is used together with the entropy coding.

The data structure of MPEG-4 VLBV algorithm is:

1. **Picture layer**. It provides the information about the picture resolution, its relative temporal positions among other pictures and the type of encoding (inter, intra).
2. **Group of blocks layer**. This layer contains a group of macroblocks (with a fixed size defined by the standard) and has a similar function as slices in the MPEG-1 and MPEG-2.
3. **Macroblock layer** consists of 4 blocks carrying information about luminance and 2 blocks with chrominance components. Therefore, its header contains information about the type of macroblock, about the motion vectors, etc.
4. **Block layer** consists of quantized and coded transform coefficients from the $8 \times 8$ blocks.

Shape coding is used to provide information about the shape of video object plane (VOP). In other words, it is used to determine whether a pixel belongs to an object or not, and thus, it defines the contours of the video object. The shape information can be coded as a binary (pixel either belongs to the object or not) or gray scale information (coded by 8 bits to provide more description about possible overlapping, pixel transparency, etc.). The binary and greyscale shape masks need to be sufficiently compressed for efficient representation.

Objects are encoded by using the $16 \times 16$ blocks. Note that all pixels within the block can completely belong to an object, but can also be on the edge of the object. For blocks that are completely inside the object plane, the motion estimation is performed similarly to the MPEG-1 and MPEG-2 algorithms. For the blocks outside the object (blocks with transparent pixels) no motion estimation is performed. For the blocks on the boundaries of the video object plane, the motion estimation is done as follows. In the reference frame, the blocks ($16 \times 16$ or $8 \times 8$) on the object boundary are padded by the pixels from the object edge, in order to fill the transparent pixels. Then the block in the current frame is compared with the blocks in the referent frame. The MSE (or SAE) is calculated only for pixels that are inside the video object plane.

Motion estimation is done for video object plane as follows:

- For the I frame, the motion estimation is not performed;
- For the P frame, the motion prediction is based on the I frame or the previous P frame;
- For the B frame, the video object plane is coded by using the motion prediction from I and P frames (backward and forward).

The MPEG-4 in its structure contains spatial and temporal scalability. We can change the resolution with spatial scaling, while the time scaling can change the time resolution for objects and background (e.g., we can display objects with more frames/s, and the background with less frames/s). Also, at the beginning of the video sequence, we can transmit larger backgrounds than the one that is actually displayed at the moment. Hence, when zooming or moving the camera, the background information already exists. This makes the compression more efficient.

### 5.3.4 VCEG Algorithms

The VCEG algorithms are used for video coding and they belong to the ITU standards. Thus, they are more related to the communication applications. Some of the algorithms belonging to this group are described in the sequel, particularly: H.261, H.263, and H.264 (which is currently the most widely used codec).

### 5.3.5 H.261

This standard was developed in the late 80s and early 90s. The main objective was to establish the standards for video conferencing via an ISDN network with a bit rate equal to p $\times$ 64 Kb/s. A typical bit rates achieved with this standard are in the range 64–384 Kb/s. The CIF and QCIF formats are used with the 4:2:0 YCbCr scheme. The coding unit is a macroblock containing 4 luminance and 2 chrominance blocks (of size $8 \times 8$). This compression approach requires relatively simple hardware and software, but has a poor quality of video signals at bit rates below 100 Kb/s.

### 5.3.6 H.263

In order to improve compression performance, the H.263 standard is developed as an extension of H.261. It can support video communication at bit rates below 20 Kb/s with a quite limited video quality that may be used, for example, in video telephony. The functionality of H.263 is identical to the MPEG-4 algorithm. It uses 4:2:0 sampling scheme. The motion prediction can be done separately for each of the four $8 \times 8$ luminance blocks or for the entire $16 \times 16$ block. The novelty of this approach can be seen in introducing an extra frame, called a PB frame, whose macroblocks contain data from the P and B frames, which increases the efficiency of compression (H.263+ optional modes).

An illustration of bit rate variations, depicted as a function of frames, is given in Fig. 5.12. Note that the compression of the P frames is up to 10 times higher than for the I frames.

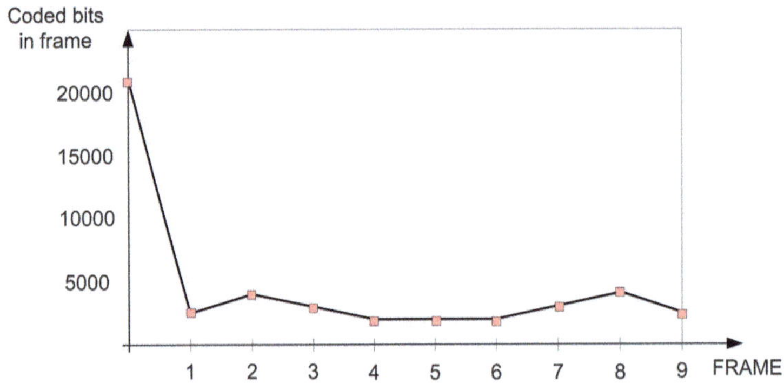


**Fig. 5.12** Illustration of the bit rate (bit rate profile)

### *5.3.7 H.264/MPEG4-AVC*

H.264/MPEG4-AVC is one of the latest standards for video encoding and has been introduced as a joint project of the ITU-T Video Coding Experts Group (VCEG) and ISO/IEC Moving Picture Experts Group (MPEG). This standard covers many current applications, including the applications for mobile phones (mobile TV), video conferencing, IPTV, HDTV, and HD video applications. H.264 replaced H.263 in telecommunications and is used for both terrestrial and satellite broadcast of digital high definition TV (HDTV). Although the broadcast of digital TV in Europe was mainly established on the basis of MPEG-2 (which was the only available at that time), there is a tendency to start broadcasting digital SDTV using H.264, due to its better compression performance over MPEG-2 (at least twice).

Some of the most important new features introduced with H.264 are listed below:

- Multiple reference pictures motion estimation and compensation
- Variable block size for motion estimation: $4 \times 4$ to $16 \times 16$
- Quarter-pixel precision for motion compensation with lower interpolation complexity
- Directional spatial prediction in intra coded macroblocks for efficient compression
- DCT is replaced by integer transform (block size $4 \times 4$ or $8 \times 8$)
- Using de-blocking filter to remove artifacts caused by motion compensation and quantization
- Two entropy coding schemes: context-adaptive variable length coding (CAVLC) and context-adaptive binary arithmetic coding (CABAC)

Some additional advantages are more error resiliency through flexible arrangement of macroblocks, transmission protection using data segmentation and

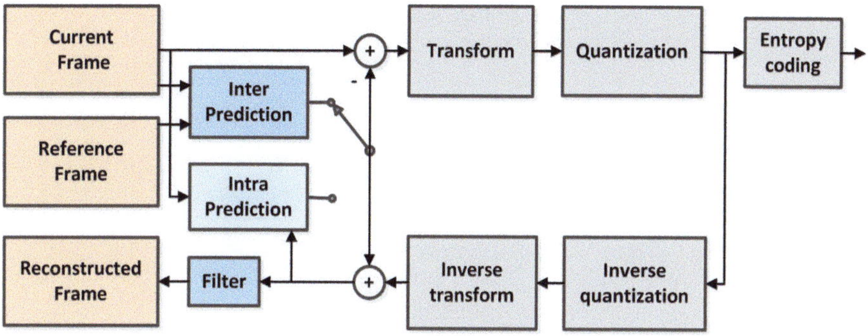


**Fig. 5.13** Block diagram of H.264/AVC encoder

packetization into three priority levels and decreasing data loss by redundant transmission of certain regions.

The H.264/AVC is a block based video standard, which generally partitions the frame using 4:2:0 sampling scheme. There are five types of frames defined by H.264 standard (I, P, B, SP, and SI). The frame is partitioned into $16 \times 16$ macroblocks. The macroblocks are further partitioned into 16 subblocks of $4 \times 4$ pixels for prediction process, and they are then intra or inter coded depending on the frame type. The H.264 algorithm is summarized in the sequel.

H.264/AVC uses aggressive spatial and temporal prediction techniques in order to improve the compression efficiency. The block diagrams of H.264/AVC encoder is presented in Fig. 5.13. The encoder consists of spatial (intra) and temporal (inter) prediction algorithms. Intra prediction is performed only within the I frames, whereas Inter prediction is processed on both P and B frames. Intra prediction computes the prediction of one pixels block by using various interpolation filters. Inter prediction compares the block pixels from the current frame with the prediction blocks in the reference frame (or multiple frames), and finds the best-matching prediction block by computing the SAE. The differences between the best prediction block and current block (residuals or prediction error) are transformed by integer transform and quantized. Quantized coefficients are further entropy encoded for efficient binary representation. Also, these quantized residuals coefficients are sent back to inverse quantization and inverse transform, then added to the best prediction block in order to be reused in the prediction process as a reference for neighboring blocks.

#### 5.3.7.1 Five Types of Frames

As mentioned earlier, the H.264/MPEG4-AVC supports five types of frames: I, P, B, SP, and SI frames. SP and SI frames are used to provide transitions from one bit rate to another.

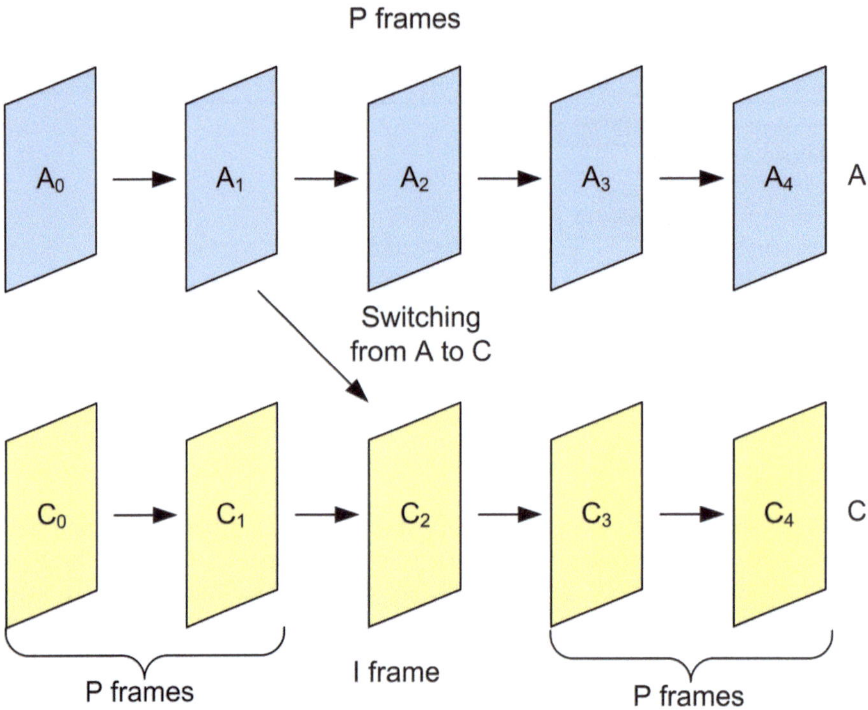


**Fig. 5.14** Switching from one to another bit rate by using I frames

***SP and SI frames***

The SP and SI frames are specially encoded and they are introduced to provide a transition between different bit rates. These frames are also used to provide other operations such as frame skipping, fast forwarding, the transition between two different video sequences, and so on. The SP and SI frames are added only if it is expected that some of these operations will be carried out. The application of these special frames can be illustrated by the following example. During the transfer of signals over the Internet, the same video is encoded for different (multiple) bit rates. The decoder attempts to decode the video with the highest bit rate, but often there is a need to automatically switch to a lower bit rate, if the incoming data stream drops.

Let us assume that during the decoding of sequence with bit rate A, we have to switch automatically to the bit rate C (Fig. 5.14). Also, assume that the P frames are predicted from one reference I frame. After decoding P frames denoted by $A_0$ and $A_1$ (sequence A), the decoder needs to switch to the bit rate C and to decode frames $C_2$, $C_3$, etc. Now, since the frames in the sequence C are predicted from other I frames, the frame in A sequence is not appropriate reference for decoding the frame in C sequence.

One solution is to determine a priori the transition points (e.g., the $C_2$ frame within the C sequence) and to insert an I frame, as shown in Fig. 5.14.

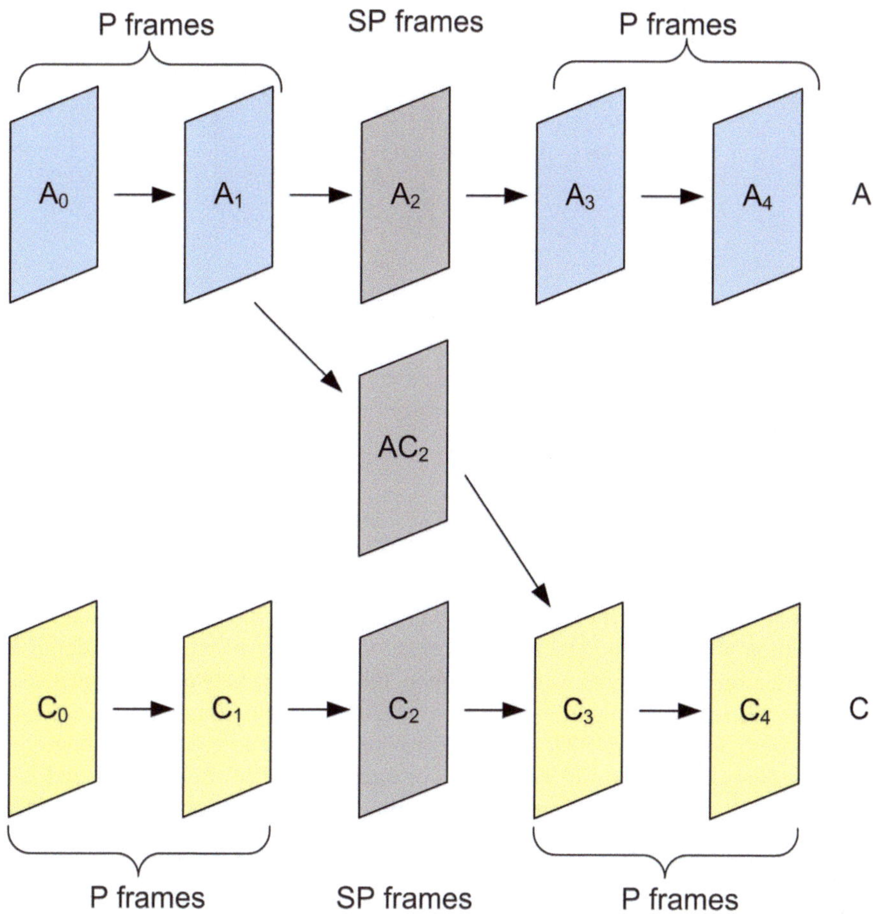


**Fig. 5.15** Switching between different bit rates by using SP frames

As a result of inserting I frames, the transitions between two video sequences would produce peaks in the bit rate. Therefore, the SP frames are designed to support the transition from one bit rate to another, without increasing the number of I frames. Transition points are defined by the SP frames (in the example these are the frames $A_2$, $C_2$, and $AC_2$ that are shown in Fig. 5.15). We can distinguish two types of SP frames: the primary ($A_2$ and $C_2$, which are the parts of the video sequences A and C) and the switching SP frame. If there is no transition, the SP frame $A_2$ is decoded by using the frame $A_1$, while the SP frame $C_2$ is decoded using $C_1$. When switching from A to C sequence, the switching secondary frame ($AC_2$) is used. This frame should provide the same reconstruction as the primary SP frame $C_2$ in order to be the reference frame for $C_3$. Also, the switching frame needs to have characteristics that provide the smooth transition between the sequences. Unlike coding of the P frames, the SP frames coding requires an additional re-quantization

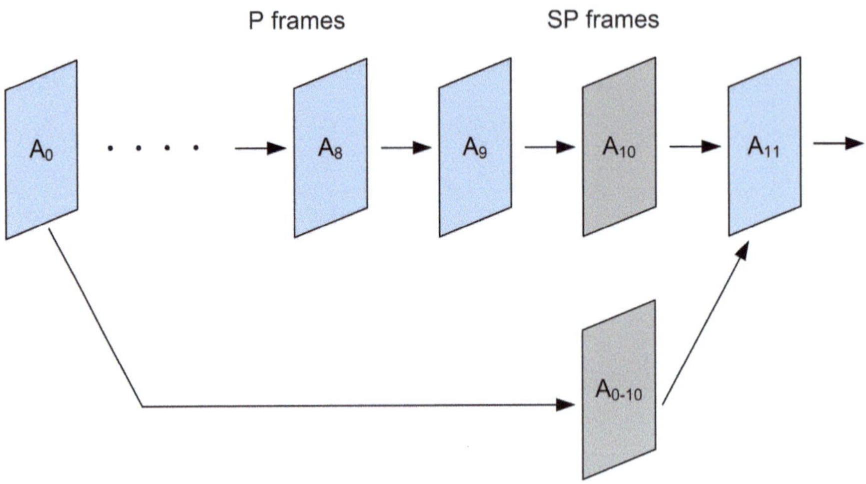


**Fig. 5.16** Illustration of the fast-forward procedure using the SP frames

procedure with a quantization step that corresponds to the step used for the switching SP frame. Obviously, the switching frame should also contain the information about the motion vectors corrections in order to provide an identical reconstruction in both cases: with and without the switching between the sequences.

In the case of switching from C to A bit rate, the switching frame $CA_2$ is needed.

Another application of SP frames is to provide arbitrary access to the frames of a video sequence, as shown in Fig. 5.16. For example, the SP frame ($A_{10}$) and the switching SP frame ($A_{0-10}$) are on the position of the 10th frame. The decoder performs a fast forward from $A_0$ frame to $A_{11}$ frame, by first decoding $A_0$, then the switching SP frame $A_{0-10}$, which will use the motion prediction from $A_0$ to decode the frame $A_{11}$.

The second type of transition frames are the SI frames. They are used in a similar way as the SP frames. These frames can be used to switch between completely different video sequences.

#### 5.3.7.2 Intra Coding in the Spatial Domain

Unlike other video encoding standards where intra coding is performed in the transform domain, the H.264/AVC intra coding is performed in the spatial domain. The pixels of the current block are predicted using the reconstructed pixels of neighboring blocks. There are proposed three types of predictions with respect to the block size: Intra $4 \times 4$ and Intra $16 \times 16$ for luminance component, while Intra $8 \times 8$ is used for chrominance component. Intra $4 \times 4$ coding is based on the prediction of $4 \times 4$ blocks, and it is used to encode the parts of pictures that contain the details. Intra $16 \times 16$ coding is based on the $16 \times 16$ blocks that are used to encode uniform (smooth) parts of the frame. Different prediction modes are

| Component | Block Size | Prediction Modes | Abbreviation |
|---|---|---|---|
| Y 16×16 | 16×16 | 0: vertical | L16_VER |
| | | 1: horizontal | L16_HOR |
| | | 2: DC | L16_DC |
| | | 3: plane | L16_PLANE |
| | 16 blocks 4×4 | 0: vertical | L4_VER |
| | | 1: horizontal | L4_HOR |
| | | 2: DC | L4_DC |
| | | 3: diagonal down-left | L4_DDL |
| | | 4: diagonal down-right | L4_DDR |
| | | 5: vertical-right | L4_VR |
| | | 6: horizontal-down | L4_HD |
| | | 7: vertical-left | L4_VL |
| | | 8: horizontal-up | L4_HU |
| Cb 8×8 | 8×8 | 0: DC | CB8_DC |
| | | 1: horizontal | CB8_HOR |
| | | 2: vertical | CB8_VER |
| | | 3: plane | CB8_PLANE |
| Cr 8×8 | 8×8 | 0: DC | CR8_DC |
| | | 1: horizontal | CR8_HOR |
| | | 2: vertical | CR8_VER |
| | | 3: plane | CR8_PLANE |

**Fig. 5.17** Different Intra prediction modes

available for each intra prediction type (Fig. 5.17). Luminance 16 × 16 and chrominance 8 × 8 predictions have 4 prediction modes (horizontal, vertical, DC, and plane mode as a linear prediction between the neighboring pixels to the left and top), whereas luminance 4 × 4 has 9 prediction modes.

For the intra coding, the prediction of each 4 × 4 block is based on the neighboring pixels. Sixteen pixels in the 4 × 4 block are denoted by a, b, …, p (Fig. 5.18a). They are coded by using the pixels: A, B, C, D, E, F, G, H, I, J, K, L, M, belonging to the neighboring blocks. One DC mode (mean value) and eight directional modes are defined in Intra 4 × 4 prediction algorithm. Fig. 5.18a shows the block that is used in the prediction and Fig. 5.18b depicts prediction directions. Fig. 5.19 illustrates the way of using some directions for block prediction.

The reference pixels used in the prediction process depend on the prediction mode. The vertical prediction, as shown in Fig. 5.19, indicates that the pixels above the current 4 × 4 block are copied to the appropriate positions according to the illustrated direction. Horizontal prediction indicates that the pixels are copied to the marked positions on the left side. DC prediction is a mean value of upper and left reference pixels. The diagonal modes are composed of 2-tap and 3-tap filters. For example, Fig. 5.20 shows the use of the filters depending on the orientation of vertical-right (VR) mode.

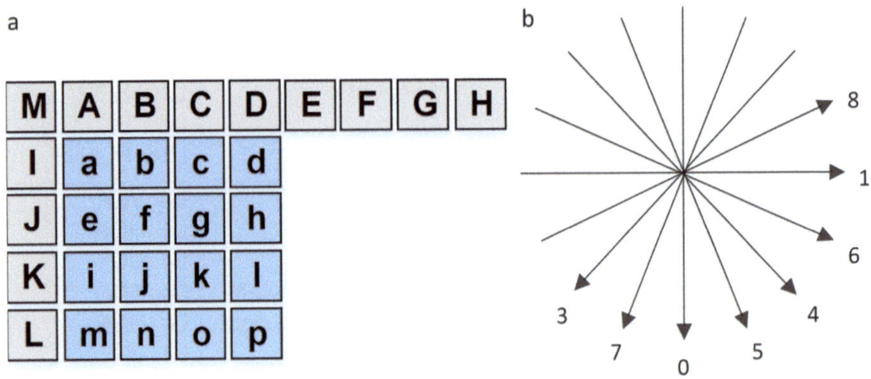


**Fig. 5.18** (**a**) Intra 4 × 4 prediction of block a-p based on the pixels A-M, (**b**) Eight prediction directions for Intra coding

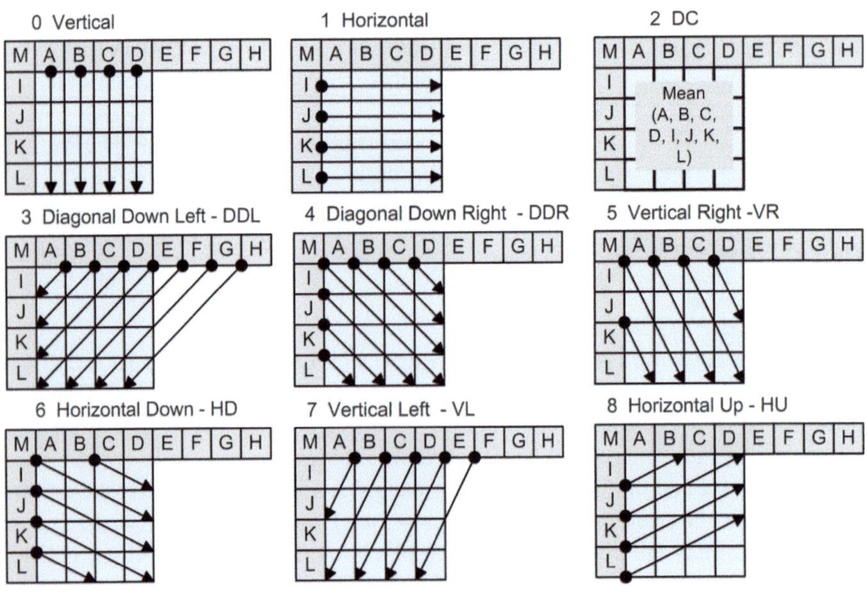


**Fig. 5.19** Nine 4 × 4 intra prediction modes

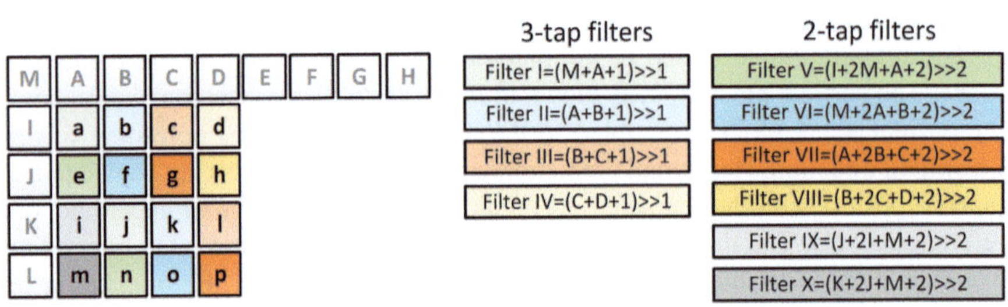


**Fig. 5.20** VR mode prediction filters

**Fig. 5.21** Slice of the reference frame used for prediction

Filter equations of VR prediction mode for each pixel within $4 \times 4$ block are defined as:

$$
\begin{aligned}
a &= (M + A + 1) >> 1 & i &= (J + 2I + M + 2) >> 2 \\
b &= (A + B + 1) >> 1 & j &= (M + A + 1) >> 1 \\
c &= (B + C + 1) >> 1 & k &= (A + B + 1) >> 1 \\
d &= (C + D + 1) >> 1 & l &= (B + C + 1) >> 1 \\
e &= (I + 2M + A + 2) >> 2 & m &= (K + 2J + I + 2) >> 2 \\
f &= (M + 2A + B + 2) >> 2 & n &= (I + 2M + A + 2) >> 2 \\
g &= (A + 2B + C + 2) >> 2 & o &= (M + 2A + B + 2) >> 2 \\
h &= (B + 2C + D + 2) >> 2 & p &= (A + 2B + C + 2) >> 2
\end{aligned}
$$

where the operator $>>$ denotes bit-shifting (to the right).

Fig. 5.21 presents an example with five $4 \times 4$ blocks within macroblock. The reference pixels in neighboring blocks used for prediction computations are marked by red.

The vertical (V) prediction pixels are copies of the pixels from the last row of the upper reference block.

$$
\begin{array}{ll}
a = A = 60 & i = A = 60 \\
b = B = 55 & j = B = 55 \\
c = C = 52 & k = C = 52 \\
d = D = 50 & l = D = 50 \\
e = A = 60 & m = A = 60 \\
f = B = 55 & n = B = 55 \\
g = C = 52 & o = C = 52 \\
h = D = 50 & p = D = 50.
\end{array}
$$

The DC prediction pixels are mean value of the reference pixels in left and upper neighboring block:

$$DC_{\text{mode}} = round\left(\frac{A + B + C + D + I + J + K + L}{8}\right) = 49.$$

The VR prediction pixels are obtained by inserting the reference pixel values into the filter functions as follows:

$$
\begin{aligned}
&a = (58 + 60 + 1)>>1 = (119)_{10}>>1 = (1110111)_2>>1 = (111011)_2 = (59)_{10} \\
&b = (60 + 55 + 1)>>1 = (116)_{10}>>1 = (1110100)_2>>1 = (111010)_2 = (58)_{10} \\
&c = (55 + 52 + 1)>>1 = (108)_{10}>>1 = (1101100)_2>>1 = (110110)_2 = (54)_{10} \\
&d = (52 + 50 + 1)>>1 = (103)_{10}>>1 = (1100111)_2>>1 = (110011)_2 = (51)_{10} \\
&e = (50 + 2 \times 58 + 60 + 2)>>2 = (228)_{10}>>2 = (11100100)_2>>2 = (57)_{10} \\
&f = (58 + 2 \times 60 + 55 + 2)>>2 = (235)_{10}>>2 = (11101011)_2>>2 = (58)_{10} \\
&g = (60 + 2 \times 55 + 52 + 2)>>2 = (224)_{10}>>2 = (11100000)_2>>2 = (56)_{10} \\
&h = (55 + 2 \times 52 + 50 + 2)>>2 = (211)_{10}>>2 = (11010011)_2>>2 = (52)_{10} \\
&i = (45 + 2 \times 50 + 58 + 2)>>2 = (205)_{10}>>2 = (11001101)_2>>2 = (51)_{10} \\
&j = a = (59)_{10} \\
&k = b = (58)_{10} \\
&l = c = (54)_{10} \\
&m = (40 + 2 \times 45 + 50 + 2)>>2 = (182)_{10}>>2 = (10110110)_2>>2 = (45)_{10} \\
&n = e = (57)_{10} \\
&o = f = (58)_{10} \\
&p = g = (56)_{10}
\end{aligned}
$$

The computed predicted pixels for VR mode within the active block are presented in Fig. 5.22.

The best mode among defined modes is found by computing the SAE between original block in the current frame and the predictions.

#### 5.3.7.3 Inter Frame Prediction with Increased Accuracy of Motion Parameters Estimation

As in the previously described standards, inter prediction is the process of predicting blocks from the reference pictures that has previously been encoded.

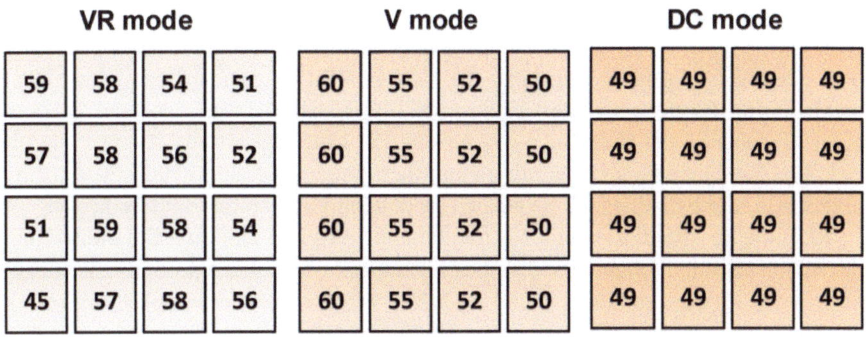


**Fig. 5.22** Prediction pixels for three modes in intra $4 \times 4$ prediction

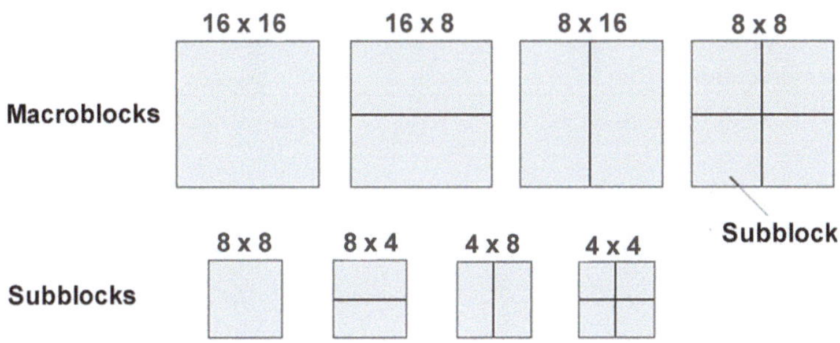


**Fig. 5.23** Macroblocks and subblocks

Recall that the search area in the reference frame is centered on the current block position and can vary in size. The searching procedure results in the best-matched block, which becomes the predictor for the current block in the current frame (motion estimation). The differences between block in the current frame and its best prediction are further encoded and transmitted (motion compensation).

Having in mind that the inter frame prediction and coding includes motion estimation and motion compensation, H.264 has three significant improvements over the other mentioned standards: variable block size motion estimation, quarter-pixel precision for determination of moving vectors, and multiple reference frames for motion compensation. Namely, the inter prediction in H.264 uses blocks of sizes $16 \times 16$, $16 \times 8$, $8 \times 16$, and $8 \times 8$. The $8 \times 8$ can be further divided into the sub-blocks of sizes $8 \times 4$, $4 \times 8$, or $4 \times 4$, as shown in Fig. 5.23.

The macroblocks are partitioned into smaller blocks in order to reduce the cost of coding. In flat areas, it is suitable to use the larger blocks since we will have less motion vectors to encode (consequently, less bits are required). In busy areas, smaller blocks sizes are used since it will significantly reduce the prediction error. Therefore, the total bit rate of encoded data will be reduced.

#### 5.3.7.4 Quarter Pixel Precision

In comparison to other algorithms, the H.264/MPEG-4 standard provides higher precision for the motion vectors estimation. Namely, its accuracy is equal to 1/4 of pixels distance in the luminance component. For other algorithms, the precision is usually 1/2 of the distance. Starting from the integer positions within the existing pixels grid, the half-distance pixels can be obtained by using the interpolation.

After the best prediction block is found in integer-pixel motion estimation within the search region, the encoder searches for half-pixel positions around that integer position. If a better match is found on half-pixel interpolated samples, the search continues on quarter-pixel level around the best half-pixel position.

A 6-tap FIR filter is used to obtain the interpolation accuracy equal to 1/2. Filter coefficients are (1, −5, 20, 20, −5, 1), which can be considered as a low-pass filter. Then the bilinear filter is applied to obtain the precision equal to 1/4 pixel.

The pixels $b$, $h$, $j$, $m$, and $s$ (Fig. 5.24) are obtained following the relations (1/2 pixel precision):

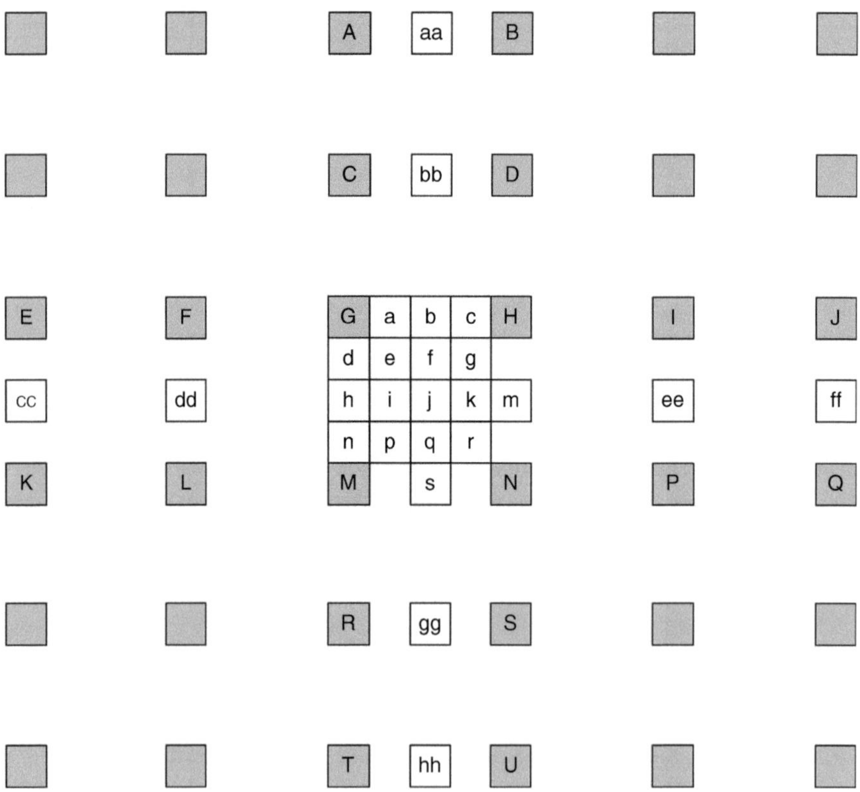


**Fig. 5.24** Interpolation method for ¼ pixel precision (luminance component)

$$\begin{aligned}
b &= ((E - 5F + 20G + 20H - 5I + J) + 16)/32 \\
h &= ((A - 5C + 20G + 20M - 5R + T) + 16)/32 \\
m &= ((B - 5D + 20H + 20N - 5S + U) + 16)/32 \\
s &= ((K - 5L + 20M + 20N - 5P + Q) + 16)/32 \\
j &= ((cc - 5dd + 20h + 20m - 5ee + ff) + 512)/1024 \\
or\ j &= ((aa - 5bb + 20b + 20s - 5gg + hh) + 512)/1024
\end{aligned} \tag{5.6}$$

To obtain a pixel $j$, it is necessary to calculate the values of pixels $cc$, $dd$, $ee$, and $ff$, or $aa$, $bb$, $gg$, and $hh$. Pixels placed at the quarter of the distance between the pixels $a, c, d, e, f, g, i, k, n, p, q$ are obtained as (1/4 pixel precision):

$$\begin{aligned}
a &= \frac{(G+b+1)}{2} & c &= \frac{(H+b+1)}{2} \\
d &= \frac{(G+h+1)}{2} & n &= \frac{(M+h+1)}{2} \\
f &= \frac{(b+j+1)}{2} & i &= \frac{(h+j+1)}{2} \\
k &= \frac{(j+m+1)}{2} & q &= \frac{(j+s+1)}{2} \\
e &= \frac{(b+h+1)}{2} & g &= \frac{(b+m+1)}{2} \\
p &= \frac{(h+s+1)}{2} & r &= \frac{(m+s+1)}{2}
\end{aligned} \tag{5.7}$$

#### 5.3.7.5 Multiple Reference Frames

The H.264 introduces the concept of multiple reference frames (Fig. 5.25). Specifically, the decoded reference frames are stored in the buffer. It allows finding the best possible references from the two sets of buffered frames (*List 0* is a set of past frames, and *List 1* is a set of future frames). Each buffer contains up to 16 frames. The prediction for the block is calculated as a weighted sum of blocks from different multiple reference frames. The motion estimation based on multiple reference frames increases computational cost, but can improve motion prediction performance, especially in the scenes where there is a change in perspective, zoom, or the scene where new objects appear.

Another novelty with the H.264 standard is a generalization of the B frames concept. B frames can be encoded using *List 0*, *List 1*, or bidirectional prediction where the macroblocks are predicted as the weighted average of different frames from the *List 0* and *List 1*.

#### 5.3.7.6 Coding in the Transform Domain Using Integer Transform

The H.264/AVC as well as other coding standards encodes the difference between the current and the reference frame. However, unlike the previous standards (such

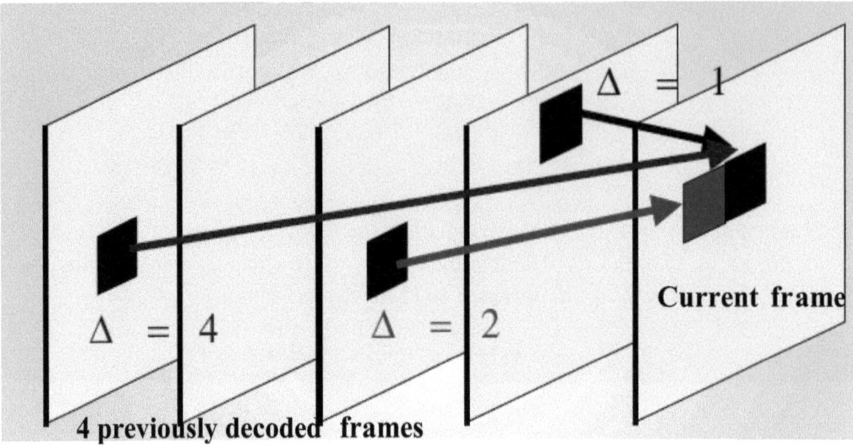


**Fig. 5.25** Multiple reference frames

as MPEG-2 and H.263) based on the DCT coefficients, the H.264/AVC uses integer transform (based on the $4 \times 4$ or $8 \times 8$ transform matrices), which is simpler to implement and allows accurate inverse transform. The commonly used $4 \times 4$ transform matrix is given by:

$$\mathbf{H} = \begin{bmatrix} 1 & 1 & 1 & 1 \\ 2 & 1 & -1 & -2 \\ 1 & -1 & -1 & 1 \\ 1 & -2 & 2 & -1 \end{bmatrix}, \qquad \mathbf{H}^{-1} = \begin{bmatrix} 1 & 1 & 1 & 1 \\ 1 & 1/2 & -1/2 & -1 \\ 1 & -1 & -1 & 1 \\ 1/2 & -1 & 1 & -1/2 \end{bmatrix} \tag{5.8}$$

where $\mathrm{H}^{-1}$ denotes the inverse transform. The inverse transform is exact (unlike in the DCT case). Also, note that the realization of the integer transform can be done using simple operations such as addition and shift operations.

In order to obtain the orthonormal transformation matrix, the 2D transform coefficients are multiplied by a scaling matrix **E**:

$$\mathbf{X} = \left(\mathbf{HBH}^T\right) \otimes \mathbf{E}, \tag{5.9}$$

where $\otimes$ denotes multiplication of elements at the same positions, **B** is block of pixels, and the scaling matrix **E** is given by:

$$\mathbf{E} = \begin{bmatrix} a^2 & \frac{ab}{2} & a^2 & \frac{ab}{2} \\ \frac{ab}{2} & \frac{b^2}{4} & \frac{ab}{2} & \frac{b^2}{4} \\ a^2 & \frac{ab}{2} & a^2 & \frac{ab}{2} \\ \frac{ab}{2} & \frac{b^2}{4} & \frac{ab}{2} & \frac{b^2}{4} \end{bmatrix}, \quad \text{and} \quad \mathbf{E}^{-1} = \begin{bmatrix} a^2 & ab & a^2 & ab \\ ab & b^2 & ab & b^2 \\ a^2 & ab & a^2 & ab \\ ab & b^2 & ab & b^2 \end{bmatrix} \tag{5.10}$$

where: $a = 1/2$, $b = \sqrt{2/5}$.

Notation $\mathbf{E}^{-1}$ is used for the inverse scaling matrix, which is used to weight the 2D transform coefficients before the inverse transform.

In the case of intra $16 \times 16$ coding, there is still high spatial redundancy (high correlation between DC values of sixteen $4 \times 4$ blocks). Therefore, the $4 \times 4$ Hadamard transform is applied to DC coefficients from $4 \times 4$ integer transform blocks:

$$\mathbf{HD}_{4\times4} = \begin{bmatrix} 1 & 1 & 1 & 1 \\ 1 & 1 & -1 & -1 \\ 1 & -1 & -1 & 1 \\ 1 & -1 & 1 & -1 \end{bmatrix}.$$

The chrominance DC values are transformed using $2 \times 2$ Hadamard transform:

$$\mathbf{HD}_{2\times2} = \begin{bmatrix} 1 & 1 \\ 1 & -1 \end{bmatrix}.$$

The high profiles of H.264 (such as *High profile*, *High 10 profile*, *High 4:2:2 profile*, *High 4:4:4 profile*), as an additional option, can apply $8 \times 8$ integer transform (instead of $4 \times 4$ defined by Eq. (5.8)) to motion-compensated residual blocks:

$$\mathbf{H} = \frac{1}{8}\begin{bmatrix} 8 & 8 & 8 & 8 & 8 & 8 & 8 & 8 \\ 12 & 10 & 6 & 3 & -3 & -6 & -10 & -12 \\ 8 & 4 & -4 & -8 & -8 & -4 & 4 & 8 \\ 10 & -3 & -12 & -6 & 6 & 12 & 3 & 10 \\ 8 & -8 & -8 & 8 & 8 & -8 & -8 & 8 \\ 6 & -12 & 3 & 10 & -10 & -3 & 12 & -6 \\ 4 & -8 & 8 & -4 & -4 & 8 & 8 & 4 \\ 3 & -6 & 10 & -12 & 12 & -10 & 6 & -3 \end{bmatrix}.$$

#### 5.3.7.7 Quantization

H.264 uses a scalar quantizer, implemented in a way to avoid division and/or floating point arithmetic. There is a total number of 52 quantization step sizes denoted by $Q_{step}$, ranging from the lowest step size 0.625, to the largest step size 224. The step sizes are addressed using the quantization parameter $QP$ with the values from 0 to 51. The values of $Q_{step}$ are given in Table 5.1. The exact relation between $Q_{step}$ and $QP$ is given by:

$$Q_{step}(QP) = Q_{step}(\mathrm{mod}(QP, 6))2^{floor(QP/6)}, \tag{5.11}$$

**Table 5.1** Quantization step sizes in H.264 standard

| *QP* | 0 | 1 | 2 | 3 | 4 | 5 | 6 | 7 | 8 |
|---|---|---|---|---|---|---|---|---|---|
| *Qstep* | 0.625 | 0.6875 | 0.8125 | 0.875 | 1 | 1.125 | 1.25 | 1.375 | 1.625 |
| *QP* | 9 | 10 | ... | 18 | ... | 24 | ... | ... | 51 |
| *Qstep* | 1.75 | 2 | ... | 5 | ... | 10 | ... | ... | 224 |

where the first six *QP* and corresponding $Q_{step}$ are known (*mod* denotes remainder after dividing *QP* by 6, and *floor* denotes rounding to the smallest integer).

*Qstep* is doubled for every increment of 6 in *QP*. The possibility of using different quantization steps (with large range of values) allows us to efficiently manage trade-off between bit rate and quality.

***Direct quantization and transform***. The weighting coefficients from the scaling matrix **E** are implemented within the quantizer using the scaling factor (*SF*). First, the input block **B** is transformed to give a block of unscaled coefficients $\mathbf{X}=\mathbf{HBH}^{\mathrm{T}}$ (i.e., coefficients that are not yet scaled by **E**). Then, each coefficient on the position (*i*,*j*), denoted by $X_{ij}$, is quantized and scaled as follows:

$$Z_{ij} = \text{round}\left(\frac{X_{ij}SF_{ij}}{Q_{step}}\right). \tag{5.12}$$

We may observe from Eq. (5.10) that the scaling factor $SF_{ij}$ in **E** can have values $\boldsymbol{a^2}$, $\boldsymbol{ab/2}$, or $\boldsymbol{b^2/4}$ depending on the position (*i*,*j*) within the $4\times4$ matrix:

| Position (*i*,*j*) | Scaling factor SF (direct transform) |
|---|---|
| (0,0), (2,0), (0,2), (2,2) | $a^2=0.25$ |
| (1,1), (1,3), (3,1), (3,3) | $b^2/4=0.1$ |
| Other | $ab/2=0.1581$ |

In the implementation of H.264, the division operation $SF_{ij}/Q_{step}$ is replaced by the multiplication with factor $MF_{ij}$:

$$MF_{ij} = \frac{SF_{ij}}{Q_{step}}2^{qbits}, \quad \text{and} \quad qbits = 15 + floor(QP/6) \tag{5.13}$$

For simplicity, consider the case: $Q_{step}=1$, $QP=4$, $qbits = 15+floor(4/6) = 15$, and position $(i,j)=(0,0)$, which means that $SF=a^2=0.25$.

Therefore: $MF_{ij}=SF_{ij}/Q_{step}2^{qbits}=0.25\cdot2^{15}=8192$.

The multiplication factors are given in Table 5.2 for $QP=\{0,\ldots,5\}$.

If $QP>5$, *MF* is the same as in the case mod(*QP*,6), but the factor $2^{qbits}$ will increase by a factor of 2 for each increment of 6 in *QP*.

Using the above defined multiplication and shifting operations, the quantization can be implemented as:

**Table 5.2** Multiplication factor *MF* for different (i,j)

| *QP* | (*i,j*) = (0,0),(0,2),(2,0),(2,2) | (*i,j*) = (1,1),(1,3),(3,1),(3,3) | Other |
|---|---|---|---|
| 0 | 13107 | 5243 | 8066 |
| 1 | 11916 | 4660 | 7490 |
| 2 | 10082 | 4194 | 6554 |
| 3 | 9362 | 3647 | 5825 |
| 4 | 8192 | 3355 | 5243 |
| 5 | 7282 | 2893 | 4559 |

$$|Z_{ij}| = \left(|X_{ij}|MF_{ij} + f\right) \gg qbits$$
$$sign(Z_{ij}) = sign(X_{ij})$$

where $f = 2^{qbits}/3$ for Intra and $f = 2^{qbits}/6$ for Inter blocks. Note that $qbits = 15$ for $QP = \{0,\ldots,5\}$, $qbits = 16$ for $QP = \{6,\ldots,11\}$, $qbits = 17$ for $QP = \{12,\ldots,17\}$, etc.

***Inverse quantization and transform***. During the inverse quantization and inverse transform, the first step is to calculate the rescaled coefficients:

$$X'_{ij} = Z_{ij} \cdot Q_{step} \cdot SF_{ij} \cdot 64,$$

where the constant scaling factor 64 is applied to avoid rounding errors, while the scaling factors *SF* corresponds to $a^2$, $b^2$, $ab$ (0.25, 0.4, and 0.3162, respectively) for the inverse scaling matrix $\mathbf{E}^{-1}$:

| Position (*i,j*) | Scaling factor SF (inverse transform) |
|---|---|
| (0,0), (2,0), (0,2),(2,2) | $a^2 = 0.25$ |
| (1,1), (1,3), (3,1), (3,3) | $b^2 = 0.4$ |
| Other | $ab = 0.3162$ |

The total re-scaling parameter is denoted as:

$$V = Q_{step} \cdot SF \cdot 64, \tag{5.14}$$

and it is defined within the standard as a table of values for $QP = \{0,\ldots,5\}$ and each coefficients position (*i,j*) (Table 5.3). If $QP > 5$, *V* is the same as in the case mod (*QP*,6).

The rescaling operation can be written as:

$$X'_{ij} = Z_{ij}V_{ij}2^{floor(QP/6)}, \tag{5.15}$$

**Table 5.3** Rescaling coefficients $V$ for different $(i,j)$

| $QP$ | $(i,j)=(0,0),(0,2),(2,0),(2,2)$ | $(i,j)=(1,1),(1,3),(3,1),(3,3)$ | Other |
|---|---|---|---|
| 0 | 10 | 16 | 13 |
| 1 | 11 | 18 | 14 |
| 2 | 13 | 20 | 16 |
| 3 | 14 | 23 | 18 |
| 4 | 16 | 25 | 20 |
| 5 | 18 | 29 | 23 |

or in a matrix form: $\mathbf{X}' = 2^{floor(QP/6)}\mathbf{Z} \otimes \mathbf{V}$, with $\otimes$ again denotes the multiplication of elements at the same positions. Note that the factor $2^{floor(QP/6)}$ increases the output by a factor of 2 for every increment of 6 in $QP$.

The block pixels are then obtained as:

$$\mathbf{B} = \mathbf{H}^{-1}\mathbf{X}'\left(\mathbf{H}^{-1}\right)^{T}. \tag{5.16}$$

#### 5.3.7.8 Arithmetic Coding

H.264 standard provides significantly better compression ratio than the existing standards. At the same time, it uses advanced entropy coding, such as CAVLC (Context Adaptive Variable Length Coding), and especially CABAC (Context-based Adaptive Binary Arithmetic Coding).

***Context-Based Adaptive Variable Length Coding (CAVLC)***

It is used to encode zig-zag scanned and quantized residuals ($4 \times 4$ blocks of transform domain coefficients). These blocks of coefficients mainly contain zero values, while the non-zero coefficients are often $\pm 1$. Furthermore, the number of nonzero coefficients in neighboring blocks is correlated (left and upper block are observed). The number of coefficients is encoded using one out of four look-up tables, where the lookup table is chosen on the basis of the number of nonzero coefficients in neighboring blocks.

For each transform block, CAVLC encodes the following.

- **Coeff_Token** which includes the number of non-zero coefficients (**TotalCoeffs**) and trailing ones $\pm 1$ (**TrailingOnes**).

  **TotalCoeffs** can be between 0 and 16. The number of **TrailingOnes** can be between 0 and 3 (even if there are more than 3, only the last three are considered, the others are coded as normal coefficients). **Coeff_token** are coded using 4 look-up tables (3 of them with variable-length codes: ***Table 1*** for small number of coefficients, ***Table 2*** for medium and ***Table 3*** for high number of coefficients), and the fourth with fixed 6-bit codes to every pair of **TotalCoeff** and **TrailingOnes** (Tables are given in the Appendix). Only one of the tables should be chosen for encoding, and the choice depends on the number of nonzero coefficients in the left and upper blocks that are already coded. The table selection parameter $nC$ for the current block is calculated as the average value of the number of nonzero coefficients from the left ($nC_A$) and upper $4 \times 4$ block

($nC_B$). If only one block is available than the parameter is equal to the number of nonzero coefficients from that block ($nC = nC_A$ or $nC = nC_B$), or it is equal to zero if both blocks are unavailable. The table selection parameter adapts VLC to the number of coded coefficients in neighboring blocks (*context adaptive*). The values $nC = 0$ and $nC = 1$ select ***Table 1***, values $nC = 2$ and $nC = 3$ selects ***Table 2***, values $nC = \{4,5,6,7\}$ use ***Table 3***, otherwise ***Table 4*** is used.

As an illustration, let us observe an example of a block with values:

| | | | |
|---|---|---|---|
| 0 | 4 | 1 | 0 |
| 0 | −1 | −1 | 0 |
| 1 | 0 | 0 | 0 |
| 0 | 0 | 0 | 0 |

**Zig-zag scanned block**: 0,4,0,1,−1,1,0, −1,0,0…

We may observe that:

**TotalCoeffs** = 5 (indexed from 4 to 0, starting from the end of sequence, such that: the last non-zero coefficients with value −1 is indexed by 4, while the first non-zero coefficients having value 4 is indexed by 0)

**TrailingOnes** = 3 (there are four **trailing ones** but only three can be encoded)

Assume that the table selection parameter $nC < 2$ selects ***Table 1*** for coding (***Table 1*** is represented as one of the columns of the table in the Appendix). For the pair **TrailingOnes** = 3 (field **TO** in the table) and **TotalCoeffs** = 5 we read the code from ***Table 1***: 0000100.

| Element | Value | Code |
|---|---|---|
| Coeff_Token | TotalCoeffs = 5 | 0000100 |
| | TrailingOnes = 3<br>(e.g., use Table 1 given in the Appendix) | |

- Encode the sign of **TrailingOnes**: + sign is encoded by "0" bit, − sign is encoded by "1" bit. The signs are encoded in reverse order (from the end of the sequence). Hence, in the example we include the signs:

| Element | Value | Code |
|---|---|---|
| Coeff_Token | TotalCoeffs = 5 | 0000100 |
| | TrailingOnes = 3<br>(use Table 1) | |
| TrailingOne sign(4) | − | 1 |
| TrailingOne sign(3) | + | 0 |
| TrailingOne sign(2) | − | 1 |

- **Levels of remaining nonzero coefficients**: starting from the end of sequence, the sign and magnitude (level) of remaining non-zero coefficients are encoded.

  Each level is encoded using a prefix and a suffix. Seven level_VLC tables are used to encode the level (ITU-T Rec., 2002). The length of the suffix is between 0 and 6 bits and it is adapted according to the each successive coded level ("context adaptive"): a small value is appropriate for levels with low

**Table 5.4** Suffix length

| Current table | Threshold | suffixLength to be set |
|---|---|---|
| Level_VLC0 | 0 | 0 |
| Level_VLC1 | 3 | 1 |
| Level_VLC2 | 6 | 2 |
| Level_VLC3 | 12 | 3 |
| Level_VLC4 | 24 | 4 |
| Level_VLC5 | 48 | 5 |
| Level_VLC6 | N/A (highest suffixLength) | 6 |

magnitudes, larger value for levels with high magnitudes. The choice of **suffixLength** is adapted as follows:

1. Initialize **suffixLength** to 0, i.e., **Level_VLC0** (if the number of nonzero coefficients is higher than 10, and there are less than 3 trailing ones, initialize **suffixLength** to 1, i.e., **Level_VLC1**).
2. Encode the last nonzero coefficient (encoding is done in reverse order).
3. If the magnitude of this coefficient is larger than a predefined threshold, increment **suffixLength** by 1 (or equivalently, move to the next VLC table) (Table 5.4).

| Element | Value | Code |
|---|---|---|
| Coeff_Token | TotalCoeffs = 5 | 0000100 |
| | TrailingOnes = 3 (use Table 1) | |
| TrailingOne sign(4) | − | 1 |
| TrailingOne sign(3) | + | 0 |
| TrailingOne sign(2) | − | 1 |
| Level (1) | +1 (use suffixLength = 0) | 1 (prefix) |
| Level (0) | +4 (use suffixLength = 1) | 0001 (prefix) 0 (suffix) |

- **Total number of zeros (totalzeros) before the last nonzero coefficient**
  In the considered example the total number of zeros is **Total zeros** = 3, and thus the coding table includes the additional line:

| Element | Value | Code |
|---|---|---|
| Coeff_Token | TotalCoeffs = 5 | 0000100 |
| | TrailingOnes = 3 (use Table 1) | |
| TrailingOne sign(4) | − | 1 |
| TrailingOne sign(3) | + | 0 |
| TrailingOne sign(2) | − | 1 |
| Level (1) | +1 (use suffixLength = 0) | 1 (prefix) |
| Level (0) | +4 (use suffixLength = 1) | 0001 (prefix) 0 (suffix) |
| Total zeros | 3 | 111 |

The code for **totalzeros** (depending on the number of non-zero coefficients) is given in Table 5.5.

- **Runs of zeros**: the last step is to encode the number of zeros preceding each nonzero coefficient (**run_before, Table 5.6**) starting from the end of sequence (from the last nonzero coefficients). The number of zeros before the first coefficients (last in reverse coding order) is not necessary to encode. Also, if there are no more zeros other than **totalzeros** (zeros before the last nonzero coefficients), then no **run_before** is encoded.

  Now, after we add **run_before** in the considered example: 0,4,0,1,−1,1,0, −1,0,0…

  run_before(4): 0,4,0,1,−1,−1, 0,**1**,0,0… Zerosleft = 3, run_before = 1
  run_before(3): 0,4,0,1,−1,−**1**,0,1,0,0… Zerosleft = 2, run_before = 0
  run_before(2): 0,4,0,1,−**1**,−1,0,1,0,0… Zerosleft = 2, run_before = 0
  run_before(1): 0,4, 0,**1**,−1,−1,0,1,0,0… Zerosleft = 2, run_before = 1
  run_before(0): 0,**4**,0,1,−1,−1,0,1,0,0… Zerosleft = 1, run_before = 1
  (no code required)

  Note that **run_before** is the number of zeros (0) preceding each observed non-zero coefficient (marked in red).

| Element | Value | Code |
|---|---|---|
| Coeff_Token | TotalCoeffs = 5 | 0000100 |
| | TrailingOnes = 3 | |
| | (use Table 1) | |
| TrailingOne sign(4) | − | 1 |
| TrailingOne sign(3) | + | 0 |
| TrailingOne sign(2) | − | 1 |
| Level (1) | +1 (use suffixLength = 0) | 1 (prefix) |
| Level (0) | +4 (use suffixLength = 1) | 0001 (prefix) 0 (suffix) |
| Total zeros | 3 | 111 |
| run_before(4) | ZerosLeft = 3; run before = 1 | 10 |
| run_before(3) | ZerosLeft = 2; run before = 0 | 1 |
| run_before(2) | ZerosLeft = 2; run before = 0 | 1 |
| run_before(1) | ZerosLeft = 2; run before = 1 | 01 |
| run_before(0) | ZerosLeft = 1; run before = 1 | No code req. (last coeff.) |

The encoded sequence for this block is obtained as: 000010010110 0010111101101

**Table 5.5** Total_zeros table for 4 × 4 blocks

| | | Number of non-zero coefficients | | | | | | | | | | | | | | |
|---|---|---|---|---|---|---|---|---|---|---|---|---|---|---|---|---|
| | | 1 | 2 | 3 | 4 | 5 | 6 | 7 | 8 | 9 | 10 | 11 | 12 | 13 | 14 | 15 |
| Total_zeros | 0 | 1 | 111 | 0101 | 00011 | 0101 | 000001 | 000001 | 000001 | 000001 | 00001 | 0000 | 0000 | 000 | 00 | 0 |
| | 1 | 011 | 110 | 111 | 111 | 0100 | 00001 | 00001 | 0001 | 000000 | 00000 | 0001 | 0001 | 001 | 01 | 1 |
| | 2 | 010 | 101 | 110 | 0101 | 0011 | 111 | 101 | 00001 | 0001 | 001 | 001 | 01 | 1 | 1 | |
| | 3 | 0011 | 100 | 101 | 0100 | 111 | 110 | 100 | 011 | 11 | 11 | 010 | 1 | 01 | | |
| | 4 | 0010 | 011 | 0100 | 110 | 110 | 101 | 011 | 11 | 10 | 10 | 1 | 001 | | | |
| | 5 | 00011 | 0101 | 0011 | 101 | 101 | 100 | 11 | 10 | 001 | 01 | 011 | | | | |
| | 6 | 00010 | 0100 | 100 | 100 | 100 | 011 | 010 | 010 | 01 | 0001 | | | | | |
| | 7 | 000011 | 0011 | 011 | 0011 | 011 | 010 | 0001 | 001 | 00001 | | | | | | |
| | 8 | 000010 | 0010 | 0010 | 011 | 0010 | 0001 | 001 | 000000 | | | | | | | |
| | 9 | 0000011 | 00011 | 00011 | 0010 | 00001 | 001 | 000000 | | | | | | | | |
| | 10 | 00000011 | 00010 | 00010 | 00010 | 0001 | 000000 | | | | | | | | | |
| | 11 | 00000011 | 000011 | 000001 | 00001 | 00000 | | | | | | | | | | |
| | 12 | 00000010 | 000010 | 00001 | 00000 | | | | | | | | | | | |
| | 13 | 000000011 | 000001 | 000000 | | | | | | | | | | | | |
| | 14 | 000000010 | 000000 | | | | | | | | | | | | | |
| | 15 | 000000001 | | | | | | | | | | | | | | |

**Table 5.6** Table for run_before

| | Zeros left | | | | | | |
|---|---|---|---|---|---|---|---|
| run_before | 1 | 2 | 3 | 4 | 5 | 6 | >6 |
| 0 | 1 | 1 | 11 | 11 | 11 | 11 | 111 |
| 1 | 0 | 01 | 10 | 10 | 10 | 000 | 110 |
| 2 | – | 00 | 01 | 01 | 011 | 001 | 101 |
| 3 | – | – | 00 | 001 | 010 | 011 | 100 |
| 4 | – | – | – | 000 | 001 | 010 | 011 |
| 5 | – | – | – | – | 000 | 101 | 010 |
| 6 | – | – | – | – | – | 100 | 001 |
| 7 | – | – | – | – | – | – | 0001 |
| 8 | – | – | – | – | – | – | 00001 |
| 9 | – | – | – | – | – | – | 000001 |
| 10 | – | – | – | – | – | – | 0000001 |
| 11 | – | – | – | – | – | – | 00000001 |
| 12 | – | – | – | – | – | – | 000000001 |
| 13 | – | – | – | – | – | – | 0000000001 |
| 14 | – | – | – | – | – | – | 00000000001 |

## Decoding

The steps of the decoding procedure are summarized below until the decoded sequence is obtained.

| Code | Element | Value | Output array |
|---|---|---|---|
| 0000100 | coeff token | TotalCoeffs = 5, TrailingOnes = 3 | Empty |
| 1 | TrailingOne sign | – | −1 |
| 0 | TrailingOne sign | + | 1, −1 |
| 1 | TrailingOne sign | – | −1,1, −1 |
| 1 | Level | +1 (suffixLength = 0; increment suffixLength after decoding) | 1, −1,1, −1 |
| 00010 | Level | +4 (suffixLength = 1) | 4, 1, −1,1,−1 |
| 111 | **Total_zeros** | **3** | |
| 10 | run before | 1 | 4, 1, −1,1, 0,−1 |
| 1 | run before | 0 | 4, 1, −1,1, 0, −1 |
| 1 | run before | 0 | 4, 1, −1,1, 0, −1 |
| 01 | run before | 1 | 4, 0, 1, −1,1, 0, −1 |
| | | **since Total_zeros = 3 ⇒** | 0, 4, 0, 1, −1, 1, 0, −1 |

The decoded sequence is: 0,4,0,1,−1,1,0,−1,...

## 5.4 Data Rate and Distortion

The video sequences, in general, have variable bit rates that depend on several factors:

- Applied algorithms—intra and inter coding techniques use different compression approaches. Hence, it is clear that different types of frames have different compression factors.
- Dynamics of videos sequence—compression will be higher in the sequences where there are fewer movements and moving objects.
- Encoding parameters—the choice of quantization steps will also influence the bit rate.

It is obvious that the video compression ratio is closely related to the degree of quality distortion, which is in turn related to the degree of quantization $Q$. Therefore, an important issue is to provide a compromise between the data rate and quality distortion, which can be described by:

$$\min\{D\} \quad \text{for} \;\; R \le R_{\max}, \tag{5.17}$$

where $D$ is a distortion, while $R$ is the data rate. The algorithm searches for an optimal combination of $D$ and $R$. It can be summarized as follows:

- Encode a video signal for a certain set of compression parameters and measure the data rate and distortion of decoded signals.
- Repeat the coding procedure for different sets of compression parameters, which will produce different compression ratio. For each compression ratio, a measure of distortion is calculated.
- As a result, different points in the R-D (rate-distortion) plane are obtained.

The optimal point in the R-D plane is obtained by using the Lagrangian optimization:

$$\min\{J = D + \lambda R\}, \tag{5.18}$$

where $\lambda$ is the Lagrange constant. It will find the nearest point on the convex optimization curve.

In practice, most applications require the constant bit rate for the video signals. For this purpose, the bit rate control system, shown in Fig. 5.26, can be used.

$R_V$ indicates the variable bit rates, while $R_C$ denotes the constant ones. The system buffers a video signal with variable bit rate obtained at the output of coder, and then transmits the buffered signal with the constant bit rate. A buffer feedback controls the quantization step size, and consequently, the compression ratio.

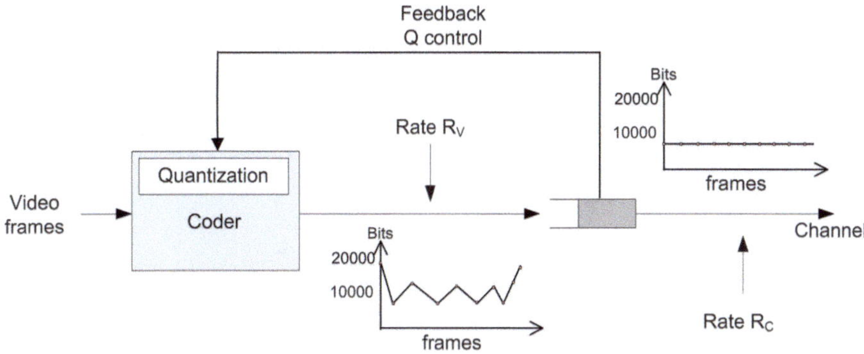


**Fig. 5.26** Rate control for video signal

Namely, when the bit rate of the input signal to the buffer is high, the buffer may overflow. Then, the quantization step should be increased to increase compression and to reduce the bit rate at the output of the encoder.

A field dealing with matching the video quality and transmission capacity of the network is called the Quality of Service (QoS). On the one side we have the QoS required by the application and on the other side, the available QoS offered by the network. QoS differs for different video applications and transmission scenarios. For example, a one-sided simplex transmission (broadcasting) requires a different QoS compared to a two-sided duplex transmission (video conferencing). In simplex, it is important to have video and audio synchronization, because the synchronization loss greater than 0.1 s becomes obvious. In duplex, delays greater than 0.4 s cause difficulties and unnatural communication.

Digital data can be carried over networks with constant or variable bit rates. Networks with constant rates are the PSTN networks (Public Switched Telephone Networks—circuit switched networks) and ISDN networks (Integrated Services Digital Networks). Networks with variable bit rates are the Asynchronous Transfer Mode networks (ATM—packet switched networks), where the bit rate depends on the traffic within the network.

Errors that can occur when transferring video material can generally be divided into spatial and temporal. The spatial error occurs in one of the macroblocks within the frame and it affects other blocks that are intra coded by using the erroneous block (Fig 5.27).

Correcting these errors is done by interpolation, using the undamaged parts of the frame. The most prominent time errors are those that occur in the initial frame and they are transmitted through the motion vectors to B and P frames. When such an error is detected, then the motion prediction is done by using the previous error-free reference frame. An illustration of removing these errors is given in Fig. 5.27. Alternatively, the error correction can be efficiently done using gradient-based signal reconstruction algorithm for images presented in Chap. 6, which is able to recover erroneous blocks by treating them as the missing ones.

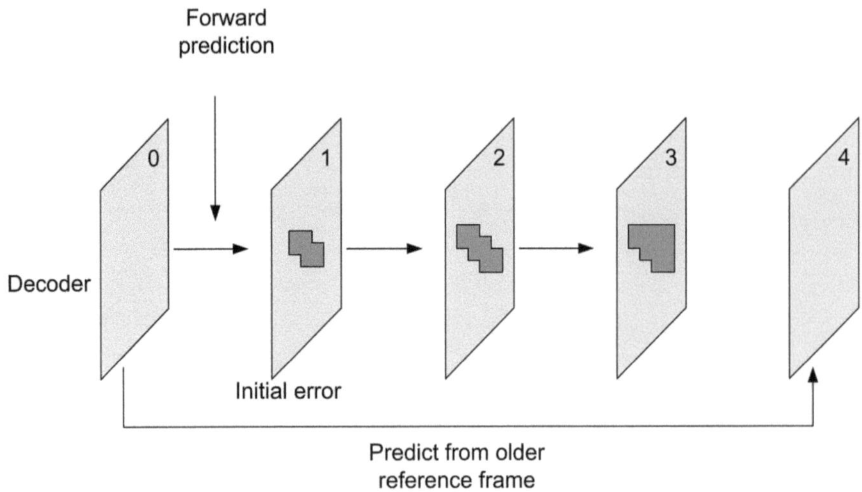


**Fig. 5.27** Error reduction by using the older reference frame

## 5.5 Communications Protocols for Multimedia Data

In this part we provide an overview of some multimedia protocols used in different networks. In the PSTN networks (typically ISDN), H.324 and H.320 protocols are used, and both have a constant bit rate. For multimedia data over the IP and LAN networks the H.323 protocol can be used, which has a variable delay and unreliable data transfer.

## 5.6 H.323 Multimedia Conference

The H.323 protocol provides the multimedia communication sessions (voice and videoconferencing in point-to-point and multipoint configurations). This standard involves call signaling, control protocol for multimedia communication, bandwidth control, etc. The H.323 network usually includes four components: the H.323 terminal, gatekeeper, gateway, and multipoint control units (MCU). The H.323 terminals are the endpoints on the LAN that provide real-time communications. The gateway provides communication between H.323 networks and other networks (PSTN or ISDN). The gatekeeper is used to translate IP addresses and to manage the bandwidth. The MCU allows communication between multiple conference units. The structure of the H.323 terminal is given in Fig. 5.28.

This protocol requires the audio coding and control protocols, while the video coding and Real-Time Transport Protocol (RTP) are optional. The audio signals are encoded using the G.711, G.723, and G.729 standards, while the video signals are

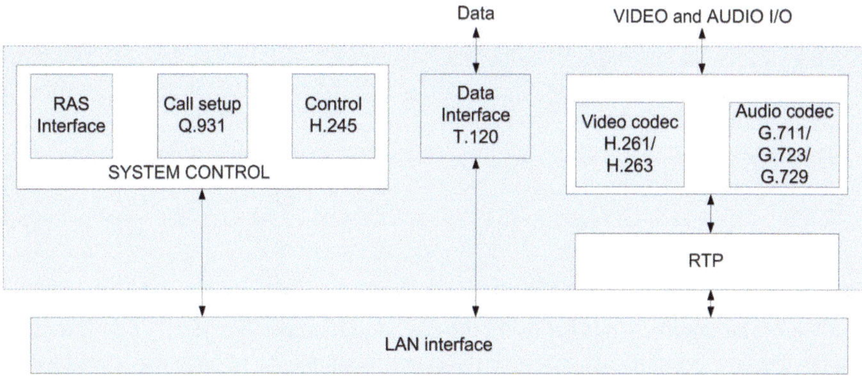


**Fig. 5.28** H.323 terminal

encoded using the H.261 and H.263 standards. The block Q.931 is used to set up the calls, the H.245 block controls the operation of the network, and the RAS block is used to communicate with the gatekeeper.

A centralized conference assumes that all connections are routed through the MCU (unicast communication). Then the MCU is very loaded. In a decentralized conference (multicast communication) each terminal sends data to all other terminals. The basic transport protocol in the H.323 is the UDP (User Datagram Protocol).

### 5.6.1 *SIP Protocol*

The Session Initiation Protocol (SIP) is a protocol designed for the session control in the multi-service networks. The software that provides real-time communications between the end-users can use SIP to establish, maintain, and terminate the communication between two or more end-points. These applications include the voice over IP (VoIP), video teleconferencing, virtual reality applications, multiplayer video games, etc. The SIP does not provide all the functions required for communication between these programs, but it is an important component that facilitates communication.

One of the major demands that the network should meet is the maintenance of QoS for the client application. The SIP is a client–server protocol, based on the protocols HTTP (HyperText Transfer Protocol) and SMTP (Simple Mail Transfer Protocol). The SIP can use either UDP or TCP (Transmission Control Protocol) as a transport protocol.

The SIP messages can be in a form of a *request* (from a client to a server) or a *response* (from a server to a client).

SIP performs five basic functions:

- Determines the location of endpoint.
- Determines the availability of endpoint, i.e., whether the endpoint is able to participate in a session.
- Determines the characteristics of users, i.e., the parameters of the medium that are essential for communication.
- Establishes a session or performs the exchange of parameters for establishing a session.
- Manages sessions.

One of the main reasons for using the SIP is to increase flexibility in multimedia data exchange. Specifically, users of these applications can change the location and use different computers, with multiple user names and user accounts, or to communicate using a combination of voice, text, and other media that require different protocols separately. The SIP uses various components of the network to identify and locate the users. The data go through a proxy server that is used to register and forward the user connection requests. Given that there are different protocols for voice, text, video, and other media, the SIP is positioned above any of these protocols.

The SIP architecture is illustrated in Fig. 5.29. The SIP is independent of network topology and can be used with different transport protocols such as the UDP, TCP, X.25, ATM AAL5, CLNP, TP4, IPX, and PPP. The SIP does not require a reliable transport protocol, and therefore, the client side can use the UDP. For servers, it is recommended to support both protocols, the UDP and TCP. The TCP connection is opened only when the UDP connection cannot be established.

The functionality of SIP is mainly based on signaling. This is its main difference in comparison to the H.323 which includes all necessary functions to carry out the

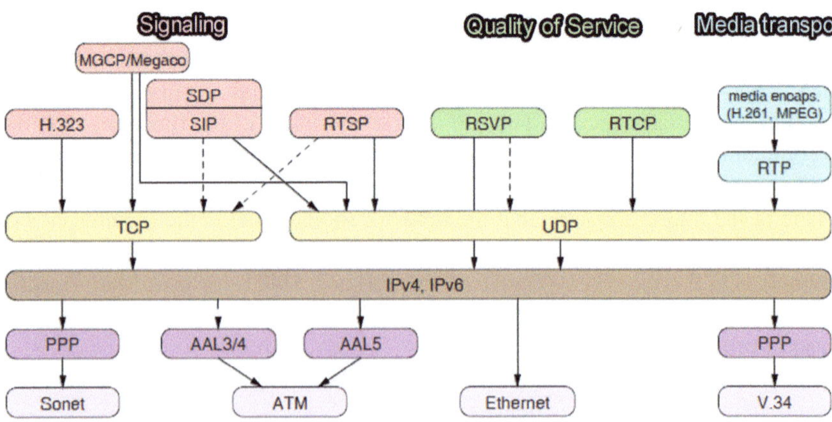


**Fig. 5.29** SIP architecture

conference. The SIP architecture is designed to be modular so that the different functions can be easily replaced. The SIP environment can implement some components of the H.323 protocol.

For a description of multimedia sessions, SIP uses the Session Description Protocol (SDP). To transfer in real time, the SIP architecture includes the RTP protocol. It also includes the Real-Time Streaming Protocol (RTSP), which is a control protocol for streaming multimedia data in real time. This protocol is suitable for audio/video on-demand streaming.

In the SIP protocol, the following methods are used:
INVITE—making the connection,
BYE—end connection,
OPTIONS—indicates information about the possibilities,
ACK—is used for reliable messaging,
CANCEL—cancels the last request,
REGISTER—SIP server provides information about the location.

## 5.7 Audio Within a TV Signal

Audio signal together with a video sequence is an integral part of the TV signal. Inserting audio in the video signal requires knowledge of many different disciplines like compression algorithms, multiplexing, standards for packetized data stream, and algorithms for signal modulation. In the case of digital TV, the compression algorithms have the main influence to the received signal quality. A system for transmission of audio and video data is an iso-synchronized system. This means that both transmitter and receiver use the data buffering to avoid asynchronous data. Video and audio data from a channel form the elementary stream (ES). Multiple program channels are combined such that the variable-length elementary stream is packetized into the fixed length transport stream packets. A simplified block diagram of this system is shown in Fig. 5.30.

The metadata provides the synchronization of audio and video data (the timing reference). It should be noted that the stream of coded audio and video data is packetized by using the PES (Packetized Elementary Stream) blocks, which have a defined structure for both video and audio data. In video compression, the frames are not included in the same order as they are generated, so that the video block must contain a part that takes care of when the frame is played. The Transport Stream (TS) is composed of the fixed length packets (188 bytes). The TS packet is made of the header and the data. Multiplexer is an important part of the system, because it combines data from various channels.

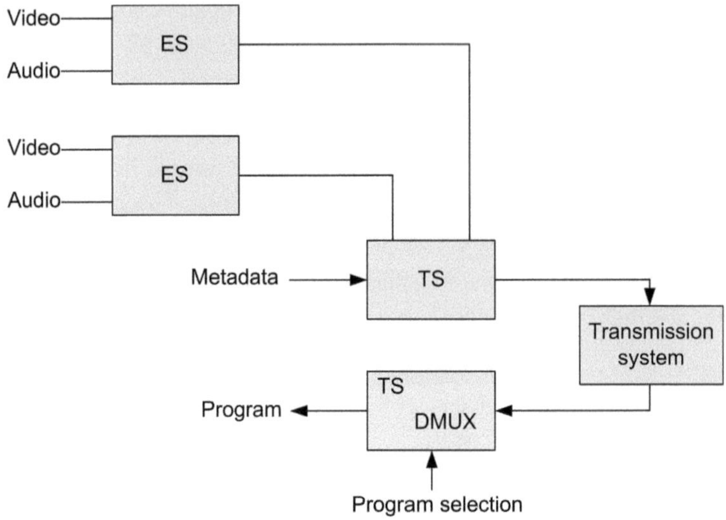


**Fig. 5.30** Transport stream multiplexing and demultiplexing

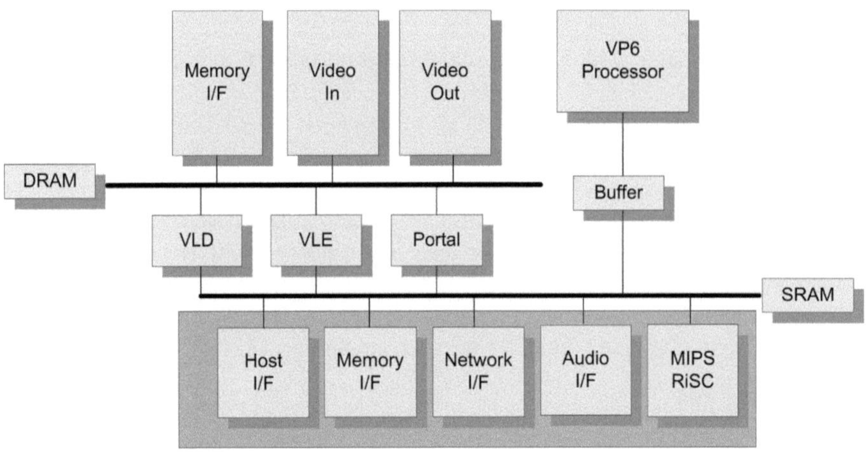


**Fig. 5.31** Video Processor

## 5.8 Video Signal Processor

An example of a simple video processor (VCPex processor) is shown in Fig. 5.31.

The RISC is the major processor. The SRAM bus is used for lower bit rates, such as the compressed data (audio and video). The DRAM bus is used for higher bit rates, as it is the case with the uncompressed material. The RISC and VP6 processor can be reprogrammed to support different coding standards. The VLE (variable-length encoding) and VLD (variable-length decoding) are used to encode and decode the signal.

## 5.9 Examples

5.1. Determine the number of bits used to represent 16 pixels, by using the following sampling schemes:

(a) 4:4:4
(b) 4:2:2
(c) 4:2:0

Solution:

(a) 4:4:4
We observe 4 blocks with 4 pixels, each having the three components (Y, Cb, Cr)
$4{\cdot}(4{\cdot}3{\cdot}8)\ \text{b} = 4{\cdot}96\ \text{b} = 384\ \text{b}$ or $16\ \text{pixel} \cdot 24\ \text{b/pixel} = 384\ \text{b}$

(b) 4:2:2
According to this scheme, 2 out of 4 pixels within the observed block are represented by using three components (Y, Cb, Cr), while the remaining 2 pixels contain just the luminance Y.
$4{\cdot}(2{\cdot}3{\cdot}8\ \text{b}) + 4{\cdot}(2{\cdot}1{\cdot}8\ \text{b}) = 256\ \text{b}$ or $16\ \text{pixel} \cdot 16\text{b/pixel} = 256\ \text{b}$

(c) 4:2:0
In this case only one pixel is represented with a full resolution (Y, Cb, Cr), while the remaining 3 pixels contains the luminance components Y. Hence, for the observed 16 pixels, we have:
$4{\cdot}(1{\cdot}3{\cdot}8\ \text{b}) + 4{\cdot}(3{\cdot}1{\cdot}8\ \text{b}) = 192\ \text{b}$ or $16\ \text{pixel} \cdot 12\ \text{b/pixel} = 192\ \text{b}$

5.2. Determine the bit rate of the PAL video sequence for the CIF format and sampling schemes:

(a) 4:4:4
(b) 4:2:2
(c) 4:2:0

Solution:
The CIF format resolution is $352 \times 288$. Hence, we obtain the following bit rates:

(a) $352{\cdot}288{\cdot}24{\cdot}25\ \text{b/s} = 60825600\ \text{b/s} = 60.8\ \text{Mb/s}$
(b) $352{\cdot}288{\cdot}16{\cdot}25\ \text{b/s} = 40550400\ \text{b/s} = 40.5\ \text{Mb/s}$
(c) $352{\cdot}288{\cdot}12{\cdot}25\ \text{b/s} = 30412800\ \text{b/s} = 30.4\ \text{Mb/s}$

5.3. How many minutes of the uncompressed video data in the CIF format with sampling scheme 4:2:2 can be stored on a DVD (capacity 4.7 GB)? The PAL system is assumed.

Solution:

$$t = \frac{4.7 \cdot 1024^3 \cdot 8\ \text{b}}{25 \cdot 352 \cdot 288 \cdot 16 \cdot 60\ \text{b/s}} \approx 16.6\ \text{min}$$

5.4. Consider a $4 \times 4$ block of pixels within a current frame and the $6 \times 6$ region cenetred at the same position in the reference frame. Determine the motion vector by using the block matching technique based on the MSE assuming that the motion vector is within the given $6 \times 6$ block.

$$\begin{matrix} 1 & 4 & 7 & 6 \\ 9 & 11 & 8 & 11 \\ 4 & 4 & 6 & 11 \\ 4 & 4 & 9 & 12 \end{matrix} \qquad \begin{matrix} 2 & 5 & 7 & 7 & 19 & 19 \\ 9 & 11 & 8 & 9 & 5 & 4 \\ 4 & 6 & 6 & 10 & 1 & 1 \\ 4 & 6 & 10 & 11 & 4 & 5 \\ 4 & 8 & 7 & 3 & 1 & 3 \\ 8 & 15 & 8 & 8 & 11 & 8 \end{matrix}$$

Solution:
The observed $4 \times 4$ block is compared with the corresponding $4 \times 4$ block (within $6 \times 6$ block) in the reference frame, centered at (0,0). The MSE is calculated. Then, the procedure is repeated for eight positions around the central one.

$$\begin{matrix} 1 & 4 & 7 & 6 \\ 9 & 11 & 8 & 11 \\ 4 & 4 & 6 & 11 \\ 4 & 4 & 9 & 12 \end{matrix} \quad \overset{MSE_{00}}{\longleftrightarrow} \quad \begin{matrix} 11 & 8 & 9 & 5 \\ 6 & 6 & 10 & 1 \\ 6 & 10 & 11 & 4 \\ 8 & 7 & 3 & 1 \end{matrix}$$

$$\begin{aligned} \mathrm{MSE}_{00} = {} & ((1-11)^2+(4-8)^2+(7-9)^2+(6-5)^2+(9-6)^2+(11-4)^2+ \\ & (8-10)^2+(11-1)^2+(4-6)^2+(4-10)^2+(6-11)^2+(11-4)^2 \\ & +(4-8)^2+(4-7)^2+(9-3)^2+(12-1)^2)/16=34.69 \end{aligned}$$

$$\begin{matrix} 1 & 4 & 7 & 6 \\ 9 & 11 & 8 & 11 \\ 4 & 4 & 6 & 11 \\ 4 & 4 & 9 & 12 \end{matrix} \quad \overset{MSE_{-10}}{\longleftrightarrow} \quad \begin{matrix} 9 & 11 & 8 & 9 \\ 4 & 6 & 6 & 10 \\ 4 & 6 & 10 & 11 \\ 4 & 8 & 7 & 3 \end{matrix}$$

$$\begin{aligned} \mathrm{MSE}_{-10} = {} & ((1-9)^2+(4-11)^2+(7-8)^2+(6-9)^2+(9-4)^2+(11-6)^2+ \\ & (8-6)^2+(11-10)^2+(4-4)^2+(4-6)^2+(6-10)^2+(11-11)^2 \\ & +(4-4)^2+(4-8)^2+(9-7)^2+(12-3)^2)/16=18.68 \end{aligned}$$

$$\begin{matrix} 1 & 4 & 7 & 6 \\ 9 & 11 & 8 & 11 \\ 4 & 4 & 6 & 11 \\ 4 & 4 & 9 & 12 \end{matrix} \quad \overset{MSE_{-11}}{\longleftrightarrow} \quad \begin{matrix} 2 & 5 & 7 & 7 \\ 9 & 11 & 8 & 9 \\ 4 & 6 & 6 & 10 \\ 4 & 6 & 10 & 11 \end{matrix}$$

$$\begin{aligned} \mathrm{MSE}_{-11} = {} & ((1-2)^2+(4-5)^2+(7-7)^2+(6-7)^2+(9-9)^2+(11-11)^2+ \\ & (8-8)^2+(11-9)^2+(4-4)^2+(4-6)^2+(6-6)^2+(11-10)^2 \\ & +(4-4)^2+(4-6)^2+(9-10)^2+(12-11)^2)/16=1.125 \end{aligned}$$

$$\begin{array}{cccc} 1 & 4 & 7 & 6 \\ 9 & 11 & 8 & 11 \\ 4 & 4 & 6 & 11 \\ 4 & 4 & 9 & 12 \end{array} \quad \xleftrightarrow{MSE_{11}} \quad \begin{array}{cccc} 7 & 7 & 19 & 19 \\ 8 & 9 & 5 & 4 \\ 6 & 10 & 1 & 1 \\ 10 & 11 & 4 & 5 \end{array}$$

$$\begin{aligned} \mathrm{MSE}_{11} = {} & ((1-7)^2+(4-7)^2+(7-19)^2+(6-19)^2+(9-8)^2+(11-9)^2+ \\ & (8-5)^2+(11-4)^2+(4-6)^2+(4-10)^2+(6-1)^2+(11-1)^2+ \\ & (4-10)^2+(4-11)^2+(9-4)^2+(12-5)^2)/16 = 46.56 \end{aligned}$$

$$\begin{array}{cccc} 1 & 4 & 7 & 6 \\ 9 & 11 & 8 & 11 \\ 4 & 4 & 6 & 11 \\ 4 & 4 & 9 & 12 \end{array} \quad \xleftrightarrow{MSE_{-1-1}} \quad \begin{array}{cccc} 4 & 6 & 6 & 10 \\ 4 & 6 & 10 & 11 \\ 4 & 8 & 7 & 3 \\ 8 & 15 & 8 & 8 \end{array}$$

$$\begin{aligned} \mathrm{MSE}_{-1-1} = {} & ((1-4)^2+(4-6)^2+(7-6)^2+(6-10)^2+(9-4)^2+(11-6)^2+- \\ & (8-10)^2+(11-11)^2+(4-4)^2+(4-8)^2+(6-7)^2+(11-3)^{2-} \\ & +(4-8)^2+(4-15)^2+(9-8)^2+(12-8)^2)/16 = 19.93 \end{aligned}$$

$$\begin{array}{cccc} 1 & 4 & 7 & 6 \\ 9 & 11 & 8 & 11 \\ 4 & 4 & 6 & 11 \\ 4 & 4 & 9 & 12 \end{array} \quad \xleftrightarrow{MSE_{0-1}} \quad \begin{array}{cccc} 6 & 6 & 10 & 11 \\ 6 & 10 & 11 & 4 \\ 8 & 7 & 3 & 1 \\ 15 & 8 & 8 & 11 \end{array}$$

$$\begin{aligned} \mathrm{MSE}_{0-1} = {} & ((1-6)^2+(4-6)^2+(7-10)^2+(6-1)^2+(9-6)^2+(11-10)^2+ \\ & (8-11)^2+(11-4)^2+(4-8)^2+(4-7)^2+(6-3)^2+(11-1)^2 \\ & +(4-15)^2+(4-8)^2+(9-8)^2+(12-11)^2)/16 = 25.25 \end{aligned}$$

$$\begin{array}{cccc} 1 & 4 & 7 & 6 \\ 9 & 11 & 8 & 11 \\ 4 & 4 & 6 & 11 \\ 4 & 4 & 9 & 12 \end{array} \quad \xleftrightarrow{MSE_{1-1}} \quad \begin{array}{cccc} 6 & 10 & 1 & 1 \\ 10 & 11 & 4 & 5 \\ 7 & 3 & 1 & 3 \\ 8 & 8 & 11 & 8 \end{array}$$

$$\begin{aligned} \mathrm{MSE}_{1-1} = {} & ((1-6)^2+(4-10)^2+(7-1)^2+(6-1)^2+(9-10)^2+(11-11)^2+ \\ & (8-4)^2+(11-5)^2+(4-7)^2+(4-3)^2+(6-1)^2+(11-3)^2+ \\ & (4-8)^2+(4-8)^2+(9-11)^2+(12-8)^2)/16 = 20.37 \end{aligned}$$

$$\begin{array}{cccc} 1 & 4 & 7 & 6 \\ 9 & 11 & 8 & 11 \\ 4 & 4 & 6 & 11 \\ 4 & 4 & 9 & 12 \end{array} \quad \xleftrightarrow{MSE_{10}} \quad \begin{array}{cccc} 8 & 9 & 5 & 4 \\ 6 & 10 & 1 & 1 \\ 10 & 11 & 4 & 5 \\ 7 & 3 & 1 & 3 \end{array}$$

$$\mathrm{MSE}_{10}=((1-8)^2+(4-9)^2+(7-5)^2+(6-4)^2+(9-6)^2+(11-10)^2+(8-1)^2+(11-1)^2+(4-10)^2+(4-11)^2+(6-4)^2+(11-5)^2+(4-7)^2+(4-3)^2+(9-1)^2+(12-3)^2)/16=32.56$$

$$\begin{matrix} 1 & 4 & 7 & 6 \\ 9 & 11 & 8 & 11 \\ 4 & 4 & 6 & 11 \\ 4 & 4 & 9 & 12 \end{matrix} \quad \overset{MSE_{01}}{\longleftrightarrow} \quad \begin{matrix} 5 & 7 & 7 & 19 \\ 11 & 8 & 9 & 5 \\ 6 & 6 & 10 & 1 \\ 6 & 10 & 11 & 4 \end{matrix}$$

$$\mathrm{MSE}_{01}=((1-5)^2+(4-7)^2+(7-7)^2+(6-19)^2+(9-11)^2+(11-8)^2+(8-9)^2+(11-5)^2+(4-6)^2+(4-6)^2+(6-10)^2+(11-1)^2+(4-6)^2+(4-10)^2+(9-11)^2+(12-4)^2)/16=29.75$$

Since $\min(\mathrm{MSE}_{nm})=\mathrm{MSE}_{-11}$, we may conclude that the position $(-1,1)$ represents the motion vector.

5.5. Consider a video sequence with $N=1200$ frames. The frames are divided into $8\times 8$ blocks, in order to analyze the stationarity of the coefficients. We assume that the stationary blocks do not vary significantly over the sequence duration. The coefficients from the stationary blocks are transmitted only once (within the first frame). The coefficients from the nonstationary blocks change significantly over time. In order to reduce the amount of data that will be sent, the nonstationary coefficients are represented by using $K$ Hermite coefficients, where $N/K=1.4$. Determine how many bits are required for encoding the considered sequence and what is the compression factor? The original video frames can be coded by using on average 256 bits per block.

| Blocks statistics | |
|---|---|
| Total number of frames | 1200 |
| Frame size | 300 × 450 |
| Stationary blocks | 40 % |
| Nonstationary blocks | 60 % |

Solution:
The stationary blocks are transmitted only for the first frame. Thus, the total number of bits used to represent the coefficients from the stationary blocks is:

$$n_s=\frac{40}{100}\left(\frac{300\cdot 450}{64}\cdot 256\right)=216\cdot 10^3\mathrm{b}.$$

In the case of nonstationary blocks, we observe the sequences of coefficients which are on the same position within different video frames. Hence, each sequence having $N=1200$ coefficients, is represented by using $K$ Hermite coefficients, where $N/K = 1.4$ holds. The total number of bits used to encode the coefficients from the nonstationary blocks is:

$$n_n = 1200 \cdot \frac{K}{N} \cdot \left( \frac{60}{100} \cdot \left( \frac{300 \cdot 450}{64} \cdot 256 \right) \right) = 2.77 \cdot 10^8 \text{ b}.$$

The number of bits that is required for sequence coding is:

$$p = 1200 \cdot 300 \cdot 450 \cdot 4 = 6.4 \cdot 10^8 \text{ b}.$$

The compression factor is: $\frac{6.4 \cdot 10^8}{216 \cdot 10^3 + 2.77 \cdot 10^8} = 2.33.$

5.6. A part of the video sequence contains 126 frames in the JPEG format (Motion JPEG—MJPEG format) and its total size is 1.38 MB. The frame resolution is $384 \times 288$, while an average number of bits per $8 \times 8$ block is $B = 51.2$. Starting from the original sequence, the DCT blocks are classified into stationary $S$ and nonstationary NS blocks. The number of the blocks are $No\{S\} = 1142$ and $No\{NS\} = 286$. The coefficients from the S blocks are almost constant over time and can be reconstructed from the first frame. The coefficients from the NS blocks are represented by using the Hermite coefficients. Namely, the each sequence of 126 coefficients is represented by 70 Hermite coefficients. Calculate the compression ratio between the algorithm based on the blocks classification and Hermite expansion, and the MJPEG algorithm.

Solution:
A set of 126 frames in the JPEG format requires $No\{S\} \cdot B \cdot 126$ bits for stationary and $No\{NS\} \cdot B \cdot 126$ bits for nonstationary blocks. In other words, the total number of bits for the original sequence in the MJPEG format is:

$$\begin{aligned} &No\{S\} \cdot B \cdot 126 + No\{NS\} \cdot B \cdot 126 = \\ &(1142 + 286) \cdot 51.2 \cdot 126 = 9.21 \cdot 10^6 \text{ b} \end{aligned}$$

The algorithm based on the classification of blocks will encode the stationary blocks from the first frame only: $No\{S\} \cdot B$.

For nonstationary blocks, instead of 126 coefficients over time, it uses 70 Hermite coefficients, with the required number of bits equal to: $No\{NS\} \cdot N \cdot B$.

The total number of bits for stationary and nonstationary blocks is:

$$No\{S\} \cdot B + No\{NS\} \cdot N \cdot B = 1142 \cdot 51.2 + 286 \cdot 70 \cdot 51.2 = 1.083 \cdot 10^6 \text{ b}$$

In this example, the achieved compression factor is approximately 8.5 times.

5.7. Consider the H.264 quantization procedure. For $QP = 4$ determine the multiplication factor $MF$ for the position (0,2).

Solution:
$Qstep = 1.0$
$(i,j) = (0,2)$, $SF = a^2 = 0.25$
$qbits = 15$, hence $2^{qbits} = 32768$
hence, $MF = (32768 \times 0.25)/1 = 8192$

5.8. Consider an image block $\mathbf{B}$ ($4 \times 4$) and a given quantization parameter $QP = 9$ (H.264 algorithm is considered). Calculate the integer transform $\mathbf{X}$, the multiplication factors $MF$ for each $(i,j)$, $i = \{0,\ldots,3\}$ and $j = \{0,\ldots,3\}$. Further, calculate the rescaling factor $V$ for the inverse quantization for each position $(i,j)$ and the rescaled transform $\mathbf{X}'$ and the output block of pixels $\mathbf{B}'$.

$$\mathbf{B} = \begin{array}{c} \\ i=0 \\ 1 \\ 2 \\ 3 \end{array} \begin{array}{c} \begin{array}{cccc} j=0 & 1 & 2 & 3 \end{array} \\ \begin{bmatrix} 11 & 14 & 18 & 11 \\ 9 & 10 & 14 & 11 \\ 1 & 4 & 8 & 1 \\ 14 & 18 & 19 & 17 \end{bmatrix} \end{array}$$

Solution:

$$\mathbf{X} = \mathbf{HBH}^T,$$

where $\mathbf{H}$ is defined by Eq. (5.8). The resulting transform matrix $\mathbf{X}$ (non-scaled version or the core transform) is given by:

$$\mathbf{X} = \begin{array}{c} \\ i=0 \\ 1 \\ 2 \\ 3 \end{array} \begin{array}{c} \begin{array}{cccc} j=0 & 1 & 2 & 3 \end{array} \\ \begin{bmatrix} 180 & -23 & -30 & 21 \\ 2 & 2 & -2 & 16 \\ 64 & 1 & -2 & -7 \\ -74 & 11 & -16 & 13 \end{bmatrix} \end{array}$$

The multiplication factors for each of the positions $(i,j)$ are given by the following matrix (the values can be found in Table 5.2 for $QP$ corresponding to $\mathrm{mod}(QP,6) = 3$):

$$MF = \begin{array}{c} \\ i=0 \\ 1 \\ 2 \\ 3 \end{array} \begin{array}{c} \begin{array}{cccc} j=0 & 1 & 2 & 3 \end{array} \\ \begin{bmatrix} 9363 & 5825 & 9363 & 5825 \\ 5825 & 3647 & 5825 & 3647 \\ 9363 & 5825 & 9363 & 5825 \\ 5825 & 3647 & 5825 & 3647 \end{bmatrix} \end{array}.$$

The quantized and scaled coefficients are calculated as:

$$\mathbf{Z} = \text{round}\left(\mathbf{X} \otimes \frac{MF}{2^{16}}\right),$$

where $\otimes$ denotes multiplication of elements on the same positions $(i,j)$, and:

$$\mathbf{X} \otimes \frac{MF}{2^{16}} = \begin{bmatrix} 25.7135 & -2.0443 & -4.2856 & 1.8665 \\ 0.1778 & 0.1113 & -0.1778 & 0.8904 \\ 9.1426 & 0.0889 & -0.2857 & -0.6222 \\ -6.5773 & 0.6121 & -1.4221 & 0.7234 \end{bmatrix}.$$

The quantized and scaled coefficients are obtained as follows:

$$\mathbf{Z} = \begin{bmatrix} 26 & -2 & -4 & 2 \\ 0 & 0 & 0 & 1 \\ 9 & 0 & 0 & -1 \\ -7 & 1 & -1 & 1 \end{bmatrix}.$$

The rescaling factors $V(i,j)$ for the inverse quantization for $QP=9$ has the following values obtained from Table 5.3 (use the row corresponding to mod $(QP,6)=3$):

$$\mathbf{V} = \begin{bmatrix} 14 & 18 & 14 & 18 \\ 18 & 23 & 18 & 23 \\ 14 & 18 & 14 & 18 \\ 18 & 23 & 18 & 23 \end{bmatrix}.$$

Prior to the inverse transform the coefficients needs to be rescaled using the rescaling factors $V(i,j)$ multiplied by $2^{\text{floor}(QP/6)}=2$:

$$\mathbf{X}' = 2\mathbf{Z} \otimes \mathbf{V} = \begin{bmatrix} 728 & -72 & -112 & 72 \\ 0 & 0 & 0 & 46 \\ 252 & 0 & 0 & -36 \\ -252 & 46 & -36 & 46 \end{bmatrix}.$$

Then the inverse transform is applied as follows:

$$\mathbf{B}' = \left(\mathbf{H}^{-1}\right)^T \mathbf{X}' \mathbf{H}^{-1},$$

while the block pixels are obtained after the post-scaling by 64:

$$\mathbf{B}'' = \text{round}(\mathbf{B}'/64) = \begin{bmatrix} 11 & 13 & 17 & 10 \\ 9 & 10 & 15 & 8 \\ 2 & 4 & 8 & 0 \\ 15 & 17 & 20 & 14 \end{bmatrix}.$$

# Appendix

**Table 5.7** Coeff-Token coding (TrailingOnes-TO, Total number of nonzero coefficients is numCoeff)

| | | Table 1 | Table 2 | Table 3 | Table 4 |
|---|---|---|---|---|---|
| TO | TotalCoeff | $0 \leq nC < 2$ | $2 \leq nC < 4$ | $4 \leq nC < 8$ | $8 \leq nC$ |
| 0 | 0 | 1 | 11 | 1111 | 000011 |
| 0 | 1 | 000101 | 001011 | 001111 | 000000 |
| 0 | 2 | 00000111 | 000111 | 001011 | 000100 |
| 0 | 3 | 000000111 | 0000111 | 001000 | 001000 |
| 0 | 4 | 0000000111 | 00000111 | 0001111 | 001100 |
| 0 | 5 | 00000000111 | 00000100 | 0001011 | 010000 |
| 0 | 6 | 0000000001111 | 000000111 | 0001001 | 010100 |
| 0 | 7 | 0000000001011 | 00000001111 | 0001000 | 011000 |
| 0 | 8 | 0000000001000 | 00000001011 | 00001111 | 011100 |
| 0 | 9 | 00000000001111 | 000000001111 | 00001011 | 100000 |
| 0 | 10 | 00000000001011 | 000000001011 | 000001111 | 100100 |
| 0 | 11 | 000000000001111 | 000000001000 | 000001011 | 101000 |
| 0 | 12 | 000000000001011 | 0000000001111 | 000001000 | 101100 |
| 0 | 13 | 0000000000001111 | 0000000001011 | 0000001101 | 110000 |
| 0 | 14 | 0000000000001011 | 0000000000111 | 0000001001 | 110100 |
| 0 | 15 | 0000000000000111 | 00000000001001 | 0000000101 | 111000 |
| 0 | 16 | 0000000000000100 | 00000000000111 | 0000000001 | 111100 |
| 1 | 1 | 01 | 10 | 1110 | 000001 |
| 1 | 2 | 000100 | 00111 | 01111 | 000101 |
| 1 | 3 | 00000110 | 001010 | 01100 | 001001 |
| 1 | 4 | 000000110 | 000110 | 01010 | 001101 |
| 1 | 5 | 0000000110 | 0000110 | 01000 | 010001 |
| 1 | 6 | 00000000110 | 00000110 | 001110 | 010101 |
| 1 | 7 | 0000000001110 | 000000110 | 001010 | 011001 |
| 1 | 8 | 0000000001010 | 00000001110 | 0001110 | 011101 |
| 1 | 9 | 00000000001110 | 00000001010 | 00001110 | 100001 |
| 1 | 10 | 00000000001010 | 000000001110 | 00001010 | 100101 |
| 1 | 11 | 000000000001110 | 000000001010 | 000001110 | 101001 |
| 1 | 12 | 000000000001010 | 0000000001110 | 000001010 | 101101 |
| 1 | 13 | 000000000000001 | 0000000001010 | 000000111 | 110001 |

(continued)

**Table 5.7** (continued)

| | | Table 1 | Table 2 | Table 3 | Table 4 |
|---|---|---|---|---|---|
| TO | TotalCoeff | 0 ≤ nC < 2 | 2 ≤ nC < 4 | 4 ≤ nC < 8 | 8 ≤ nC |
| 1 | 14 | 0000000000001110 | 00000000001011 | 0000001100 | 110101 |
| 1 | 15 | 0000000000001010 | 00000000001000 | 0000001000 | 111001 |
| 1 | 16 | 0000000000000110 | 00000000000110 | 0000000100 | 111101 |
| 2 | 2 | 001 | 011 | 1101 | 000110 |
| 2 | 3 | 0000101 | 001001 | 01110 | 001010 |
| 2 | 4 | 00000101 | 000101 | 01011 | 001110 |
| 2 | 5 | 000000101 | 0000101 | 01001 | 010010 |
| 2 | 6 | 0000000101 | 00000101 | 001101 | 010110 |
| 2 | 7 | 00000000101 | 000000101 | 001001 | 011010 |
| 2 | 8 | 0000000001101 | 00000001101 | 0001101 | 011110 |
| 2 | 9 | 0000000001001 | 00000001001 | 0001010 | 100010 |
| 2 | 10 | 00000000001101 | 000000001101 | 00001101 | 100110 |
| 2 | 11 | 00000000001001 | 000000001001 | 00001001 | 101010 |
| 2 | 12 | 000000000001101 | 0000000001101 | 000001101 | 101110 |
| 2 | 13 | 000000000001001 | 0000000001001 | 000001001 | 110010 |
| 2 | 14 | 0000000000001101 | 0000000000110 | 0000001011 | 110110 |
| 2 | 15 | 0000000000001001 | 00000000001010 | 0000000111 | 111010 |
| 2 | 16 | 0000000000000101 | 00000000000101 | 0000000011 | 111110 |
| 3 | 3 | 00011 | 0101 | 1100 | 001011 |
| 3 | 4 | 000011 | 0100 | 1011 | 001111 |
| 3 | 5 | 0000100 | 00110 | 1010 | 010011 |
| 3 | 6 | 00000100 | 001000 | 1001 | 010111 |
| 3 | 7 | 000000100 | 000100 | 1000 | 011011 |
| 3 | 8 | 0000000100 | 0000100 | 01101 | 011111 |
| 3 | 9 | 00000000100 | 000000100 | 001100 | 100011 |
| 3 | 10 | 0000000001100 | 00000001100 | 0001100 | 100111 |
| 3 | 11 | 00000000001100 | 00000001000 | 00001100 | 101011 |
| 3 | 12 | 00000000001000 | 000000001100 | 00001000 | 101111 |
| 3 | 13 | 000000000001100 | 0000000001100 | 000001100 | 110011 |
| 3 | 14 | 000000000001000 | 0000000001000 | 0000001010 | 110111 |
| 3 | 15 | 0000000000001100 | 0000000000001 | 0000000110 | 111011 |
| 3 | 16 | 0000000000001000 | 00000000000100 | 0000000010 | 111111 |

## References


1. Akbulut O, Urhan O, Ertürk S (2006) Fast sub-pixel motion estimation by means of one-bit transform. In: Proc ISCIS. Springer, Berlin, pp 503–510
2. Angelides MC, Agius H (2011) The handbook of MPEg applications: standards in practice. Wiley, New York, NY
3. Djurović I, Stanković S (2003) Estimation of time-varying velocities of moving objects in video-sequences by using time-frequency representations. IEEE Trans Image Process 12 (5):550–562

4. Djurović I, Stanković S, Oshumi A, Ijima H (2004) Motion parameters estimation by new propagation approach and time-frequency representations. Signal Process 19(8):755–770
5. Fuhrt B (2008) Encyclopaedia of multimedia, 2nd edn. Springer, New York, NY
6. Furth B, Smoliar S, Zhang H (1996) Video and image processing in Multimedia systems. Kluwer Academic Publishers, Boston, MA
7. Ghanbar M (2008) Standard codecs: image compression to advanced video coding, 3rd edn. IET, London
8. Grob B, Herdon C (1999) Basic television and video systems. McGraw-Hill, New York, NY
9. ISO/IEC 13818-2 (MPEG-2) www.comp.nus.edu.sg/~cs5248/l04/IEC-13818-2_Specs.pdf
10. Haskell BG, Puri A (2012) The MPEG representation of digital media. Springer, New York, NY
11. http://iris.ee.iisc.ernet.in/web/Courses/mm_2012/pdf/CAVLC_Example.pdf
12. Karczewicz M, Kurceren R (2001) A proposal for SP-frames. ITU-T video coding experts group meeting, Eibsee, Germany, Doc. VCEG-L-27
13. Karczewicz M, Kurceren R (2003) The SP- and SI-frames design for H.264/AVC. IEEE Trans Circ Syst Video Technol 13(7):637–644
14. Kaup A (1999) Object-based texture coding of moving video in MPEG-4. IEEE Trans Circ Syst Video Technol 9(1):5–15
15. Lie WN, Yeh HC, Lin TCI, Chen CF (2005) Hardware-efficient computing architecture for motion compensation interpolation in H.264 video coding. In: IEEE International Symposium on Circuits and Systems. IEEE, Washington, DC, pp 2136–2139
16. Lin YLS, Kao CY, Kuo HC, Chen JW (2010) VLSI design for video coding: H.264/AVC encoding form standard specification to CHIP. Springer, New York, NY
17. Malvar HS, Hallapuro A, Karczevicz M, Kerofsky L (2003) Low-complexity transform and quantization in H.264/AVC. IEEE Trans Circ Syst Video Technol 13(7):598–603
18. Mandal M (2003) Multimedia signals and systems. Springer, New York, NY
19. Marpe D, Wiegand T, Sullivan GJ (2006) The H.264/MPEG-4 advanced video coding standard and its applications. IEEE Commun Mag 44(8):134–143
20. Nisar H, Choi TS (2009) Fast and efficient fractional pixel motion estimation for H.264/AVC video coding. In: International Conference on Image Processing (ICIP 2009). IEEE, Washington, DC, pp 1561–1564
21. Richardson I (2010) The H.264 advanced video compression standard. John Wiley and Sons, Chichester
22. Richardson I (2003) H.264 and MPEG-4 video compression: video coding for next-generation multimedia. John Wiley & Sons Ltd., Chichester
23. Richardson I (2002) Video codec design. Wiley, Chichester
24. Salomon D, Motta G, Bryant D (2009) Handbook of data compression. Springer, New York, NY
25. Stanković S, Djurović I (2001) Motion parameter estimation by using time frequency representations. Electron Lett 37(24):1446–1448
26. Stanković S, Orović I, Krylov A (2010) Video frames reconstruction based on time-frequency analysis and hermite projection method. EURASIP J Adv Signal Process, Article ID 970105, 11 pages
27. Steinmetz R (2000) Multimedia systems. McGraw Hill, New York, NY
28. Sullivan GJ, Wiegand T (2005) Video compression - from concepts to the H.264/AVC standard. Proc IEEE 93(1):18–31
29. Sullivan GJ, Topiwala P, Luthra A (2004) The H264/AVC advanced video coding standard: overview and introduction to the fidelity range extensions. Proc SPIE 5558:454. doi:10.1117/12.564457
30. Wiegand T, Sullivan GJ, Bjontegaard G, Luthra A (2003) Overview of the H.264/AVC video coding standard. IEEE Trans Circ Syst Video Technol 13(7):560–576

# Chapter 6
# Compressive Sensing

In the era of digital technology expansion, the acquisition/sensing devices are producing a large amount of data that need to be stored, processed, or transmitted. This is a common issue in sensing systems dealing with multimedia signals, medical and biomedical data, radar signals, etc. The required amount of digital information given as the number of digital samples, measurements or observations per time unit is defined by the fundamental theorem in the communications which has been known as the Shannon–Nyquist sampling theorem. Accordingly, a signal can be reconstructed if the sampling frequency is at least twice higher than the maximal signal frequency ($2f_{\max}$). Obviously the sampling procedure results in a large number of samples for signals with considerably high maximal frequency. Until recently, the signal acquisition process in the real applications was mainly done according to the sampling theorem and then, in order to respond to the storage, transmission, and computational challenges, the data are compressed up to the acceptable quality by applying complex and demanding algorithms for data compression. As discussed in the previous chapters, the lossy compression algorithms are mainly based on the fact that signals actually contain a large amount of redundant information not needed for perceiving good signal quality. For lossy compression, we used two basic assumptions: the imperfection of human perception (sense), and the specific signals properties in a certain transform domain. For instance, in the case of images, a large energy compaction in the low-frequency region is achieved by using the DCT transform. Hence, a large number of coefficients can be omitted without introducing visible image quality degradation. In summary, this means that the entire information is firstly sensed and then most of it is thrown away through compression process, which seems to be a waste of resources. The data acquisition process has still remained demanding in terms of resources (e.g., sensors technology) and acquisition time. The question is whether it is possible to significantly reduce the amount of data during the acquisition process? Is it always necessary to sample the signals according to the Shannon–Nyquist criterion?

 
S. Stanković et al., *Multimedia Signals and Systems*,
DOI 10.1007/978-3-319-23950-7_6

In recent years, the compressive sensing approaches have been intensively developed to overcome the limits of traditional sampling theory by applying a concept of compression during the sensing procedure. Compressive sensing aims to provide the possibility to acquire much smaller amount of data, but still achieving the same quality (or almost the same) of the final representation as if the physical phenomenon is sensed according to the conventional sampling theory. In that sense, significant efforts have been done toward the development of methods that would allow to sample data in the compressed form using much lower number of samples. Compressive sensing opens the possibility to simplify very expensive devices and apparatus for data recording, imaging, sensing (for instance MRI scanners, PET scanners for computed tomography, high-resolution cameras, etc.). Furthermore, the data acquisition time can be significantly reduced, and in some applications even to almost 10 or 20 % of the current needs.

The compressive sensing theory states that the signal can be reconstructed using just a small set of randomly acquired samples if it has a sparse (concise) representation in a certain transform domain. Sparsity means that the signal, in certain domain (usually transform domain), can be represented with small number of nonzero samples, which implies its compressibility, decreasing numerical computations and memory usage. Another important condition, apart from the signal sparsity, is incoherence between the measurement matrix and the sparsity basis. Higher incoherence will allow fewer measurements to be used for successful signal reconstruction.

Compressive sensing is a field dealing with the above problem of interest and provides a solution that differs from the classical signal theory (Fig. 6.1). If the samples acquisition process is linear, than the problem of data reconstruction from

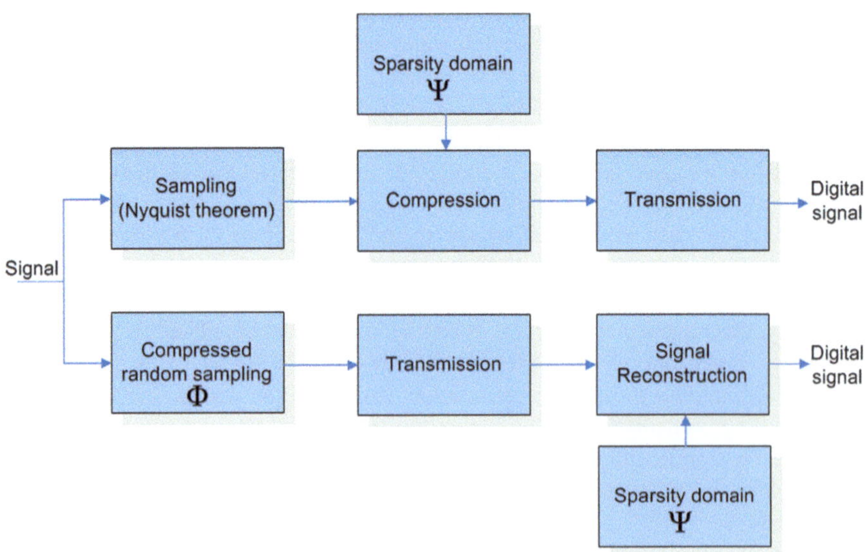


**Fig. 6.1** A classical approach for signal sampling and its compressive sensing alternative

acquired measurements can be done by solving a linear system of equations. Assume that a measurement process can be modelled by the measurement matrix $\mathbf{\Phi}$. Hence, the linear measurement process of certain signal $\mathbf{f}$ with $N$ samples can be represented as a signal reconstruction problem using a set of $M$ measurements obtained by the measurement matrix $\mathbf{\Phi}$ as follows:

$$\mathbf{\Phi f} = \mathbf{y}, \tag{6.1}$$

where $\mathbf{y}$ represents the acquired measurements.

Note that that in the traditional sampling systems, the number of measurements $M$ is set to be at least as large as the signal length $N$. However, compressive sensing systems allows us to selects $M$ measurements in a random manner, and to reconstruct the signal even though the number of measurement can be much smaller than the signal length, $M << N$. Hence, it is necessary to reconstruct the original signal by using just a small set of samples. Recall that the sparsity is one of the main requirements that should be satisfied, in order to efficiently perform the signal reconstruction. In compressive sensing, the full signal reconstruction is actually formulated as a problem of solving undetermined system of linear equations using sparsity constraint. Properly chosen basis can provide a sparse signal representation. If the signal is not sparse, then it cannot be accurately reconstructed from compressive measurements.

Compressive sensing is based on powerful mathematical algorithms for error minimization. There are several standard algorithms that are commonly employed for this purpose. For instance, the constrained $\ell_1$-norm minimization has been used as one of the first approaches for finding the sparse solutions and it is known as basis pursuit. Alternative approaches are called greedy algorithms and among them the most popular is the iterative orthogonal matching pursuit (with a variety of modifications). In this chapter, the focus is made on different reconstruction algorithms and its applications to multimedia signals. Here, we also mention the construction of a linear measurement process. It is defined as a problem of designing a measurement matrix, which would be adequate and optimal for the observed compressive sensing scenario. In most compressive sensing applications, the commonly used are random matrices such as the Gaussian matrix, with random variables distributed according to the normal distribution, and Bernoulli matrices with random variables $+1$ and $-1$ having equal probability of appearing. Also, partial random Fourier transform matrices, as well as partial random Toeplitz and circular matrices have been widely used.

## 6.1 The Compressive Sensing Requirements

### 6.1.1 Sparsity Property

Sparsity means that the signal in a transform domain contains only a small number of nonzero coefficients comparing to the signal length. Most of the real signals can be considered as sparse or almost sparse if they are represented in the proper basis. The signal $f(n)$ with $N$ samples can be represented as a linear combination of the orthonormal basis vectors:

$$f(n) = \sum_{i=1}^{N} x_i \psi_i(n), \text{ or : } \mathbf{f} = \mathbf{\Psi x}. \tag{6.2}$$

Now we can firstly define the support of vector $\mathbf{x}$ as the set of positions having non-zero entries:

$$\text{supp}(\mathbf{x}) = \{n \in (1, \ldots, N) |\ x(n) \neq 0\}. \tag{6.3}$$

If the number of nonzero coefficients in $\mathbf{x}$ is $K << N$, i.e., $\text{card}\{\text{supp}(\mathbf{x})\} \leq K$, then we can say that the signal is sparse with the sparsity level (index) $K$.

The advantages of signal sparsity have been exploited in compression algorithms: coding is done for the $K$ most significant coefficients in the transform domain. Namely, the signals in real applications are approximately sparse (not strictly sparse), meaning that the remaining $N - K$ coefficients are negligible and can be considered as zeros. Note that different basis can be used, such as: the Fourier basis (DFT), the DCT basis, the wavelet basis, etc.

In compressive sensing theory, the sparsity property is usually written by using the $\ell_0$-norm, defined as the number of nonzero elements in a certain vector:

$$\|\mathbf{x}\|_0 = \text{card}\{\text{supp}(\mathbf{x})\} \leq K, \tag{6.4}$$

where $\|\mathbf{x}\|_0 = \lim_{p\to 0} \|\mathbf{x}\|_p^p = \lim_{p\to 0} \sum_{n=1}^{N} |x(n)|^p = \sum_{n=1;\, x(n)\neq 0}^{N} 1 = K.$

Now, we can summarize the assumptions we have introduced so far (Fig. 6.2):

- The set of random measurements are selected from the signal $\mathbf{f}$ ($N \times 1$), which can be written by using the random measurement matrix $\mathbf{\Phi}$ ($M \times N$) as follows:

$$\mathbf{y} = \mathbf{\Phi f}. \tag{6.5}$$

- In order to reconstruct $\mathbf{f}$ from $\mathbf{y}$, the transform domain representation of $\mathbf{f}$ (defined by the orthogonal basis matrix $\mathbf{\Psi}$ ($N \times N$)), should be sparse:

$$\mathbf{f} = \mathbf{\Psi x}, \quad \text{where} \quad \|\mathbf{x}\|_0 \leq K,\ K \ll N. \tag{6.6}$$

Using Eqs. (6.5) and (6.6) we have:

$$\mathbf{y} = \mathbf{\Phi \Psi x} = \mathbf{Ax}. \tag{6.7}$$

Now, we may observe that the measurements vector is $M$ samples long and the matrix $\mathbf{A}$ is of size ($M \times N$). Thus, we have $M$ linear equations with

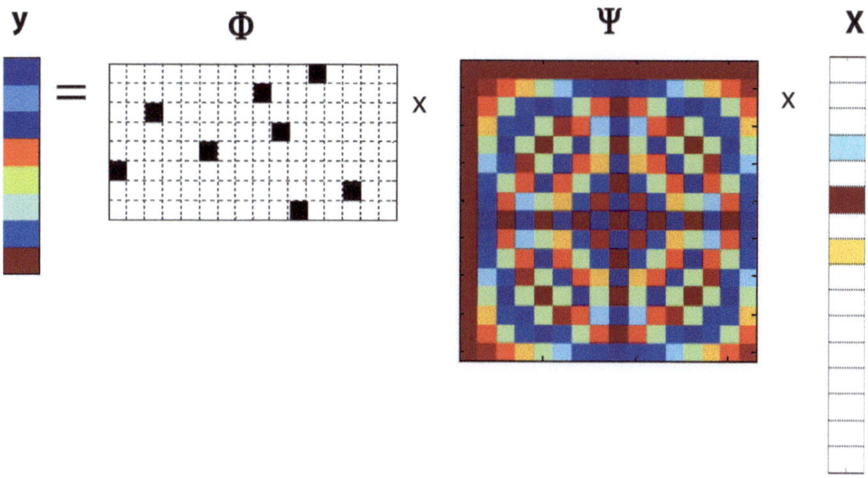


**Fig. 6.2** An illustration of the compressive sensing concept (*white fields* represent zero values)

$N$ unknowns in $\mathbf{x}$. Hence, the system is undetermined and may have infinitely many solutions. However, there is an important fact we know about $\mathbf{x}$ and that is the sparsity property.

Hence, most of the entries in $\mathbf{x}$ are zeros, and consequently, we need to determine only the non-zero components. It will reduce the problem dimension.

Let us observe the systems from Eqs. (6.6) and (6.7) in a matrix form. The signal $\mathbf{f}$ can be represented using a transform domain vector $\mathbf{x}$ in a basis $\mathbf{\Psi}$ as follows:

$$\underbrace{\begin{bmatrix} f_{(1)} \\ f_{(2)} \\ f_{(3)} \\ \cdots \\ f_{(i-1)} \\ f_{(i)} \\ \cdots \\ f_{(N)} \end{bmatrix}}_{\mathbf{f}} = \underbrace{\begin{bmatrix} \psi_{1,1} & \psi_{1,2} & \psi_{1,3} & \cdots & \psi_{1,i-1} & \psi_{1,i} & \cdots & \psi_{1,N} \\ \psi_{2,1} & \psi_{2,2} & \psi_{2,3} & \cdots & \psi_{2,i-1} & \psi_{2,i} & \cdots & \psi_{2,N} \\ \psi_{3,1} & \psi_{3,2} & \psi_{3,3} & \cdots & \psi_{3,i-1} & \psi_{3,i} & \cdots & \psi_{3,N} \\ \cdots & & \cdots & & & & & \cdots \\ \psi_{i-1,1} & \psi_{i-1,2} & \psi_{i-1,3} & \cdots & \psi_{i-1,i-1} & \psi_{i-1,i} & \cdots & \psi_{i-1,N} \\ \psi_{i,1} & \psi_{i,2} & \psi_{i,3} & \cdots & \psi_{i,i-1} & \psi_{i,i} & \cdots & \psi_{i,N} \\ \cdots & & \cdots & & & & & \cdots \\ \psi_{N,1} & \psi_{N,2} & \psi_{N,3} & \cdots & \psi_{N,i-1} & \psi_{N,i} & \cdots & \psi_{N,N} \end{bmatrix}}_{\mathbf{\Psi}} \cdot \underbrace{\begin{bmatrix} x_{(1)} \\ x_{(2)} \\ x_{(3)} \\ \cdots \\ x_{(i-1)} \\ x_{(i)} \\ \cdots \\ x_{(N)} \end{bmatrix}}_{\mathbf{x}}. \tag{6.8}$$

However due to the compressive sensing process we will miss some of the signal values, for example those marked in red, and this can be modelled using the multiplication with a measurement matrix. Now the measurement vector can be represented in matrix form (compliant with Eq. (6.7)) as follows:

$$\underbrace{\begin{bmatrix} y_{(1)} \\ \cdots \\ y_{(j)} \\ \cdots \\ y_{(M)} \end{bmatrix}}_{\mathbf{y}} = \begin{bmatrix} f_{(2)} \\ \cdots \\ f_{(i)} \\ \cdots \\ f_{(N)} \end{bmatrix} = \underbrace{\begin{bmatrix} \psi_{2,1} & \psi_{2,2} & \psi_{2,3} & \cdots & \psi_{2,i-1} & \psi_{2,i} & \cdots & \psi_{2,N} \\ \cdots & & & \cdots & & & & \cdots \\ \psi_{i,1} & \psi_{i,2} & \psi_{i,3} & \cdots & \psi_{i,i-1} & \psi_{i,i} & \cdots & \psi_{i,N} \\ \cdots & & & \cdots & & & & \cdots \\ \psi_{N,1} & \psi_{N,2} & \psi_{N,3} & \cdots & \psi_{N,i-1} & \psi_{N,i} & \cdots & \psi_{N,N} \end{bmatrix}}_{\mathbf{A}} \underbrace{\begin{bmatrix} x_{(1)} \\ x_{(2)} \\ x_{(3)} \\ \cdots \\ x_{(i-1)} \\ x_{(i)} \\ \cdots \\ x_{(N)} \end{bmatrix}}_{\mathbf{x}} \tag{6.9}$$

Note that the matrix **A** is obtained as a partial random transform matrix, by keeping only the rows of $\mathbf{\Psi}$ that correspond to the positions of measurements. If we use the notation with both $\mathbf{\Phi}$ and $\mathbf{\Psi}$ matrices, then $\mathbf{\Phi}$ should be modelled as a matrix of size $M \times N$ having only one entry per row equal to 1 at the position of available measurements, and zeros otherwise:

$$A = \underbrace{\begin{bmatrix} 0 & 1 & 0 & \ldots & 0 & 0 & \ldots & 0 \\ \cdots & & & \cdots & & & & \cdots \\ 0 & 0 & 0 & \ldots & 0 & 1 & \ldots & 0 \\ \cdots & & & \cdots & & & & \cdots \\ 0 & 0 & 0 & \ldots & 0 & 0 & \ldots & 1 \end{bmatrix}}_{\mathbf{\Phi}}$$

$$\times \underbrace{\begin{bmatrix} \psi_{1,1} & \psi_{1,2} & \psi_{1,3} & \cdots & \psi_{1,i-1} & \psi_{1,i} & \cdots & \psi_{1,N} \\ \psi_{2,1} & \psi_{2,2} & \psi_{2.3} & \cdots & \psi_{2,i-1} & \psi_{2,i} & \cdots & \psi_{2,N} \\ \psi_{3,1} & \psi_{3,2} & \psi_{3,3} & \cdots & \psi_{3,i-1} & \psi_{3,i} & \cdots & \psi_{3,N} \\ \cdots & & & \cdots & & & & \cdots \\ \psi_{i-1,1} & \psi_{i-1,2} & \psi_{i-1,3} & \cdots & \psi_{i-1,i-1} & \psi_{i-1,i} & \cdots & \psi_{i-1,N} \\ \psi_{i,1} & \psi_{i,2} & \psi_{i,3} & \cdots & \psi_{i,i-1} & \psi_{i,i} & \cdots & \psi_{i,N} \\ \cdots & & & \cdots & & & & \cdots \\ \psi_{N,1} & \psi_{N,2} & \psi_{N,3} & \cdots & \psi_{N,i-1} & \psi_{N,i} & \cdots & \psi_{N,N} \end{bmatrix}}_{\mathbf{\Psi}}$$

In order to generalize the notations, we use the set of available samples given by the indices: $\mathbf{N}_M = \{n_1, n_2, \ldots, n_M\}$, and then we can write:

$$\underbrace{\begin{bmatrix} y_{(1)} \\ \cdots \\ y_{(j)} \\ \cdots \\ y_{(M)} \end{bmatrix}}_{\mathbf{y}} = \underbrace{\begin{bmatrix} \psi_{n_1,1} & \psi_{n_1,2} & \psi_{n_1,3} & \cdots & \psi_{n_1,i-1} & \psi_{n_1,i} & \cdots & \psi_{n_1,N} \\ \cdots & & & \cdots & & & & \cdots \\ \psi_{n_i,1} & \psi_{n_i,2} & \psi_{n_i,3} & \cdots & \psi_{n_i,i-1} & \psi_{n_i,i} & \cdots & \psi_{n_i,N} \\ \cdots & & & \cdots & & & & \cdots \\ \psi_{n_M,1} & \psi_{n_M,2} & \psi_{n_M,3} & \cdots & \psi_{n_M,i-1} & \psi_{n_M,i} & \cdots & \psi_{n_M,N} \end{bmatrix}}_{\mathbf{A}} \underbrace{\begin{bmatrix} x_{(1)} \\ \cdots \\ x_{(i-1)} \\ x_{(i)} \\ \cdots \\ x_{(N)} \end{bmatrix}}_{\mathbf{x}}. \tag{6.10}$$

The measurement procedure and the measurement matrix should be properly created to provide the reconstruction of signal **f** (of length $N$) by using $M<<N$ measurements. The reconstructed signal is obtained as a solution of $M$ linear equations with $N$ unknowns. Having in mind that this system is undetermined and can have infinitely many solutions, the optimization based mathematical algorithms should be used to search for the sparsest solution, consistent with the linear measurements.

Consider a simplified case when a signal has three possible values in the transform domain vector **x**: $x_{(1)}$, $x_{(2)}$, $x_{(3)}$ (for $N=3$). Also, we assume that **x** is sparse with $K=1$ component, meaning that only one coefficient is nonzero: $x_{(1)}=0$, $x_{(2)}\neq 0$, $x_{(3)}=0$. Now, in the case when we have only one measurement, there is only one equation:

$$y_{(1)} = \begin{bmatrix} \psi_{n_1,1} & \psi_{n_1,2} & \psi_{n_1,3} \end{bmatrix} \begin{bmatrix} x_{(1)} \\ x_{(2)} \\ x_{(3)} \end{bmatrix}. \tag{6.11}$$

The measurement $y_{(1)} = x_{(1)}\psi_{n_1,1} + x_{(2)}\psi_{n_1,2} + x_{(3)}\psi_{n_1,3}$ represents the three-dimensional plane, which has three intersection points with the coordinate axis, meaning that we have three different solutions. Consequently, we need to include more measurements (equations) to obtain the unique solution of the problem. Now, we observe the system obtained for two measurements $y_{(1)}$ and $y_{(2)}$:

$$\begin{bmatrix} y_{(1)} \\ y_{(2)} \end{bmatrix} = \begin{bmatrix} \psi_{n_1,1} & \psi_{n_1,2} & \psi_{n_1,3} \\ \psi_{n_2,1} & \psi_{n_2,2} & \psi_{n_2,3} \end{bmatrix} \begin{bmatrix} x_{(1)} \\ x_{(2)} \\ x_{(3)} \end{bmatrix}, \tag{6.12}$$

or equivalently,

$$\begin{aligned} \textit{plane A}: \quad & y_{(1)} = x_{(1)}\psi_{n_1,1} + x_{(2)}\psi_{n_1,2} + x_{(3)}\psi_{n_1,3} \\ \textit{plane B}: \quad & y_{(2)} = x_{(1)}\psi_{n_2,1} + x_{(2)}\psi_{n_2,2} + x_{(3)}\psi_{n_2,3} \end{aligned}. \tag{6.13}$$

In this case, we have two planes **A** and **B** (Fig. 6.3). The intersection of the planes is line ***p***. If ***p*** intersects with only one coordinate axis in only one point then this is the unique solution of the problem. In other words, the direction vector of ***p*** does not have any zero coordinate.

Next, we discuss a few additional interesting cases. For instance, if ***p*** intersects with two coordinate axes (***p*** lies in one of the coordinate planes), then there will be two possible solutions. Hence, the two planes are not sufficient to find the unique solution. If ***p*** belongs to one of the axes, the solution cannot be determined.

The direction vector of ***p*** is normal to the vectors of the planes **A** and **B**. The coordinates of the direction vector of ***p*** are:

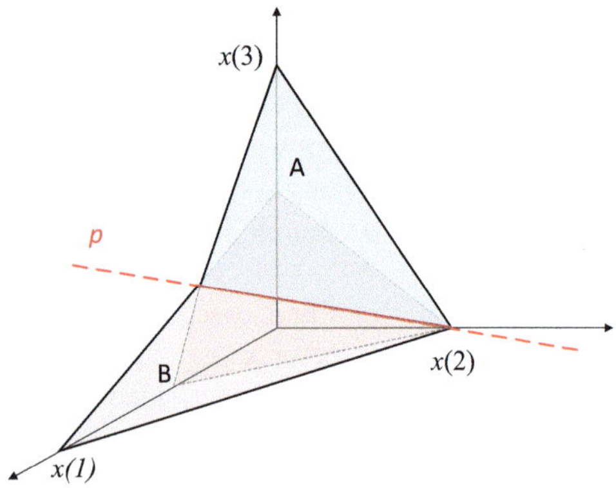


**Fig. 6.3** An illustration of the two planes A and B, intersection line p, and the common point on the coordinate axis

$$\vec{d} = (d_1, d_2, d_3) = \begin{vmatrix} \vec{i} & \vec{j} & \vec{k} \\ \psi_{n_1,1} & \psi_{n_1,2} & \psi_{n_1,3} \\ \psi_{n_2,1} & \psi_{n_2,2} & \psi_{n_2,3} \end{vmatrix}$$

where,

$$\begin{aligned} d_1 &= \psi_{n_1,2}\psi_{n_2,3} - \psi_{n_1,3}\psi_{n_2,2}, \\ d_2 &= \psi_{n_1,1}\psi_{n_2,3} - \psi_{n_1,3}\psi_{n_2,1}, \\ d_3 &= \psi_{n_1,1}\psi_{n_2,2} - \psi_{n_1,2}\psi_{n_2,1}. \end{aligned} \tag{6.14}$$

For $K = 1$, the solution is unique if:

$$\min\{|d_1|, |d_2|, |d_3|\} > 0, \tag{6.15}$$

meaning that $\boldsymbol{p}$ does not belong to any of the coordinate planes.

### 6.1.2 *Restricted Isometry Property*

There are two important conditions that should be met in order to achieve successful reconstruction for a wider range of sparsity level. One is the Restricted Isometry Property (RIP). For a proper isometry constant, RIP ensures that any subset of columns in **A** with cardinality less than sparsity level *K*, is nearly orthogonal. This results in better guarantees for the successful signal reconstruction from small set of measurements. Another condition is a small mutual coherence between the measurement matrix and the transform representation matrix (incoherence property). The two conditions will be analyzed in the sequel.

A matrix **A** satisfies the isometry property if it preserves the vector intensity in the *N*-dimensional space:

$$\|\mathbf{Ax}\|_2^2 = \|\mathbf{x}\|_2^2. \tag{6.16}$$

For instance in the case when **A** is the full Fourier transform matrix:
$\mathbf{A} = N\boldsymbol{\Psi}$, we can write:

$$N\|\boldsymbol{\Psi}\mathbf{x}\|_2^2 = \|\mathbf{x}\|_2^2, \tag{6.17}$$

which corresponds to the Parseval's theorem in the Fourier domain analysis stating that the total energy of signal calculated over all time instants is equal to the total energy of its Fourier transform calculated over all frequency components (the unitary property of the Fourier transform). The relation Eq. (6.17) can be also written as:

$$\frac{N\|\boldsymbol{\Psi}\mathbf{x}\|_2^2 - \|\mathbf{x}\|_2^2}{\|\mathbf{x}\|_2^2} = 0. \tag{6.18}$$

Now, we can also define the restricted isometry property of certain matrix **A**, where **A** will refer to the compressive sensing matrix (of size $M \times N, M < N$) in the sequel. For each integer number $K$ (representing the sparsity index), the isometry constant $\delta_K$ of the matrix **A** is the smallest number for which the relation:

$$(1 - \delta_K)\|\mathbf{x}\|_2^2 \le \|\mathbf{Ax}\|_2^2 \le (1 + \delta_K)\|\mathbf{x}\|_2^2, \tag{6.19}$$

holds for all $K$-sparse vectors, where $\mathbf{A} = \boldsymbol{\Phi\Psi}$. Equivalently, the restricted isometry property can be rewritten as:

$$\left|\frac{\|\mathbf{Ax}\|_2^2 - \|\mathbf{x}\|_2^2}{\|\mathbf{x}\|_2^2}\right| \le \delta_K, \tag{6.20}$$

where $0 < \delta_K < 1$ is the restricted isometry constant. Note that for the restricted isometry constants holds:

$$\delta_1 \le \delta_2 \le \ldots \le \delta_K \le \ldots \le \delta_N. \tag{6.21}$$

If the matrix **A** satisfies RIP we may say that it acts as a near-isometry on sparse vectors and approximately preserves the Euclidian length of sparse vectors. Hence, for the RIP matrix **A** with $(2K, \delta_K)$ and $\delta_K < 1$ we can say that all subsets of $2K$ columns are linearly independent. In other words:

$$spark(\mathbf{A}) > 2K, \tag{6.22}$$

where *spark* represents the smallest number of dependent columns, and $spark(\mathbf{A}) \leq rank(\mathbf{A}) + 1$. For the matrix $\mathbf{A}$ of size $M \times N$ ($M$ columns, $N$ rows) we can write:

$$2 \leq spark(\mathbf{A}) \leq M + 1. \tag{6.23}$$

Only in the case when one of the columns has all zero values, then $spark(\mathbf{A}) = 1$. If there are no dependent columns: $spark(\mathbf{A}) = M + 1$. Then according to Eq. (6.22):

$$K < \frac{1}{2} spark(\mathbf{A}) \leq \frac{1}{2}(M + 1). \tag{6.24}$$

It means that the number of measurements should be at least twice the number of components $K$: $M \geq 2K$, which can be proven on a previous simple example: two measurements $\mathbf{y} = [y_{(1)} y_{(2)}]^{\mathrm{T}}$, $N = 3$ and $\mathbf{x} = [x_{(1)} x_{(2)} \ldots x_{(N)}]^{\mathrm{T}} = [x_{(1)} x_{(2)} x_{(3)}]^{\mathrm{T}}$. Then the observed system of equations $\mathbf{y} = \mathbf{A}\mathbf{x}$ can be written as:

$$\begin{bmatrix} y_{(1)} \\ y_{(2)} \end{bmatrix} = \begin{bmatrix} A_{n_1,1} & A_{n_1,2} & A_{n_1,3} \\ A_{n_2,1} & A_{n_2,2} & A_{n_2,3} \end{bmatrix} \begin{bmatrix} x_{(1)} \\ x_{(2)} \\ x_{(3)} \end{bmatrix},$$

where $\mathbf{A}$ can be random partial Fourier matrix as discussed before (or any other compressive sensing matrix):

$$\begin{bmatrix} A_{1,1} & A_{1,2} & A_{1,3} \\ A_{2,1} & A_{2,2} & A_{2,3} \end{bmatrix} = \begin{bmatrix} \Psi_{n_1,1} & \Psi_{n_1,2} & \Psi_{n_1,3} \\ \Psi_{n_2,1} & \Psi_{n_2,2} & \Psi_{n_2,3} \end{bmatrix}.$$

If every pair of columns in $\mathbf{A}$ is independent, i.e.,

$$\begin{aligned} A_{1,1}A_{2,2} - A_{2,1}A_{1,2} \neq 0 \quad & \Psi_{n_1,1}\Psi_{n_2,2} - \Psi_{n_2,1}\Psi_{n_1,2} \neq 0 \\ A_{1,2}A_{2,3} - A_{2,2}A_{1,3} \neq 0 \quad & \Psi_{n_1,2}\Psi_{n_2,3} - \Psi_{n_2,2}\Psi_{n_1,3} \neq 0 \\ A_{1,1}A_{2,3} - A_{2,1}A_{1,3} \neq 0 \quad & \Psi_{n_1,1}\Psi_{n_2,3} - \Psi_{n_2,1}\Psi_{n_1,3} \neq 0 \end{aligned}$$

and there are no zero columns, then we can conclude:

$$spark(\mathbf{A}) = 3,$$

meaning that the signal can be recovered if $K < \frac{3}{2} = 1$ (since $spark(\mathbf{A}) > 2K$ according to Eq. (6.22)).

The RIP property has been used to establish sufficient conditions for sparse signal recovery. For instance, for any sparse vector $\mathbf{x}$, the exact recovery is possible by the $\ell_1$-minimization, if the CS matrix satisfies the RIP of order $2K$ with $\delta_{2K} < 0.414$ (or even for $\delta_{2K} < 0.465$). Namely, if $\delta_{2K} < 1$ then the $\ell_0$-minimization has a unique solution with sparsity index $K$, while if $\delta_{2K} < 0.414$ the solution of the $\ell_1$-minimization problem corresponds exactly to the solution of $\ell_0$-minimization problem.

Finally, the RIP has been established for some commonly used matrices in compressive sensing such as random Gaussian, Bernoulli, and partial random Fourier matrices. Generally, for random matrices, the columns are linearly independent, i.e., $spark(\mathbf{A}) = M + 1$ with high probability.

#### 6.1.2.1 Restricted Isometry Property of Some Common Matrices

The commonly used measurement matrices that satisfy the RIP include random Gaussian, Bernoulli, partial random Fourier matrices, etc. Each type of measurement matrix has a different lower bound for the number of measurements. Hence, in the case of Gaussian and Bernoulli random matrix with zero mean and variance $1/M$, the lower bound:

$$M \geq C \cdot K \cdot \log(N/K), \tag{6.25}$$

is achievable and can guarantee the exact reconstruction of $\mathbf{x}$ with overwhelming probability ($C$ is a positive constant). The same condition can be applied to binary matrices with independent entries taking values $\pm 1/\sqrt{M}$. As another example, consider the partial random Fourier matrix obtained by selecting $M$ normalized rows from full DFT matrix. The lower bound of $M$ will be:

$$M \geq CK(\log N)^4 . \tag{6.26}$$

In the case of general orthogonal matrix $\mathbf{A}$ obtained by randomly selecting $M$ rows from $N \times N$ orthonormal matrix and renormalizing the columns so that they are unit-normed, the lower bound is given by:

$$M \geq C\mu K(\log N)^6 . \tag{6.27}$$

The recovery condition depends on the mutual coherence denoted by $\mu$, which measures the similarity between $\mathbf{\Phi}$ and $\mathbf{\Psi}$, and is further discussed below.

### 6.1.3 *Incoherence*

The incoherence is an important condition that matrices $\mathbf{\Psi}$ and $\mathbf{\Phi}$ should satisfy to make compressive sensing possible. It is related to the property that signals, having sparse representation in the transform domain $\mathbf{\Psi}$, should be dense in the domain where the acquisition is performed (e.g., time domain). For instance, it is well-know that the signal, represented by the Dirac pulse in one domain, is spread in an another (inverse) domain. Hence, the compressive sensing approach assumes that a signal is acquired in the domain where it is rich of samples, so that by using random sampling we can collect enough information about the signal.

The relation between the number of nonzero samples in the transform domain $\boldsymbol{\Psi}$ and the number of measurements (required to reconstruct the signal) depends on the coherence between the matrices $\boldsymbol{\Psi}$ and $\boldsymbol{\Phi}$. For example, if $\boldsymbol{\Psi}$ and $\boldsymbol{\Phi}$ are maximally coherent, then all coefficients would be required for signal reconstruction. The matrices $\boldsymbol{\Phi}$ and $\boldsymbol{\Psi}$ are incoherent if the rows of $\boldsymbol{\Phi}$ are spread out in the domain $\boldsymbol{\Psi}$ (rows of $\boldsymbol{\Phi}$ cannot provide sparse representation of the columns of $\boldsymbol{\Psi}$, and vice versa). The mutual coherence between two matrices $\boldsymbol{\Psi}$ and $\boldsymbol{\Phi}$ measures the maximal absolute value of correlation between two elements from $\boldsymbol{\Psi}$ and $\boldsymbol{\Phi}$, and it is defined as[1]:

$$\mu(\boldsymbol{\Phi},\boldsymbol{\Psi}) = \max_{i\neq j}\left|\frac{\langle\phi_i,\psi_j\rangle}{\|\phi_i\|^2\|\psi_j\|^2}\right|, \tag{6.28}$$

where $\phi_i$ and $\psi_j$ are rows of $\boldsymbol{\Phi}$ and columns of $\boldsymbol{\Psi}$, respectively.

Since, $\mathbf{A}=\boldsymbol{\Phi}\boldsymbol{\Psi}$, the mutual coherence can be also defined as the maximum absolute value of normalized inner product between all columns in $\mathbf{A}$:

$$\mu(\mathbf{A}) = \max_{i\neq j,\,1\leq i,\,j\leq M}\left|\frac{\langle A_i,A_j\rangle}{\|A_i\|^2\|A_j\|^2}\right|, \tag{6.29}$$

where $A_i$ and $A_j$ denote columns of matrix $\mathbf{A}$. The maximal mutual coherence will have the value 1 in the case when certain pair of columns coincides.

*Proposition* If the matrix $\mathbf{A}$ has $\ell_2$-normalized columns $A_0,\ldots,A_{N-1}$ or equivalently $\|A_n\|_2^2 = 1$ for all columns then[2]:

$$\delta_K = (K-1)\mu.$$

---

[1] Depending whether the elements of matrices are normalized or not, different forms of the coherence can be found in the literature.

[2] *Proof*:

$$\|\mathbf{Ax}\|_2^2 = \left|\sum_{k=0}^{N-1} x(k)A_{1,k}\right|^2 + \ldots + \left|\sum_{k=0}^{N-1} x(k)A_{M,k}\right|^2$$

$$= \sum_{k_1=0}^{N-1} x(k_1)A_{1,k_1}\sum_{k_2=0}^{N-1} x^*(k_2)A^*_{1,k_2} + \ldots + \sum_{k_1=0}^{N-1} x(k_1)A_{M,k_1}\sum_{k_2=0}^{N-1} x^*(k_2)A^*_{M,k_2}$$

$$= \sum_{i=1}^{M}\sum_{k_1=0}^{N-1}|x(k_1)|^2|A_{i,k_1}|^2 + \sum_{k_1=0}^{N-2}\sum_{k_2=k_1+1}^{N-1} 2\mathrm{Re}\{x(k_1)x^*(k_2)\}\sum_{i=1}^{M} A_{i,k_1}A^*_{i,k_2}$$

The coherence of a matrix **A** of size $M \times N$ with $\ell_2$-normalized columns satisfies:

$$\mu \geq \sqrt{\frac{N-M}{M(N-1)}}, \tag{6.30}$$

that represents a fundamental lower bound called Welch bound. Therefore we have: $\mu \in \left[\sqrt{\frac{N-M}{M(N-1)}}, 1\right]$. Note that for large $N$, $\mu \geq 1/\sqrt{M}$. For proving this lower bound, we start from two matrices derived from **A**:

$\mathbf{G}=\mathbf{A}^*\mathbf{A}$ (of size $N \times N$), and $\mathbf{H}=\mathbf{A}\mathbf{A}^*$ (of size $M \times M$), where $\mathbf{A}^*$ denotes conjugate transpose of **A**.

---

$$\|\mathbf{Ax}\|_2^2 = |x(0)|^2 \sum_{i=1}^{M} |A_{i,0}|^2 + \ldots + |x(N-1)|^2 \sum_{i=1}^{M} |A_{i,N-1}|^2$$
$$+ \sum_{k_1=0}^{N-2} \sum_{k_2=k_1+1}^{N-1} 2\mathrm{Re}\{x(k_1)x^*(k_2)\} \sum_{i=1}^{M} A_{i,k_1} A^*_{i,k_2}$$
$$\|\mathbf{Ax}\|_2^2 < \sum_{k_1=0}^{N-1} |x(k_1)|^2 + \sum_{k_1=0}^{N-2} \sum_{k_2=k_1+1}^{N-1} 2|x(k_1)x^*(k_2)|\mu(k_1,k_2)$$

Here, we have used the condition: $\sum_{i=1}^{M} |A_{i,0}|^2 = \cdots = \sum_{i=1}^{M} |A_{i,N-1}|^2 = \| A_n \|^2_{2(n=0,\ldots,N-1)} = 1.$
The previous relation can be rewritten as follows:

$$\frac{\| \mathbf{Ax} \|_2^2 - \| \mathbf{x} \|_2^2}{\| \mathbf{x} \|_2^2} < \sum_{k_1=0}^{N-2} \sum_{k_2=k_1+1}^{N-1} \frac{2|x(k_1)x^*(k_2)|\mu(k_1,k_2)}{\| \mathbf{x} \|_2^2}$$

$$\delta = 2\mu\max\left\{ \sum_{k_1=0}^{N-2} \sum_{k_2=k_1+1}^{N-1} \frac{|x(k_1)x^*(k_2)|}{\| \mathbf{x} \|_2^2} \right\}.$$

Assuming that we have $K$ components in $\mathbf{x}$: $\mathbf{x} = \left[x(k_{p_1}),\ \ldots, x(k_{p_K})\right]$, the RIP constant $\delta_K$ can be derived as follows:

$$\delta_K = \mu\max\left\{ \frac{2\{|x(k_{p_1})||x(k_{p_2})| + |x(k_{p_1})||x(k_{p_3})| + \ldots + |x(k_{p_{K-1}})||x(k_{p_K})|\}}{|x(k_{p_1})|^2 + |x(k_{p_2})|^2 + \cdots + |x(k_{p_K})|^2} \right\}$$
$$= \mu\max\left\{ \frac{\left(|x(k_{p_1})| + |x(k_{p_2})| + \cdots + |x(k_{p_K})|\right)^2}{|x(k_{p_1})|^2 + |x(k_{p_2})|^2 + \cdots + |x(k_{p_K})|^2} - 1 \right\}$$
$$= \mu\left\{ \frac{K^2|x(k_p)|^2}{K|x(k_p)|^2} - 1 \right\} \text{ for } |x(k_{p_1})| = \cdots = |x(k_{p_K})|$$

Finally, we can say that: $\delta_K = \mu(K-1)$.

Then the trace of matrix **G** and matrix **H** is used:

$$\mathrm{tr}(\mathbf{G}) = \sum_{i=1}^{N} \|A_i\|_2^2 = N \text{ and } \mathrm{tr}(\mathbf{H}) = \mathrm{tr}(\mathbf{G}). \tag{6.31}$$

Furthermore, the following inequality holds:

$$\mathrm{tr}(\mathbf{H}) \leq \sqrt{M}\sqrt{\mathrm{tr}(\mathbf{G}\mathbf{G}^*)},$$

or equivalently,

$$\mathrm{tr}(\mathbf{G}) \leq \sqrt{M}\sqrt{\mathrm{tr}(\mathbf{G}\mathbf{G}^*)},$$

which leads to:

$$\begin{aligned} N &\leq \sqrt{M}\sqrt{\sum_{i=1}^{N}\sum_{j=1}^{N}\left(\langle A_i, A_j\rangle\right)^2}, \\ N^2 &\leq M\left(\sum_{i=1}^{N}(A_i)^2 + \sum_{i=1}^{N}\sum_{j=1,\, j\neq i}^{N}\left(\langle A_i, A_j\rangle\right)^2\right). \end{aligned} \tag{6.32}$$

The $\ell_2$-normalized columns of **A** are denoted as $A_i$ and $A_j$. Since $\langle A_i, A_j\rangle \leq \mu$ and there are $N^2 - N$ of such terms, we obtain:

$$N^2 \leq M\left(N + (N^2 - N)\mu^2\right)\Big). \tag{6.33}$$

Therefore, from Eq. (6.33) follows:

$$\mu \leq \sqrt{\frac{M - N}{M(N-1)}}. \tag{6.34}$$

If the number of measurements $M$ is of order:

$$M \geq C \cdot K \cdot \mu(\mathbf{\Phi}, \mathbf{\Psi}) \cdot \log N, \tag{6.35}$$

then the sparsest solution is exact with a high probability ($C$ is a constant). It is assumed that the original signal $f \in \mathbb{R}^N$ is K-sparse in $\mathbf{\Psi}$. Lower the coherence between $\mathbf{\Phi}$ and $\mathbf{\Psi}$, a smaller number of random measurements is required for signal reconstruction.

## 6.2 Signal Reconstruction Approaches

The main challenge of the CS reconstruction is to solve an underdetermined system of linear equations using sparsity assumption. However, it has been shown that in the case of CS matrices that satisfy RIP, instead of $\ell_0$-minimization problem, it is much easier to use the $\ell_1$-optimization based on linear programming methods to provide signal reconstruction with high accuracy. Due to their complexity, the linear programming techniques are not always suitable in practical applications, and thus, many alternative approaches have been proposed in the literature, including the greedy and threshold based algorithms.

Before we start describing some commonly used signal reconstruction algorithms, we will observe one simplified case of the direct search method for the reconstruction of missing signal samples.

### 6.2.1 *Direct (Exhaustive) Search Method*

A simple and intuitive approach based on a direct parameter search is considered to provide a better insight into the problem of recovering missing data by exploiting the sparsity property. Let assume that the original vector with 20 entries (original signal samples) is given below:

$\mathbf{f}$ = [0, 0.5878, 0.9511, 0.9511, 0.5878, 0, **−0.5878**, −0.9511, −0.9511, −0.5878, 0, 0.5878, 0.9511, 0.9511, 0.5878, 0, −0.5878, −0.9511, **−0.9511**, −0.5878];

Now, assume that we have a vector of measurements $\mathbf{y}$ with two missing samples compared to the original vector $\mathbf{f}$, and the 7th and the 19th samples are missing. Hence:

$\mathbf{y}$ = [0, 0.5878, 0.9511, 0.9511, 0.5878, 0, −0.9511, −0.9511, −0.5878, 0, 0.5878, 0.9511, 0.9511, 0.5878, 0, −0.5878, −0.9511, −0.5878];

Furthermore, let us use the Fourier transform domain as a domain of signal sparsity (the samples in $\mathbf{f}$ belongs to a single sinusoidal signal producing two symmetric components in the Fourier domain). It means that the matrix $\mathbf{A}$ will be partial random Fourier transform matrix $\boldsymbol{\Psi}$ obtained by omitting the 7th and the 19th row of the full Fourier transform matrix. The signal with missing samples is plotted in Fig. 6.4a, while the corresponding Fourier transform, which is referred to as the initial Fourier transform, is plotted in Fig. 6.4b. Note that the omitted rows in the Fourier transform matrix produce the same result as if zeros are inserted at the

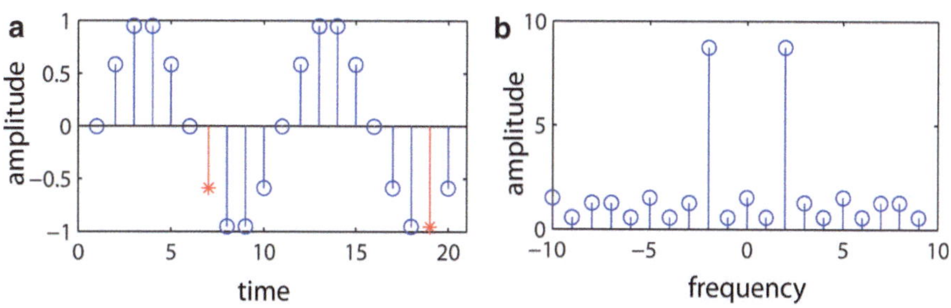


**Fig. 6.4** (**a**) Time domain samples (missing samples are marked by *red star*), (**b**) Fourier transform vector for the case when two measurements are missing

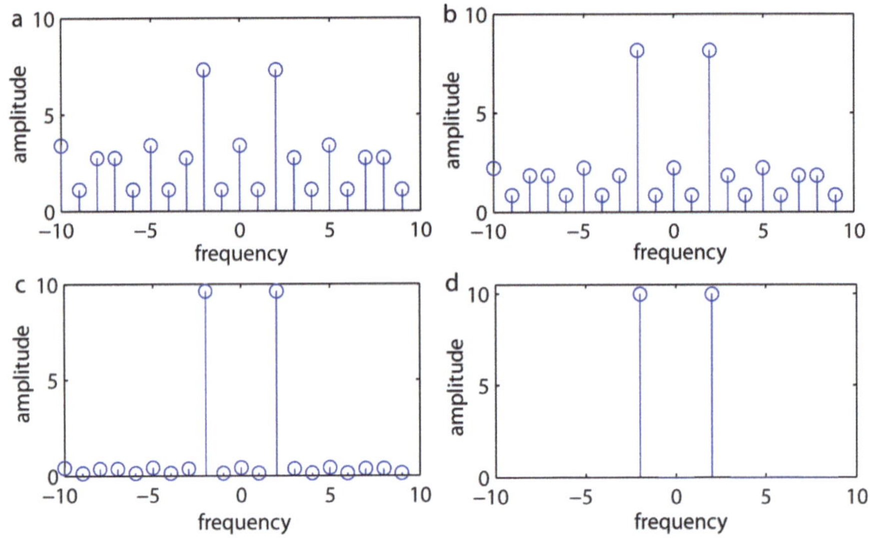


**Fig. 6.5** The discrete Fourier transform of **f** for different values of (**a**,**b**): (**a**) (0.95,0.92), (**b**) (0.25,0.44), (**c**) close to exact values (−0.4,−0.8), (**d**) exact values (−0.5878,−0.9511)

positions of missing samples. It is obvious that the initial transform vector is not sparse, due to the effects caused by missing samples. These effects are analyzed later in this chapter.

If we search for missing samples by a direct exhaustive search method taking:

**a** = −1:0.0001:1 and **b** = −1:0.0001:1.
**f** = [0, 0.5878, 0.9511, 0.9511, 0.5878, 0, **a**, −0.9511, −0.9511, −0.5878, 0, 0.5878, 0.9511, 0.9511, 0.5878, 0, −0.5878, −0.9511, **b**, −0.5878];

we would obtain the exact samples as a pair of values from **a** and **b** that provides the sparsest solution, i.e., the solution with the lowest number of non-zero components in the Fourier transform vector (Fig. 6.5).

In this case, for given ranges of **a** and **b**, the number of search iterations is approximately $10^8$. If one additional sample is missing and the search is performed in the same range, we would need more than $10^{12}$ iterations. It is obvious that for larger number of missing samples this procedure would be computationally exhaustive and impossible. In practice, we cannot use the direct search method to reconstruct missing information, but much more sophisticated reconstruction algorithms are needed.

### *6.2.2 Signal Recovering via Solving Norm Minimization Problems*

The method of solving the undetermined system $\mathbf{y}=\mathbf{\Phi\Psi x}=\mathbf{Ax}$, by searching for the sparsest solution can be described as:

$$\min \|\mathbf{x}\|_0 \ \textit{subject to} \ \mathbf{y} = \mathbf{A}\mathbf{x}, \tag{6.36}$$

where $\|\mathbf{x}\|_0$ represents the $\ell_0$-norm defined as the number of nonzero elements in $\mathbf{x}$. This is a non-convex combinatorial optimization problem and the solution requires exhaustive searches over subsets of columns of $\mathbf{A}$ with exponential complexity. Namely, in order to solve this system:

$$\mathbf{A}^T\mathbf{A}\mathbf{x} = \mathbf{A}^T\mathbf{y} \quad \text{or} \quad \mathbf{x} = (\mathbf{A}^T\mathbf{A})^{-1}\mathbf{A}^T\mathbf{y},$$

we need to search over all possible sparse vectors $\mathbf{x}$ with $K$ entries, where the subset of $K$-positions of entries are from the set $\{1,\ldots,N\}$. The total number of possible $K$-position subsets is $\binom{N}{K}$, which can mean a large number of combinations, and thus this method is not efficient for the applications (e.g., for $N = 512$ and $K = 8$ we would have more than $10^{17}$ systems to solve).

A more efficient approach uses the near optimal solution based on the $\ell_1$-norm which is defined as:

$$\|\mathbf{x}\|_1 = \sum_{i=1}^{N} |x_i|. \tag{6.37}$$

Minimization based on the $\ell_1$norm is given by:

$$\min \|\mathbf{x}\|_1 \ \textit{subject to} \ \mathbf{y} = \mathbf{A}\mathbf{x}. \tag{6.38}$$

The $\ell_1$-norm is convex and thus the linear programming can be used for solving the above optimization problem.

In real applications, we deal with noisy signals. Thus, the previous relation should be modified to include the influence of noise. Namely, it is assumed that in the presence of noise the observations contain error:

$$\mathbf{y} = \mathbf{\Phi}\mathbf{\Psi}\mathbf{x} + \mathbf{e} = \mathbf{A}\mathbf{x} + \mathbf{e}, \tag{6.39}$$

where $\mathbf{e}$ represents the error with the energy limited by the level of noise: $\|\mathbf{e}\|_2 = \varepsilon$. The optimization problem Eq. (6.38) can now be reformulated as follows:

$$\min \|\mathbf{x}\|_1 \ \textit{subject to} \ \|\mathbf{y} - \mathbf{A}\mathbf{x}\|_2 \le \varepsilon. \tag{6.40}$$

The $\ell_2$-norm is defined as: $\|a\|_2 = \sqrt{\sum_{i=1}^{P} (a_i)^2}$, with $P$ being the total number of samples in the vector $a$.

The reconstructed signal will be consistent with the original one in the sense that $\mathbf{y} - \mathbf{A}\mathbf{x}$ will remain within the noise level.

### 6.2.3 Different Formulations of CS Reconstruction Problem

In practice, most of the numerical methods and algorithms used for signal reconstruction fall into one of the following categories: $\ell_1$ minimization, greedy algorithms or total variation (TV) minimization. Among the $\ell_1$-norm minimization problems, here we focus to a few commonly used formulations: Basis Pursuit (BP), Least Absolute Shrinkage and Selection Operator (LASSO) and Basis Pursuit Denoising (BPDN).

According to the Eqs. (6.38) and (6.40), the general system of equations that should be solved in compressive sensing approach is:

$$\begin{aligned} &\min \|\mathbf{x}\|_1 \ \ s.t. \ \mathbf{A}\mathbf{x} = \mathbf{y}, \\ \text{or,} \quad &\min \|\mathbf{x}\|_1 \ \ s.t. \ \|\mathbf{A}\mathbf{x} - \mathbf{y}\|_2 < \varepsilon \end{aligned} \tag{6.41}$$

where s.t. stands for subject to. This approach is known as BP, which was introduced in computational harmonic analysis to extract sparse signal representation from highly overcomplete dictionaries. The optimization problems are solved using some of the known solvers such as simplex and interior point methods (e.g., primal-dual interior point method).

A modification of Eq. (6.41) has been known as LASSO and it is defined as:

$$\min_x \frac{1}{2} \|\mathbf{y} - \mathbf{A}\mathbf{x}\|_2^2 \ \ s.t. \ \|\mathbf{x}\|_1 < \tau, \tag{6.42}$$

where $\tau$ is a nonnegative real parameter.

Another frequently used approach is the Basis Pursuit denoising (BPDN) which considers solving this problem in Lagrangian form:

$$\min_x \frac{1}{2} \|\mathbf{y} - \mathbf{A}\mathbf{x}\|_2^2 + \lambda \|\mathbf{x}\|_1, \tag{6.43}$$

where $\lambda > 0$ is a regularization parameter.

The most commonly used greedy algorithms are Orthogonal Matching Pursuit (OMP) and Iterative thresholding. The OMP provides a sparse solution by using an iterative procedure to approximate vector $\mathbf{y}$ as a linear combination of a few columns of $\mathbf{A}$. In each iteration, the algorithm selects the column of $\mathbf{A}$ that best correlates with the residual signal. The residual signal is obtained by subtracting the contribution of a partial signal estimate from the measurement vector.

The iterative hard thresholding algorithm starts from an initial signal estimate $\widetilde{\mathbf{x}} = 0$ and then iterates a gradient descent step followed by hard thresholding until a convergence criterion is met.

### 6.2.4 An Example of Using Compressive Sensing Principles

In order to provide a better understanding of the compressive sensing, we consider one simple example, which aims to demonstrate some of the concepts introduced in this chapter (such as vector of measurements, sensing matrices, etc.).

For the purpose of signal visualization, it is a good idea to choose the sample rate several times faster than is required by the sampling theorem. Hence, a given signal $f_x(n)$ has $N = 21$ samples and it is defined as:

$$f_x(n) = \sin(2 \cdot \pi \cdot (2/N) \cdot n) \quad \text{for } n = 0, .., 20. \tag{6.44}$$

The values of the signal samples are given in the form of vector:

$$\begin{aligned}\mathbf{f}_x = [&0 \;\; 0.5633 \;\; 0.9309 \;\; 0.9749 \;\; 0.6802 \;\; 0.149 \;\; -0.4339 \;\; -0.866 \\ &-0.9972 \;\; -0.7818 \;\; -0.2948 \;\; 0.2948 \;\; 0.7818 \;\; 0.9972 \;\; 0.866 \\ &0.4339 \;\; -0.149 \;\; -0.6802 \;\; -0.9749 \;\; -0.9309 \;\; -0.5633]\end{aligned}$$

The Fourier transform of the observed signal consists of two frequency peaks: one belonging to the positive and the other belonging to the negative frequencies. The vector with the DFT coefficients of $\mathbf{f}_x$ is denoted as $\mathbf{F}_x$ ($\mathbf{F}_x$ corresponds to $\mathbf{x}$ from the previously presented theory):

$$\mathbf{F}_x = [0\;0\;0\;0\;0\;0\;0\;0\;10.5j\;0\;0\;0\;-10.5j\;0\;0\;0\;0\;0\;0\;0\;0];$$

where parameter $j$ is used a imaginary unit. The signal $\mathbf{f}_x$ and its DFT $\mathbf{F}_x$ are given in Fig. 6.6.

Note that $\mathbf{f}_x$ is sparse in the frequency domain. Hence, we may consider the signal reconstruction based on the small set of randomly selected signal samples. For this purpose, we have to define the sensing matrix.

First we calculate the elements of inverse and direct Fourier transform matrices, denoted by $\mathbf{\Psi}$ and $\mathbf{\Psi}^{-1}$ respectively:

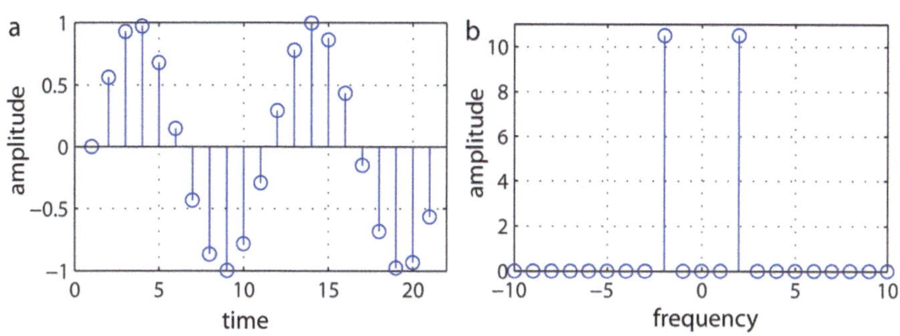


**Fig. 6.6** (**a**) The signal $\mathbf{f}_x$, (**b**) the absolute value of DFT vector $\mathbf{F}_x$

$$\boldsymbol{\Psi} = \frac{1}{21}\begin{bmatrix} 1 & 1 & 1 & 1 & \dots & 1 \\ 1 & e^{j\frac{2\pi}{21}} & e^{2j\frac{2\pi}{21}} & e^{3j\frac{2\pi}{21}} & \dots & e^{20j\frac{2\pi}{21}} \\ 1 & e^{2j\frac{2\pi}{21}} & e^{4j\frac{2\pi}{21}} & e^{6j\frac{2\pi}{21}} & \dots & e^{40j\frac{2\pi}{21}} \\ \dots & \dots & \dots & \dots & \dots & \dots \\ 1 & e^{19j\frac{2\pi}{21}} & e^{38j\frac{2\pi}{21}} & e^{57j\frac{2\pi}{21}} & \dots & e^{380j\frac{2\pi}{21}} \\ 1 & e^{20j\frac{2\pi}{21}} & e^{40j\frac{2\pi}{21}} & e^{60j\frac{2\pi}{21}} & \dots & e^{400j\frac{2\pi}{21}} \end{bmatrix}$$

$$\boldsymbol{\Psi}^{-1} = \begin{bmatrix} 1 & 1 & 1 & 1 & \dots & 1 \\ 1 & e^{-j\frac{2\pi}{21}} & e^{-2j\frac{2\pi}{21}} & e^{-3j\frac{2\pi}{21}} & \dots & e^{-20j\frac{2\pi}{21}} \\ 1 & e^{-2j\frac{2\pi}{21}} & e^{-4j\frac{2\pi}{21}} & e^{-6j\frac{2\pi}{21}} & \dots & e^{-40j\frac{2\pi}{21}} \\ \dots & \dots & \dots & \dots & \dots & \dots \\ 1 & e^{-19j\frac{2\pi}{21}} & e^{-38j\frac{2\pi}{21}} & e^{-57j\frac{2\pi}{21}} & \dots & e^{-380j\frac{2\pi}{21}} \\ 1 & e^{-20j\frac{2\pi}{21}} & e^{-40j\frac{2\pi}{21}} & e^{-60j\frac{2\pi}{21}} & \dots & e^{-400j\frac{2\pi}{21}} \end{bmatrix}$$

The matrices are of size $N \times N$. The relation between $\mathbf{f}_x$ and $\mathbf{F}_x$ is:

$$\mathbf{f}_x = \boldsymbol{\Psi}\mathbf{F}_x. \tag{6.45}$$

Now, we would like to select $M = 8$ random samples/measurements in the time domain, which will be used to reconstruct the entire signal $\mathbf{f}_x$ by using the compressive sensing approach. In other words, we should define the measurement matrix $\boldsymbol{\Phi}$ of size $M \times N$, which is used to represent a measurement vector as:

$$\mathbf{y} = \boldsymbol{\Phi}\mathbf{f}_x. \tag{6.46}$$

The measurement matrix $\boldsymbol{\Phi}$ can be defined as a random permutation matrix, and thus $\mathbf{y}$ is obtained by taking the first $M$ permuted elements of $\mathbf{f}_x$. For instance, the vector $\mathbf{y}$ can be given by:

$$\mathbf{y} = [0.7818 \ \ -0.7818 \ \ 0.8660 \ \ 0.2948 \ \ -0.9972 \ \ -0.6802 \ \ 0.6802 \ \ -0.9309],$$

where the random permutation of $N = 21$ elements is done according to:

$$perm = [13,\ 10,\ 15,\ 12,\ 9,\ 18,\ 5,\ 20,\ 16,\ 4,\ 8,\ 1,\ 2,\ 11,\ 6,\ 21,\ 17,\ 19,\ 7,\ 14,\ 3],$$

or equivalently by taking only $M = 8$ first elements we have:

$$perm(1:M) = [13,\ 10,\ 15,\ 12,\ 9,\ 18,\ 5,\ 20].$$

The $N$ points Fourier transform that corresponds to the vector $\mathbf{y}$ is obtained as:

$$\mathbf{F}_y = \mathbf{\Psi}^{-1}\mathbf{y}, \tag{6.47}$$

$$\begin{aligned}\mathbf{F}_y = [\, & (-0.0793 - 0.0997j)\ (0.0533 - 0.0739\,j)\ (-0.0642 - 0.0514\,j)\ (0.0183 + 0.0110\,j) \\ & (0.1005 + 0.0284\,j)\ (-0.0263 - 0.0178\,j)\ (0.0704 + 0.0174\,j)\ (0.0089 - 0.1007\,j) \\ & (-0.0270 + 0.2307\,j)\ (-0.0363 - 0.0171\,j)\ (-0.0365 + 0.0000\,j)\ (-0.0363 + 0.0171\,j) \\ & (-0.0270 - 0.2307\,j)\ (0.0089 + 0.1007\,j)\ (0.0704 - 0.0174\,j)\ (-0.0263 + 0.0178\,j) \\ & (0.1005 - 0.0284\,j)\ (0.0183 - 0.0110\,j)\ (-0.0642 + 0.0514\,j)\ (0.0533 + 0.0739\,j) \\ & (-0.0793 + 0.0997\,j)]\end{aligned}$$

The starting Fourier transform vector $\mathbf{F}_y$ significantly differs from $\mathbf{F}_x$ which we aim to reconstruct (Fig. 6.6). Based on equations Eqs. (6.45) and (6.46) we have:

$$\mathbf{y} = \mathbf{\Phi\Psi F}_x = \mathbf{AF}_x. \tag{6.48}$$

In analogy with the random measurements vector $\mathbf{y}$, the matrix $\mathbf{A}=\mathbf{\Phi\Psi}$ can be obtained by using the permutation of rows in $\mathbf{\Psi}$ and then selecting the first $M=8$ permuted rows. The matrix $\mathbf{A}$ for this example is:

$$\mathbf{A}_{M\times N} = \mathbf{\Phi\Psi} = \frac{1}{21}\begin{bmatrix} 1 & e^{12j\frac{2\pi}{21}} & e^{24j\frac{2\pi}{21}} & \dots & e^{240j\frac{2\pi}{21}} \\ 1 & e^{9j\frac{2\pi}{21}} & e^{18j\frac{2\pi}{21}} & \dots & e^{180j\frac{2\pi}{21}} \\ 1 & e^{14j\frac{2\pi}{21}} & e^{28j\frac{2\pi}{21}} & \dots & e^{280j\frac{2\pi}{21}} \\ 1 & e^{11j\frac{2\pi}{21}} & e^{22j\frac{2\pi}{21}} & \dots & e^{220j\frac{2\pi}{21}} \\ 1 & e^{8j\frac{2\pi}{21}} & e^{16j\frac{2\pi}{21}} & \dots & e^{160j\frac{2\pi}{21}} \\ 1 & e^{17j\frac{2\pi}{21}} & e^{34j\frac{2\pi}{21}} & \dots & e^{340j\frac{2\pi}{21}} \\ 1 & e^{4j\frac{2\pi}{21}} & e^{8j\frac{2\pi}{21}} & \dots & e^{80j\frac{2\pi}{21}} \\ 1 & e^{19j\frac{2\pi}{21}} & e^{38j\frac{2\pi}{21}} & \dots & e^{380j\frac{2\pi}{21}} \end{bmatrix}$$

Finally, we obtain the system with 8 equations and 21 unknowns.

The vector of measurements is depicted in Fig. 6.7a, where the available samples are denoted by the blue dots, while the missing samples are denoted by the red dots. The initial Fourier transform is shown in Fig. 6.7b. It is important to note that due to the missing samples, the initial Fourier transform vector is not sparse and contains noise-like components on the non-signal positions that should be zeros. The level of this kind of noise becomes higher as the number of missing samples increases. Also, note that the components on the signal frequencies do not have the true values shown earlier in Fig. 6.6.

Now assume that the signal and non-signal components differ in amplitudes, so that we can set the threshold that will separate these components in the initial

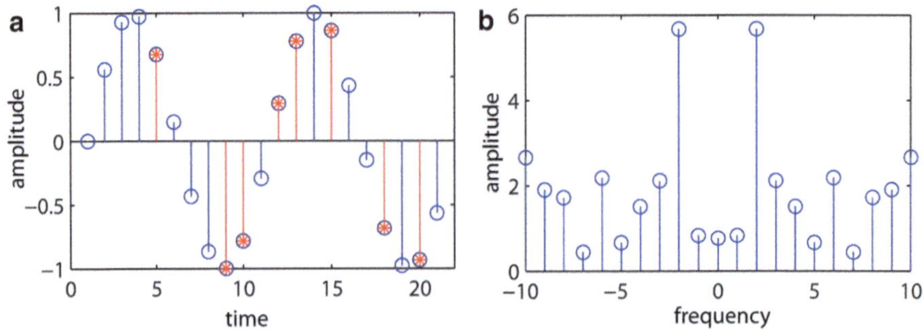


**Fig. 6.7** (**a**) The original signal and randomly selected samples denoted by the *red dots*, (**b**) $N$ point discrete Fourier transform $\mathbf{F}_y$

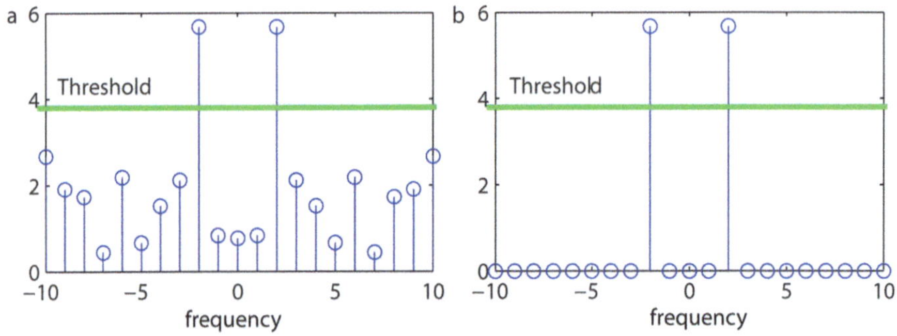


**Fig. 6.8** (**a**) The $N$ point Fourier transform $\mathbf{F}_y$, (**b**) the discrete Fourier transform of reconstructed signal. In order to center the spectrum at zero frequency in Fig. 6.7 and in this figure, the shifting operation is performed as follows: $\mathbf{F}_x = [\mathbf{F}_x(11:21)\ \ \mathbf{F}_x(1:10)]$. It means that the illustrated components positions $-2$ and 2, actually correspond to positions 19 and 2

Fourier transform vector (Fig. 6.8). The components above the threshold belongs to the signal, while the components below the threshold are non-signal components (belonging to noise) and these should be actually zeros. The threshold is derived analytically, later in this chapter.

The support of the detected $K$ signal components ($K = 2$ in our case) can be denoted as $\Omega = \{k_1, \mathrm{k}_2, \ldots, k_K\}$ and it contains the positions of components in the sparse transform vector. Thus instead of $N$ components in $\mathbf{F}_x$, we are looking for the $K$ components whose support is defined by $\Omega$.

It further means that we should select only those columns from $\mathbf{A}$ that belong to the support $\Omega$:

$$\mathbf{A}_\Omega = \mathbf{A}\{\Omega\}, \tag{6.49}$$

where matrix $\mathbf{A}_\Omega$ is of size $M \times K$.

In the considered example the support is defined by the two-component set of frequency instants:

$$\Omega = \{2, 19\}. \tag{6.50}$$

The corresponding matrix $\mathbf{A}_\Omega$ is obtained by choosing the 2nd and the 19th column of $\mathbf{A}$:

$$\mathbf{A}_\Omega = \frac{1}{21}\begin{bmatrix} e^{12j\frac{2\pi}{21}} & e^{216j\frac{2\pi}{21}} \\ e^{9j\frac{2\pi}{21}} & e^{162j\frac{2\pi}{21}} \\ e^{14j\frac{2\pi}{21}} & e^{252j\frac{2\pi}{21}} \\ e^{11j\frac{2\pi}{21}} & e^{198j\frac{2\pi}{21}} \\ e^{8j\frac{2\pi}{21}} & e^{144j\frac{2\pi}{21}} \\ e^{17j\frac{2\pi}{21}} & e^{306j\frac{2\pi}{21}} \\ e^{4j\frac{2\pi}{21}} & e^{72j\frac{2\pi}{21}} \\ e^{19j\frac{2\pi}{21}} & e^{342j\frac{2\pi}{21}} \end{bmatrix}$$

Now, the following system can be observed:

$$\mathbf{A}_\Omega \widetilde{\mathbf{F}}_x = \mathbf{y}, \tag{6.51}$$

and can be solved in the least square sense to find the exact amplitudes of the components (in vector $\widetilde{\mathbf{F}}_x$):

$$\begin{gathered} \mathbf{A}_\Omega^T \mathbf{A}_\Omega \widetilde{\mathbf{F}}_x = \mathbf{A}_\Omega^T \mathbf{y} \\ \widetilde{\mathbf{F}}_x = (\mathbf{A}_\Omega^T \mathbf{A}_\Omega)^{-1} \mathbf{A}_\Omega^T \mathbf{y} \end{gathered}. \tag{6.52}$$

Note that the resulting vector $\widetilde{\mathbf{F}}_x$ has $K$ elements and thus:

$$\begin{gathered} \mathbf{F}_x(\Omega) = \widetilde{\mathbf{F}}_x, \\ \mathbf{F}_x(\mathbf{N} \backslash \Omega) = 0, \text{ where } \mathbf{N} = \{1, \ldots, N\}. \end{gathered} \tag{6.53}$$

The signal is accurately reconstructed and it is shown in Fig. 6.9.

Note that the previously presented least square solution can be derived as follows. Define an error $\varepsilon$ and minimize the error as:

$$\varepsilon = \left\| \mathbf{y} - \mathbf{A}_\Omega \widetilde{\mathbf{F}}_x \right\|_2^2 \quad \text{and} \quad \frac{\partial \varepsilon}{\partial \widetilde{\mathbf{F}}_x} = 0, \tag{6.54}$$

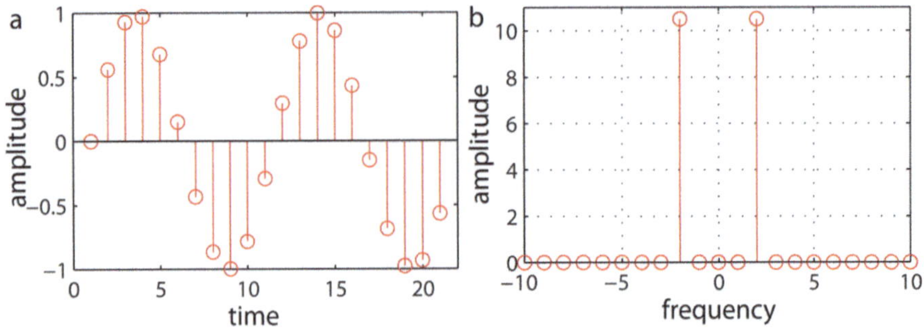


**Fig. 6.9** (**a**) The reconstructed signal (time domain samples), (**b**) the Fourier transform of the reconstructed signal

$$2\mathbf{A}_{\Omega}^{T}\left(\mathbf{y}-\mathbf{A}_{\Omega}\widetilde{\mathbf{F}}_{x}\right)=0\ \Rightarrow\ \mathbf{A}_{\Omega}^{T}\mathbf{y}=\mathbf{A}_{\Omega}^{T}\mathbf{A}_{\Omega}\widetilde{\mathbf{F}}_{x}, \tag{6.55}$$

Finally, we have $\widetilde{\mathbf{F}}_{x}=\left(\mathbf{A}_{\Omega}^{T}\mathbf{A}_{\Omega}\right)^{-1}\mathbf{A}_{\Omega}^{T}\mathbf{y}$ as in Eq. (6.52).

## 6.3 Algorithms for Signal Reconstruction

### *6.3.1 Orthogonal Matching Pursuit: OMP*

OMP is a greedy signal reconstruction algorithm that in each iteration searches for the maximum correlation between the measurements and the matrix $\mathbf{A}=\mathbf{\Phi\Psi}$. Particularly, if $\mathbf{A}$ is a partial random Fourier transform matrix, then the columns of $\mathbf{A}$ correspond to the randomly selected Fourier transform basis vector. Sometimes the matrix $\mathbf{A}$ is called dictionary, while its columns are called atoms. Thus, through the iterations it selects a certain number of columns from $\mathbf{A}$, where this number is defined by a given number of iterations.

**Algorithm: Orthogonal Matching Pursuit**
**Input**:

- Transform matrix $\mathbf{\Psi}$, Measurement matrix $\mathbf{\Phi}$
- Matrix $\mathbf{A}$: $\mathbf{A}=\mathbf{\Phi\Psi}$ (usually $\mathbf{A}$ is obtained as a random partial Fourier transform matrix: $\mathbf{A}=\mathbf{\Psi}\{\vartheta\}$, $\vartheta=\left\{\psi_{n_1},\psi_{n_2},\ldots,\psi_{n_M}\right\}$)
- Atoms are columns of $\mathbf{A}=\left\{A_j|j=1,\ \ldots,M\right\}=\left\{\psi_{n_1},\psi_{n_2},\ldots,\psi_{n_M}\right\}$
- Measurement vector $\mathbf{y}=\mathbf{\Phi f}$

**Output**:

1: $\mathbf{r}_0\leftarrow\mathbf{y}$ (*Set initial residual*)
2: $\Omega_0\leftarrow\varnothing$ (*Set initial indices*)

$\mathbf{\Theta}_0 \leftarrow [\,]$ *(Set matrix of chosen atoms)*
***for*** $i = 1, \ldots, K$
    $\omega_i \leftarrow \underset{j=1,\ldots,M}{\operatorname{argmax}} |\langle \mathbf{r}_{i-1}, A_j \rangle|$ *(Maximum correlation column)*
    $\Omega_i \leftarrow \Omega_{i-1} \cup \omega_i$ *(Update set of indices)*
    $\mathbf{\Theta}_i \leftarrow [\mathbf{\Theta}_{i-1}\ A_{\omega_i}]$ *(Update set of chosen atoms)*
    $\mathbf{x}_i = \arg\min_{\mathbf{x}} \| \mathbf{r}_{i-1} - \mathbf{\Theta}_i \mathbf{x} \|_2^2$ *(Solve the least squares)*
    $\mathbf{a}_i \leftarrow \mathbf{\Theta}_i \mathbf{x}_i$ *(New data approximation)*
    $\mathbf{r}_i \leftarrow \mathbf{y} - \mathbf{a}_i$ *(Update the residual)*
***end for***
**return** $\mathbf{x}_K$, $\mathbf{a}_K$, $\mathbf{r}_K$, $\Omega_K$

In the first step of the algorithm, the initial residual vector $\mathbf{r}_0$ is equal to the measurements vector $\mathbf{y}$. The atom from dictionary $\mathbf{A}$ having maximum correlation with residual is determined next. This atom (a column from $\mathbf{A}$) is added to the matrix of chosen atoms $\mathbf{\Theta}_i$ (initially this matrix is empty). The least square optimization is performed afterwards in the subspace spanned by all previously picked columns to determine a new signal estimate. The new approximation of the data and the new residual are calculated by removing the contribution of already chosen atoms. The final signal estimate has nonzero values at the positions given in $\Omega_K$. The values of these nonzero components are equals to the components in $\mathbf{x}_K$.

### 6.3.2 Adaptive Gradient Based Signal Reconstruction Method

This recently defined algorithm belongs to the group of gradient-based minimization approaches. The missing samples are considered as variables with the zero initial value (minimum energy solution). Then the algorithm performs an iterative procedure, where in each iteration the initial state of missing samples is changed toward their exact values. Namely, the missing samples are altered iteratively for value $+\Delta$ and $-\Delta$, where $\Delta$ is an appropriately chosen step. Then, we need to measure the influence of the changes $\pm\,\Delta$ (for each missing sample separately) and to determine the value of the gradient which is used to update the values of missing samples. Here, the measure of concentration in the transform domain representation is used to determine the value of the gradient.

A measure of concentration for a certain vector $\mathbf{x}$ of transform coefficients is defined by:

$$\eta(\mathbf{x}) = \sum_{k=0}^{N-1} |x(k)|. \tag{6.56}$$

Note that the transform domain vector $\mathbf{x}$ can be DFT, DCT, Hermite transform vector, etc. Through the iterations, the missing samples are changed to decrease the values of sparsity measure, in order to achieve the minimum of the $\ell_1$-norm based sparsity measure.

Let us consider now how the algorithm works. Assume that the initial set of $M$ measurements is modified by embedding zero values on the positions of missing samples:

$$y(n) = \begin{cases} f(n), & \text{for } n \in \mathbf{N}_M \\ 0, & \text{for } n \in \mathbf{N} \backslash \mathbf{N}_M, \end{cases} \tag{6.57}$$

where $f$ is the original signal, the set of available samples is $\mathbf{N}_M$, while the total set of samples is $\mathbf{N} = \{1,2,\ldots,N\}$. In the $i$-th iteration, the values of vector $\mathbf{y}$ on the positions of missing samples are changed as follows:

$$\begin{aligned} y^+(n) &= y_i(n) + \Delta\delta(n - n_s), \\ y^-(n) &= y_i(n) - \Delta\delta(n - n_s). \end{aligned} \tag{6.58}$$

The initial set up assumes $y_i(n) = y(n)$ for $i = 0$ and $n \in \mathbf{N}$. The missing samples positions are denoted as $n_s \in \mathbf{N} \backslash \mathbf{N}_M$, while $\delta$ is the Kronecker delta function. The initial value of the step is calculated as follows:

$$\Delta = \max_n \{|y(n)|\}. \tag{6.59}$$

Next, it is necessary to calculate the transform domain representations of both vectors $\mathbf{y}^+$ and $\mathbf{y}^-$ which will be used to determine the concentration measure in the transform domain and therefore to determine the corresponding gradient. The transform domain vectors corresponding to $\mathbf{y}^+$ and $\mathbf{y}^-$ are given by:

$$\begin{aligned} \mathbf{x}^+ &= \boldsymbol{\Psi}\mathbf{y}^+, \\ \mathbf{x}^- &= \boldsymbol{\Psi}\mathbf{y}^-, \end{aligned} \tag{6.60}$$

respectively, while $\boldsymbol{\Psi}$ is a transform matrix. Furthermore, the gradient vector is calculated as follows:

$$G^i(n_s) = \frac{\eta(\mathbf{x}^+) - \eta(\mathbf{x}^-)}{N}, \quad n_s \in \mathbf{N} \backslash \mathbf{N}_M, \tag{6.61}$$

where the concentration measure $\eta$ is defined by Eq. (6.56).

Note that, the gradient vector on the positions of available samples is equal to zero: $G^i(n) = 0$ for $n \in \mathbf{N}_M$.

An iteration is completed when all values of the gradient vector for the positions of missing samples $n_s \in \mathbf{N} \backslash \mathbf{N}_M$ are calculated. The measurement vector is then updated as:

$$y^{i+1}(n) = y^i(n) - G^i(n). \tag{6.62}$$

The procedure is repeated through the iterations. The missing values will converge toward the exact values, producing the minimal concentration measure in a chosen transformation domain.

Note that a difference of concentration measures is used to estimate the gradient. Once the gradient becomes constant, it means that the optimal point is reached. However, in practice, instead of having constant gradient at the optimal point, there will be some oscillations of the gradient direction within consecutive iterations. In this situation, we can reduce the step value $\Delta$ (for example: $\Delta/\sqrt{10}$) in order to approach to the stationary zone.

The oscillations of the gradient vector can be detected by measuring an angle between successive gradient vectors $\mathbf{G}^{i-1}$ and $\mathbf{G}^i$. The angle between two gradient vectors corresponding to iteration $(i-1)$ and $i$, can be calculated as follows:

$$\beta = \arccos \frac{\langle \mathbf{G}^{i-1}\mathbf{G}^i \rangle}{\|\mathbf{G}^{i-1}\|_2^2 \|\mathbf{G}^i\|_2^2}, \tag{6.63}$$

where a scalar (dot) product of two vectors is calculated in the brackets $\langle\rangle$. When the angle $\beta$ becomes higher than 170°, we can assume that the vector oscillation state is achieved, and the value of $\Delta$ should be decreased.

Various stopping criteria can be used to stop the iteration process. For example, when the value of $\Delta$ becomes sufficiently small, we might say that the algorithm achieved sufficient precision. Also, we can calculate the ratio between the reconstruction error and the signal in the current iteration as follows:

$$\Xi = 10\log_{10} \frac{\sum_{n \in \mathbf{N} \backslash \mathbf{N}_M} |y^p(n) - y^i(n)|^2}{\sum_{n \in \mathbf{N} \backslash \mathbf{N}_M} |y^i(n)|^2}, \tag{6.64}$$

where $y^p(n)$ is the reconstructed signal before decreasing $\Delta$, while $y^i(n)$ is the reconstructed signal in the current iteration. If $\Xi$ is larger than the required precision, the value of $\Delta$ should be decreased.

In the sequel, the special case of the gradient-based algorithm using the DFT is summarized.

**Algorithm: Adaptive Gradient-Based Signal Reconstruction**
**Input**:

- Available samples position $\mathbf{N}_M = \{n_1, n_2, \ldots n_M\}$
- Missing samples position $\mathbf{N}\backslash\mathbf{N}_M$, where $\mathbf{N} = \{1, 2, \ldots, N\}$
- Available samples $f(n), n \in \mathbf{N}_M$
- Measurement vector: $y(n) = \begin{cases} f(n), & \text{for } n \in \mathbf{N}_M \\ 0, & \text{for } n \in \mathbf{N} \backslash \mathbf{N}_M \end{cases}$

**Output**:

Set $i$=0 and $y^{(0)}(n) \leftarrow y(n)$
Set $\Delta \leftarrow \max\left|y^{(0)}(n)\right|$
**repeat**
  Set $y_p(n) = y^{(i)}(n)$
  **repeat**
    $i \leftarrow i+1$
    **for** $n_s \leftarrow 0$ to $N-1$ **do**
      **if** $n_s \in \mathbf{N} \setminus \mathbf{N}_M$ **then**
        $x^{+}(k) \leftarrow \mathrm{DFT}\left\{y^{(i)}(n) + \Delta\delta(n-n_s)\right\}$
        $x^{-}(k) \leftarrow \mathrm{DFT}\left\{y^{(i)}(n) - \Delta\delta(n-n_s)\right\}$
        $G^{(i)}(n_s) \leftarrow \frac{1}{N}\sum_{k=0}^{N-1}\left|x^{+}(k)\right| - \left|x^{-}(k)\right|$
      **else**
        $G^{(i)}(n_s) \leftarrow 0$
      **end if**
      $y^{(i+1)}(n_s) \leftarrow y^{(i)}(n_s) - G^{(i)}(n_s)$
    **end for**
    $\beta_i = \arccos \frac{\left\langle \mathbf{G}^{i-1}\mathbf{G}^{i}\right\rangle}{\left\|\mathbf{G}^{i-1}\right\|_2^2 \left\|\mathbf{G}^{i}\right\|_2^2}$
  **until** $\beta_i < 170°$
  $\Delta \leftarrow \Delta / \sqrt{10}$
  $\Xi = 10\log_{10} \frac{\sum_{n \in \mathbf{N}\setminus\mathbf{N}_M} \left|y_p(n) - y^{(i)}(n)\right|^2}{\sum_{n \in \mathbf{N}\setminus\mathbf{N}_M} \left|y^{(i)}(n)\right|^2}$
**until** $\Xi < \Xi_{max}$
**return** $y^{(i)}(n)$

**return**: reconstructed signal $y^{(i)}(n)$

### 6.3.3 Primal-Dual Interior Point Method

Let us briefly consider the primal-dual interior point method, which has been widely used in literature. The optimization problem:

$$\min\|x\|_1 \ s.t. \ Ax = y, \tag{6.65}$$

can be recast as:

$$\min_{u} \sum u \quad s.t. \;\; Ax = y,\; f_{u_1} = x - u,\; f_{u_2} = -x - u, \tag{6.66}$$

Generally, the minimization problem can be observed by forming the Lagrangian:

$$\Lambda(x, u, v, \lambda_{u_1}, \lambda_{u_2}) = f(u) + v(Ax - y) + \lambda_{u_1} f_{u_1} + \lambda_{u_2} f_{u_2}. \tag{6.67}$$

Finding its first derivatives in terms of $x$, $u$, $v$, $\lambda_{u_1}$ and $\lambda_{u_2}$ the following relations are obtained:

$$R^{u}_{dual} = \mathbf{1} - \lambda_{u_1} - \lambda_{u_2},\;\; R^{x}_{dual} = \lambda_{u_1} - \lambda_{u_2} + A^T v,\;\; R^{v}_{prim} = Ax - y, \tag{6.68}$$

$$R^{\lambda u_1}_{cent} = \lambda_{u_1} f_{u_1} + \frac{1}{\tau},\;\; R^{\lambda u_2}_{cent} = \lambda_{u_2} f_{u_2} + \frac{1}{\tau}. \tag{6.69}$$

Note that beside $A$ and $b$ which are known, we should initialize the following variables: $x = x_0$, $u = u_0$ (e.g., which is obtained by using $x_0$), $\lambda_{u_1}$ and $\lambda_{u_2}$, $v = -A(\lambda_{u_1} - \lambda_{u_2})$ and $\tau$.

In order to compute the Newton steps, the following system of equations is solved:

$$\begin{aligned}
&\frac{\partial R^{u}_{dual}}{\partial x}\Delta x + \frac{\partial R^{u}_{dual}}{\partial u}\Delta u + \frac{\partial R^{u}_{dual}}{\partial v}\Delta v = -R^{u}_{dual},\\
&\frac{\partial R^{x}_{dual}}{\partial x}\Delta x + \frac{\partial R^{x}_{dual}}{\partial u}\Delta u + \frac{\partial R^{x}_{dual}}{\partial v}\Delta v = -R^{x}_{dual},\\
&\frac{\partial R^{v}_{prim}}{\partial x}\Delta x + \frac{\partial R^{v}_{prim}}{\partial u}\Delta u + \frac{\partial R^{v}_{prim}}{\partial v}\Delta v = -R^{v}_{prim},
\end{aligned} \tag{6.70}$$

$$\begin{aligned}
&\frac{\partial R^{\lambda u_1}_{cent}}{\partial x}\Delta x + \frac{\partial R^{\lambda u_1}_{cent}}{\partial u}\Delta u + \frac{\partial R^{\lambda u_1}_{cent}}{\partial \lambda_{u_1}}\Delta \lambda_{u_1} = -\partial R^{\lambda u_1}_{cent},\\
&\frac{\partial R^{\lambda u_2}_{cent}}{\partial x}\Delta x + \frac{\partial R^{\lambda u_2}_{cent}}{\partial u}\Delta u + \frac{\partial R^{\lambda u_2}_{cent}}{\partial \lambda_{u_2}}\Delta \lambda_{u_2} = -\partial R^{\lambda u_2}_{cent}.
\end{aligned} \tag{6.71}$$

From Eq. (6.70), we have:

$$\begin{gathered}
\left(-\frac{1}{\tau f^2_{u_1}} + \frac{1}{\tau f^2_{u_2}}\right)\Delta x + \left(\frac{1}{\tau f^2_{u_1}} + \frac{1}{\tau f^2_{u_2}}\right)\Delta u = -\mathbf{1} - \frac{1}{\tau}\left(\frac{1}{f_{u_1}} + \frac{1}{f_{u_2}}\right)\\
\left(\frac{1}{\tau f^2_{u_1}} + \frac{1}{\tau f^2_{u_2}}\right)\Delta x + \left(-\frac{1}{\tau f^2_{u_1}} + \frac{1}{\tau f^2_{u_2}}\right)\Delta u + A^T \Delta v = \frac{1}{\tau}\left(\frac{1}{f_{u_1}} - \frac{1}{f_{u_2}}\right) - A^T v\\
A\Delta x = -Ax + b
\end{gathered}$$

After calculating $\Delta x$, $\Delta u$, and $\Delta v$ (e.g., by using *linsolve* in Matlab) we compute $\Delta\lambda_{u_1}$ and $\Delta\lambda_{u_2}$ by using:

$$\Delta\lambda_{u_1} = \lambda_{u_1} f_{u_1}^{-1}(-\Delta x + \Delta u) - \lambda_{u_1} - \frac{1}{\tau f_{u_1}},$$
$$\Delta\lambda_{u_2} = \lambda_{u_2} f_{u_2}^{-1}(\Delta x + \Delta u) - \lambda_{u_2} - \frac{1}{\tau f_{u_2}},$$

which are derived from Eq. (6.71). The Newton step actually represents the step direction. In order to update the values of variables for the next iteration:

$$x = x + s\Delta x,\;\; u = u + s\Delta u,\;\; v = v + s\Delta v,\; \lambda_{u_1} = \lambda_{u_1} + s\Delta\lambda_{u_1},\; \lambda_{u_2} = \lambda_{u_2} + s\Delta\lambda_{u_2},$$

the step length $s$ should be calculated. For this purpose, the backtracking line search can be applied. The general backtracking method is explained in the sequel.

*Backtracking method*
Let assume that $f(x)$ is the function that should be minimized. One solution would be to use the step length $s$ ($s \geq 0$), which minimizes the function $f(x_k + s\Delta x_k)$:

$$\underset{s\geq 0}{\operatorname{argmin}}\, f(x_k + s\Delta x_k).$$

This method can be computationally demanding, and thus the *backtracking line search* has been used:

$$\begin{aligned}&for\ given\ s\ (s>0)\\&while\ f(x_{k+1}) > f(x_k) + \alpha\cdot s\cdot f'(x_k)\cdot\Delta x_k\\&s = \beta\cdot s\\&end\end{aligned}$$

The constants $\alpha$ and $\beta$ can take values in the range $\alpha \in (0, 0.5), \beta \in (0, 1)$. Since we have five variables that should be updated, the condition in *while* loop can be modified as follows:

$$\|r_{k+1}\|_2 > (1 - \alpha\cdot s)\|r_k\|_2,$$

where $r$ is a vector that contains the elements of $R_{dual}^{u}$, $R_{dual}^{x}$, $R_{dual}^{v}$, $R_{cent}^{\lambda u_1}$, $R_{cent}^{\lambda u_2}$.

## 6.4 Analysis of Missing Samples in the Fourier Transform Domain

Observe a signal that consists of $K$ sinusoidal components in the form:

$$f(n) = \sum_{i=1}^{K} a_i e^{j2\pi k_i n/N} \tag{6.72}$$

where $a_i$ and $k_i$ denote amplitudes and frequencies of the $i$-th signal components, respectively. The DFT of such signal can be written as:

$$F(k) = N \cdot \sum_{n=1}^{N} \sum_{i=1}^{K} a_i e^{-j2\pi(k-k_i)n/N}. \tag{6.73}$$

In the compressive sensing scenario we are dealing with just a small subset of samples from $f(n)$ taken at the random positions defined by the following set:

$$\mathbf{N}_M = \{n_1, n_2, \ \ldots, \ n_M\} \subset \mathbf{N} = \{1, 2, \ \ldots \ , N\}. \tag{6.74}$$

Therefore, $\mathbf{N}_M$ represents the positions of measurements. Following the definition of the DFT in Eq. (6.73), let us observe the following sets:

$$\mathbf{f}^{\dagger} = \left\{ f^{\dagger}(n) \mid f^{\dagger}(n) = \sum_{i=1}^{K} a_i e^{-j2\pi(k-k_i)n/N}, \ n = 1, 2, \ldots, N \right\}. \tag{6.75}$$

In other words, instead of signal samples, we consider a product of samples and the Fourier basis functions. Accordingly, for a set of available samples on the positions defined by Eq. (6.74), the vector of $M$ measurements out of $N$ from the full vector $\mathbf{f}^{\dagger}$ can be written as:

$$\mathbf{y} = \left\{ y(m) \mid \ y(m) = \sum_{i=1}^{K} a_i e^{-j2\pi(k-k_i)n_m/N}, \ n_m \in \mathbf{N}_M \right\}. \tag{6.76}$$

The Fourier transform of randomly chosen set of samples (initial Fourier transform) is not sparse, although the Fourier transform of the original (full) signal is sparse. Namely, missing samples in the time domain will cause a noisy spectral representation, which will influence the signal sparsity. By analyzing the statistical properties of the Fourier coefficients (on the positions of signal and non-signal components), we can characterize the variance of the noise.

***Monocomponent signal***: Let us start the analysis by the simplest signal case having only one component $a_1 = 1$, $k_i = k_1$. The sets of samples given by Eqs. (6.75) and (6.76) become:

$$\begin{aligned} f^{\dagger}(n) &= e^{-j2\pi(k-k_1)n/N}, \; n \in \mathbf{N} \\ y(m) &= e^{-j2\pi(k-k_1)n_m/N}, \; n_m \in \mathbf{N}_M \end{aligned}, \tag{6.77}$$

where it is assumed that $f^{\dagger}(1) + f^{\dagger}(2) + \ldots + f^{\dagger}(N) = 0$ for $k \neq k_1$ since,

$$E\left\{\sum_{n=1}^{N} e^{-j2\pi(k-k_1)n/N}\right\} = E\left\{\frac{1-\left(e^{-j2\pi(k-k_1)/N}\right)^N}{1-e^{-j2\pi(k-k_1)/N}}\right\} = 0,$$

i.e., all samples are not statistically independent for $k \neq k_1$. The Fourier transform of measurements vector $\mathbf{y}$ can be written as follows:

$$Y(k) = \sum\nolimits_{m=1}^{M} y(m) = \sum\nolimits_{n=1}^{N} \{f^{\dagger}(n) - \varepsilon(n)\}, \tag{6.78}$$

where the noise can be modelled as:

$$\varepsilon(n) = \begin{cases} e^{-j2\pi(k-k_1)n/N}, & n \in \{\mathbf{N} \backslash \mathbf{N}_M\} \\ 0, & n \in \mathbf{N}_M \end{cases} \tag{6.79}$$

In other words, on the positions of available samples $\mathbf{N}_M$, we will have $y(m)$, while on the remaining positions we will have zeros. It can be observed that the expected value of the Fourier coefficients on the position of signal component $k = k_1$ is:

$$E\{Y_{k=k_1}\} = E\left\{\sum_{m=1}^{M} e^{-j2\pi(k_1-k_1)n_m/N}\right\}_{n_m \in \mathbf{N}_M} = E\left\{\sum_{m=1}^{M} e^0\right\} = M.$$

On the positions of non-signal components $k - k_1 = k_x$, we have:

$$E\{Y_{k \neq k_1,\; k-k_1=k_x}\} = 0,$$

since $E\{y(m)\} = 0$ with high probability ($y(m)$ takes values from the original sinusoid at random positions and the expectation for different realizations will tend to zero with high probability).

Now, the variance of noise in $F$ can be calculated as follows[3]:

$$\begin{aligned}\sigma^2(Y_{k\neq k_1}) &= E\left\{[y(1)+\ldots+y(M)]\cdot[y(1)+\ldots+y(M)]^*\right\}\\ &= M\cdot E\{y(n)y^*(n)\}+M(M-1)\cdot E\{y(n)y^*(m)\}\\ &= M\cdot 1+M(M-1)\frac{-1}{N-1}\end{aligned}$$

where $(\cdot)^*$ denotes the complex conjugate. Therefore, we obtain the variance in the form:

$$\sigma^2(Y_{k\neq k_1}) = M\frac{N-M}{N-1}$$

[3] A detailed derivation of variance at non-signal positions is given below:

$$\begin{aligned}\sigma^2(Y_{k\neq k_1}) &= E\left\{[y(1)+\ldots+y(M)]\cdot[y(1)+\ldots+y(M)]^*\right\}\\ &= E\left\{|y(1)|^2\right\}+\ldots+E\left\{|y(M)|^2\right\}+E\left\{y(1)y(2)^*\right\}+..+E\left\{y(1)y(M)^*\right\}+\ldots\\ &\quad +E\left\{y(M)y(1)^*\right\}+\ldots+E\left\{y(M)y(M-1)^*\right\}\end{aligned}$$

(1) Since $|y(m)| = \left|e^{-j2\pi(k-k_1)n_m/N}\right| = 1 \Rightarrow E\left\{|y(1)|^2\right\}+\ldots+E\left\{|y(M)|^2\right\} = M\cdot 1 = M$

(2) Assuming that: $f^\dagger(1)+f^\dagger(2)+\ldots+f^\dagger(N)=0$ for $k\neq k_1$

$$\Rightarrow f^\dagger(i)\left(f^\dagger(1)+\ldots+f^\dagger(N)\right)^* = 0 \Rightarrow \sum_{j=1}^{N} E\left\{f^\dagger(i)\left(f^\dagger(j)\right)^*\right\} = 0$$

(3) Set: $E\left\{f^\dagger(i)f^{\dagger*}(j)\right\} = B$ for $i\neq j$,

while from (1) we have: $E\left\{|y(i)|^2\right\} = 1 \; for \; i=j$.

Therefore: $\sum_{j=1}^{N} E\left\{f^\dagger(i)\left(f^\dagger(j)\right)^*\right\} = (N-1)B+1 = 0 \;\Rightarrow B = -\frac{1}{N-1}$

(4) Since $y(i) = f^\dagger(i)$ for available samples $i\in \mathbf{N}_M$, we have:

$$E\{y(i)y^*(j)\} = E\left\{f^\dagger(i)f^{\dagger*}(j)\right\} = B = -\frac{1}{N-1}.$$

Using (1) and (4), it can be concluded that:

$$\sigma^2(Y_{k\neq k_1}) = M+(M-1)B = M+M(M-1)\left(-\frac{1}{N-1}\right) = M\frac{N-M}{N-1}$$

Here, we have applied the following equalities:

$$\sum_{m=1}^{N} E\left\{f^{\dagger}(n)f^{\dagger^*}(m)\right\} = 0, \tag{6.80}$$

$$E\{y(n)y^*(n)\} = E\left\{f^{\dagger}(n)f^{\dagger^*}(n)\right\} = 1, \tag{6.81}$$

$$E\{y(n)y^*(m)\}_{n\neq m} = E\left\{f^{\dagger}(n)f^{\dagger^*}(m)\right\}_{n\neq m} = \frac{-1}{N-1}. \tag{6.82}$$

According to the previous analysis, we can conclude that in the case of a sparse $K$-component signal, the variance of noise that appears in the spectral domain as a consequence of missing samples is:

$$\sigma_{MS}^2 = \sigma^2\{Y_{k\neq k_i}\} = M\frac{N-M}{N-1}\sum_{i=1}^{K} a_i^2. \tag{6.83}$$

The variance of noise produced in the spectral domain depends on the number of missing samples $(N-M)$ (or alternatively on the number of available samples $M$). Observe that for $M=N$ the variance of spectral noise will be zero.

However, for $M \ll N$, we have $\frac{N-M}{N-1} \to 1$, or equivalently $\sigma^2{}_{MS} \approx M\sum_{i=1}^{K} a_i{}^2$. Thus, for low values of $M$, the noise level exceeds the values of some (or all) signal component.

The DFT variance at the positions of the $i$-th signal components is:

$$\sigma_i^2 = \sigma^2\{Y_{k=k_i}\} = M\frac{N-M}{N-1}\sum_{l=1,\, l\neq i}^{K} a_l^2. \tag{6.84}$$

In order to detect the signal components from the noisy DFT representation, it would be important to define the probability of having all signals components above the noise. In that sense, let us observe the two types of DFT values:

(a) DFT values at the position of signal components (signal positions),
(b) DFT values at the positions of non-signal components (noise-alone positions).

The error during the components detection (false components detection) may appear in the situation when the DFT values at noise-alone positions are higher than the DFT value of the signal component. Hence, with a certain probability, we need to assure the case when all $(N-K)$ noise-alone components are below the signal components values in the DFT domain.

The probability that all $(N-K)$ non-signal components are below a certain threshold value $T$ is[4]:

$$P(T) = \left(1 - \exp\left(-\frac{T^2}{\sigma_{MS}{}^2}\right)\right)^{N-K}. \tag{6.85}$$

Accordingly, we can define the probability of error (false components detection) as follows:

$$P_{err}(T) = 1 - P(T). \tag{6.86}$$

In practical applications, the approximate expression for probability $P$ can be used. Namely, assume that the DFT of the $i$-th signal component is not random but equal to $Ma_i$ (positioned at the mean value of the signals DFT). It means that the influence of noise to amplitude is symmetric and equally increases and decreases the DFT signal value. Then the approximate expression is obtained in the form:

$$P_i \cong \left(1 - \exp\left(-\frac{M^2 a_i{}^2}{\sigma_{MS}{}^2}\right)\right)^{N-K}, \tag{6.87}$$

where $i = 1,\ldots,K$. For a given $P_i$, we can calculate the number of available samples $M$, which will ensure detection of all signal components.

### 6.4.1 Threshold Based Single Iteration Algorithm

The threshold based signal reconstruction algorithm is based on the idea of separating signal components from spectral noise. Consequently, we first assume a fixed

[4] According to the central limit theorem, the real and imaginary parts of non-signal DFT values can be described by Gaussian distribution: $N(0,\sigma_{MS}^2/2)$.

The probability density function for the absolute DFT values at non-signal positions is Rayleigh distributed: $p(\xi) = \frac{2\xi}{\sigma_{MS}^2} e^{-\xi^2/\sigma_{MS}^2}$, $\xi \geq 0$.

Consequently, the DFT coefficients at non-signal (noise-alone) position takes a value greater than $T$, with the following probability: $p(T) = \int\limits_T^\infty \frac{2\xi}{\sigma_{MS}^2} e^{-\xi^2/\sigma_{MS}^2} = e^{-T^2/\sigma_{MS}^2}$

Then, the probability that only one non-signal component in DFT is below threshold $T$ is:

$Q(T) = 1 - p(T) = 1 - \int\limits_T^\infty \frac{2\zeta}{\sigma^2} e^{-\zeta^2/\sigma_{MS}{}^2} d\zeta = 1 - e^{-T^2/\sigma_{MS}{}^2}$.

Consequently, when all $N$-$K$ non-signal components are below $T$ we have: $P(T) = (1-Q(T))^{N-K}$, which is given by (6.85).

value for the probability $P(T)$ defined by Eq. (6.85) (for example $P(T) = 0.99$) in order to calculate threshold according to:

$$T = \sqrt{-\sigma_{MS}^2 \log\left(1 - P(T)^{\frac{1}{N-K}}\right)} \approx \sqrt{-\sigma_{MS}^2 \log\left(1 - P(T)^{\frac{1}{N}}\right)}, \tag{6.88}$$

where the sparsity assumption assures that the number of components $K$ is $K << N$, so that $K$ can be neglected in Eq. (6.88). Now based on the derived threshold we can define a blind signal reconstruction procedure, which does not require a priori knowledge of sparsity level.

In the case that all signal components are above the noise level in DFT, the component detection and reconstruction is achieved by using a simple single-iteration reconstruction algorithm summarized below.

**Algorithm: Automated Single-Pass Solution**
**Input**:

- Transform matrix $\mathbf{\Psi}$, Measurement matrix $\mathbf{\Phi}$
- $\mathbf{N}_M = \{n_1, n_2, \ldots, n_M\}$, $M$—number of available samples, $N$—original signal length
- Matrix $\mathbf{A}$: $\mathbf{A} = \mathbf{\Phi\Psi}$ ($\mathbf{A}$ is obtained as random partial (inverse) Fourier transform: $\mathbf{A} = \mathbf{\Phi\Psi} = \mathbf{\Psi}\{\vartheta\}$, $\vartheta = \{\psi_{n_1}, \psi_{n_2}, \ldots, \psi_{n_M}\}$)
- Measurement vector $\mathbf{y} = \mathbf{\Phi f}$

**Output**:

1: $P \leftarrow 0.99$ (*Set desired probability P*)

2: $\sigma_{MS}^2 \leftarrow M\frac{N-M}{N-1}\sum_{i=1}^{M}\frac{|y(m)|^2}{M}$ (*Calculate variance*)

3: $T \leftarrow \sqrt{-\sigma_{MS}^2 \log\left(1 - P(T)^{\frac{1}{N}}\right)}$ (*Calculate threshold*)

4: $\mathbf{X_0} \leftarrow \mathbf{y}(\mathbf{\Phi\Psi}^{-1})$ (*Calculate initial DFT of* $\mathbf{y}$)

5: $\mathbf{k} \leftarrow \arg\{|\mathbf{X}_0| > \frac{T}{N}\}$ (*Find positions of components in* $\mathbf{X}$ *higher than normalized threshold T*)

6: $\mathbf{A_{cs}} \leftarrow \mathbf{A}(\mathbf{k})$ (*Form CS matrix by using only* $\mathbf{k}$ *columns from* $\mathbf{A}$)

7: $\tilde{\mathbf{X}} = \left(\mathbf{A}_{cs}{}^{T}\mathbf{A}_{cs}\right)^{-1}\mathbf{A}_{cs}{}^{T}\mathbf{y}$

**return** $\tilde{\mathbf{X}}$, $\mathbf{k}$

*The resulting vector* $\mathbf{X}$ *will contain the values* $\tilde{\mathbf{X}}$ *at positions* $\mathbf{k}$, *while the rest of the DFT coefficients are zero.*

First we calculate the variance $\sigma_{MS}^2$ for a known number of available samples $M$ and total number of samples $N$. Since, the signal amplitudes may not be known in advance, we use:

$$\sum_{i=1}^{K} a_i{}^2 = \sum_{i=1}^{M} \frac{y(m)}{M},$$

to obtain the variance from (6.83). Then the threshold is calculated in step 3 of the algorithm and the initial Fourier transform $\mathbf{X_0}$ with $N$ elements (step 4), that corresponds to the available measurements $\mathbf{y}$. The vector $\mathbf{k}$ in step 5 will contain all positions (frequencies) of components in DFT vector $\mathbf{X_0}$ that are above the normalized threshold. The CS matrix $\mathbf{A}_{cs}$ is obtained from the matrix $\mathbf{A}$, using columns that correspond to the frequencies $\mathbf{k}$. Note that $\mathbf{A}$ is primarily obtained from the inverse DFT matrix by using only rows that correspond to $M$ available measurements. Generally, it is assumed that $K = length(\mathbf{k}) < M$. The exact DFT values at positions $\mathbf{k}$ are obtained by solving the system in step 7 in the least square sense.

### 6.4.2 Approximate Error Probability and the Optimal Number of Available Measurements

According to the approximate expression Eq. (6.87), we can write the error probability as follows:

$$P_{err}^{i} = 1 - P_i \cong 1 - \left(1 - \exp\left(-\frac{M^2 a_i^2}{\sigma_{MS}^2}\right)\right)^{N-K}, \tag{6.89}$$

which can be used as a rough approximation. Since the DFT amplitudes lower than $Ma_i$ contribute more to the error than those above $Ma_i$, a more precise approximation can be obtained by correcting the mean value $Ma_i$ for the value of one standard deviation[5]:

$$P_{err}^{i} \cong 1 - \left(1 - \exp\left(-\frac{(Ma_i - \sigma_i)^2}{\sigma_{MS}^2}\right)\right)^{N-K}, \tag{6.90}$$

where $\sigma_i^2 = \frac{M(N-M)}{(N-1)} \sum_{l=1, l\neq i}^{K} a_l^2$ is the variance at a frequency point of the observed signal component with the expected amplitude $a_i$.

Based on the previous analysis, we can define an optimal number of available samples $M$ that would allow detecting all signal components. For a fixed (desired)

[5] Note that real and imaginary parts of the DFT values at signal component position can be described by Gaussian distribution: $N(Ma_i, \sigma_i^2/2)$ and $N(0, \sigma_i^2/2)$, respectively. $\sigma_i^2$ is defined for a missing amplitude $a_i$, and real-valued $a_i$ is assumed without loss of generality. The pdf for the absolute DFT values at the position of the $i$-th signal component, is Rice-distributed: $p(\xi) = \frac{2\xi}{\sigma_i^2} e^{-(\xi^2 + M^2 a_i^2)/\sigma_i^2} I_0(Ma_i 2\xi/\sigma_i^2)$, $\xi \geq 0$, where $I_0$ is zero order Bessel function. The mean value of Rice distributed absolute DFT values can be approximated as $Ma_i$ and the variance is $\sigma_i^2$.

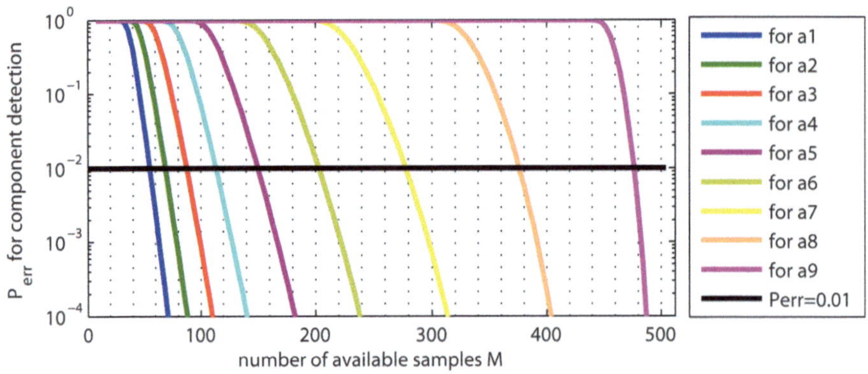


**Fig. 6.10** Probability of the component misdetection in terms of the number of available measurements

$P^i_{err} = P_{err}$ for $\forall i$, the optimal number of available samples can be obtained as a solution of the minimization problem:

$$M_{opt} \geq \underset{M}{\operatorname{argmin}} \{P_{err}\}, \quad i = 1, \ldots, K. \tag{6.91}$$

In other words, for each chosen value of $P_{err}$ and the expected value of the minimal amplitude $a_i$, there is an optimal value of $M$ that will assure the components detection. The relation between the $P_{err}$ and $M$ for different values $a_i$ are illustrated in Fig. 6.10: $a_1 = 4.5$, $a_2 = 4$, $a_3 = 3.5$, $a_4 = 3$, $a_5 = 2.5$, $a_6 = 2$, $a_7 = 1.5$, $a_8 = 1$, $a_9 = 0.5$ (additional Gaussian noise is added with $\sigma_N^2 = 4$). In the considered case, $a_{min} = 1.75$ and from Fig. 6.10, we can observe that for $P_{err} = 10^{-2}$ we need at least $M = 180$ measurements.

### 6.4.3 Algorithm 2: Threshold Based Iterative Solution

In the case when the number of available samples is such that some of the signal components are masked by noise in the spectral domain (the initial Fourier transform), we can employ the iterative threshold based algorithm, summarized below. In each iteration, we firstly detect some of the signal components that are above the threshold. Hence, in each iteration we obtain a set of components on positions $\mathbf{k}^i$ that are above the threshold. Then it is necessary to remove the contribution of detected components and to recalculate the threshold. It will reveal the remaining components that are below the noise level. Since the algorithm detects a block of components in each iteration, it needs just a few iterations to recover all signal components.

**Algorithm: Automated Threshold Based Iterative Solution**
**Input**:

- Transform matrix $\mathbf{\Psi}$, Measurement matrix $\mathbf{\Phi}$
- $\mathbf{N}_M = \{n_1, n_2, \ldots, n_M\}$, $M$—number of available samples, $N$—original signal length
- Matrix **A**: $\mathbf{A} = \mathbf{\Phi\Psi}$ (**A** is obtained as a random partial Fourier transform matrix: $\mathbf{A} = \mathbf{\Phi\Psi} = \mathbf{\Psi}\{\vartheta\}$, $\vartheta = \{\psi_{n_1}, \psi_{n_2}, \ldots, \psi_{n_M}\}$)
- Measurement vector $\mathbf{y} = \mathbf{\Phi f}$

**Output**:

1: ***Set*** $i = 1$, $\mathbf{N}_M = \{n_1, n_2, \ldots, n_M\}$
2: $\mathbf{p} = \varnothing$.
3: $P \leftarrow 0.99$ (*Set desired probability P*)
4: $\sigma_{MS}^2 \leftarrow M\frac{N-M}{N-1}\sum_{i=1}^{M}\frac{|y(m)|^2}{M}$ (*Calculate variance*)
5: $T^i \leftarrow \sqrt{-\sigma_{MS}^2 \log\left(1 - P(T)^{\frac{1}{N}}\right)}$ (*Calculate threshold*)
6: $\mathbf{X}^i \leftarrow \mathbf{A}^{-1}\mathbf{y}$ (*Calculate initial DFT of* **y**)
7: $\mathbf{k}^i \leftarrow \arg\left\{|\mathbf{X}^i| > \frac{T^i}{N}\right\}$ (*Find components in* $\mathbf{X}^i$ *above the threshold*)
8: **Update** $\mathbf{p} \leftarrow \mathbf{p} \cup \mathbf{k}_i$
9: $\mathbf{A_{cs}}^i \leftarrow \mathbf{A}(\mathbf{k}^i)$ (*CS matrix*)
10: $\tilde{\mathbf{X}}^i \leftarrow \left(\mathbf{A_{cs}}^{iT}\mathbf{A_{cs}}^i\right)^{-1}\mathbf{A_{cs}}^{iT}\mathbf{y}$
11: **for** $\forall p \in \mathbf{p}$: $\mathbf{y} \leftarrow \mathbf{y} - \frac{M}{N}\mathbf{X}^i(p)\exp(j2\pi p\mathbf{N}_M/N)$; (*Update* **y**)
13: **Update** $\sigma_{MS}^2 \leftarrow \frac{M(N-M)}{N-1}\sum|\mathbf{y}|^2/M$
14: **If** $\sum|\mathbf{y}|^2/M < \delta$ **break; Else**
15: **Set** $i = i + 1$, and go to 5.
**return** $\tilde{\mathbf{X}}^i$, **k**

A signal reconstruction example based on the iterative algorithm is illustrated in Fig. 6.11. The signal contains seven components with amplitudes: {4.5, 4, 2, 1.75, 2, 1.5, 3.5}. The total number of signal samples is $N = 512$, while the available number of samples is $M = 128$. The stopping parameter, $\delta = 2$ is used in the example. The reconstruction is done within two iterations. The first iteration is shown in Fig. 6.11a and b (selected and reconstructed DFT components), while the second iteration is shown in Fig. 6.11c and d (the entire reconstructed signal is given in Fig. 6.11d). The reconstructed components amplitudes are equal to the original ones.

#### 6.4.3.1 External Noise Influence

The previous analysis assumed a case when no external noise is present. However, in practical applications the measurements can be affected by noise. We further

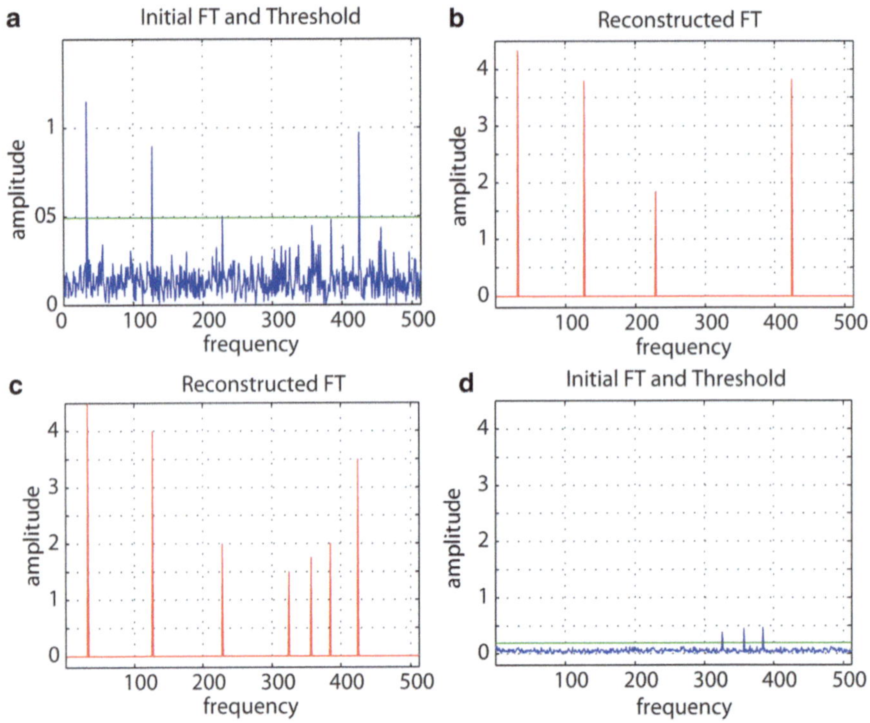


**Fig. 6.11** (**a**) Selecting components above the threshold in the first iteration, (**b**) the reconstructed components in the first iteration, (**c**) selecting components in the second iteration, (**d**) the reconstructed signal

assume that the signal is corrupted by Gaussian noise with the variance $\sigma_N^2$. Therefore, in the spectral domain, the signal components will be disturbed by two types of noise: noise that appears as a consequence of missing samples characterized by $\sigma_{MS}^2$, and external noise with variance $\sigma_N^2$. The total noise variance can be calculated as a sum of variances of external noise and noise due to the missing samples (we assume that the involved random variables are uncorrelated):

$$\sigma^2 = \sigma_{MS}^2 + M\sigma_N^2 = M\frac{N-M}{N-1}\sum_{i=1}^{K} a_i^2 + M\sigma_N^2. \tag{6.92}$$

In order to achieve the same probability of error as in the case with no external noise, it is necessary to increase the number of available samples from $M$ to some new value $M_N$. In this way, we would be able to keep the value of resulting noise variance $\sigma^2$ at the level of variance in the noiseless case which is equal to $\sigma_{MS}^2$. Consequently, the following requirement needs to be satisfied:

$$\frac{\sigma_{MS}^2}{\sigma^2} = \frac{M\frac{N-M}{N-1}\sum_{i=1}^{K} a_i^2}{M_N\frac{N-M_N}{N-1}\sum_{i=1}^{K} a_i^2 + M_N\sigma_N^2} = 1. \tag{6.93}$$

Here, $M_N$ denotes an increased number of measurements required to achieve successful components detection, as in the case with no external noise. Having in mind that:

$$SNR = \sum\nolimits_{i=1}^{K} a_i^2/\sigma_N^2, \tag{6.94}$$

we have:

$$\frac{M(N-M)}{M_N}\frac{SNR}{SNR(N-M_N)+(N-1)} = 1. \tag{6.95}$$

Now, for given $N$, $M$ and $SNR$, the required number of available samples $M_N$ in the presence of external Gaussian noise follows as a solution of the equation:

$$M_N{}^2 \cdot SNR - M_N(SNR \cdot N + N - 1) + SNR \cdot \left(MN - M^2\right) = 0. \tag{6.96}$$

For example, let us consider the case $N = 256$ and $M = 192$, while $SNR = 10$ dB is assumed. According to the previous relation, the number of available samples must be increased to $M_N = 227$.

#### 6.4.3.2 The Influence of Signal Reconstruction on the Resulting SNR

Here, it will be shown that the threshold based reconstruction method in the Fourier transform domain significantly reduces the input SNR. For a $K$ component signal with the signal amplitudes denoted by $a_i$, the input SNR of the full data set of length $N$ is by definition:

$$SNR_i = 10\log_{10}\frac{\sum_{n=0}^{N-1}|x(n)|^2}{\sum_{n=0}^{N-1}|v(n)|^2}. \tag{6.97}$$

However, we are dealing with $M$ measurements instead of $N$. Note that the amplitude of $i$-th signal components in the DFT domain for a full set of samples is $Na_i$, while in the case of CS signal with $M$ measurements the DFT amplitude is $Ma_i$. Therefore, during the signal reconstruction, the amplitudes of components are scaled from $Ma_i$ to $Na_i$ using the constant factor $N/M$. However, the noise on the observed components will be scaled with the same factor as well:

$$SNR = 10\log_{10}\frac{\sum\limits_{n=0}^{N-1}|x(n)|^2}{\sum\limits_{m=0}^{M-1}|(N/M)v(m)|^2} = 10\log_{10}\frac{\sum\limits_{n=0}^{N-1}|x(n)|^2}{\frac{N^2}{M^2}\sum\limits_{m=0}^{M-1}|v(m)|^2}. \tag{6.98}$$

Furthermore, all components detected by thresholding in the DFT domain represent the signal components. Other DFT values are set to zero. Thus, the noise remains only on $K$ out of $N$ components and the reconstruction error is decreased by the factor $K/N$:

$$SNR_{out} = 10\log_{10}\frac{\sum\limits_{n=0}^{N-1}|x(n)|^2}{\frac{K}{N}\frac{N^2}{M^2}\sum\limits_{m=0}^{M-1}|v(m)|^2} = 10\log_{10}\frac{\sum\limits_{n=0}^{N-1}|x(n)|^2}{\frac{KN}{M^2}\sum\limits_{m=0}^{M-1}|v(m)|^2}, \tag{6.99}$$

where $SNR_{out}$ is the output SNR.

Since the variance of noise is the same whether it is calculated using $N$ or only $M$ available samples: $\frac{1}{N}\sum\limits_{n=0}^{N-1}\nu(n) = \frac{1}{M}\sum\limits_{m=0}^{M-1}\nu(m)$, the resulting SNR can be written as follows:

$$SNR_{out} = 10\log_{10}\frac{\sum\limits_{n=0}^{N-1}|x(n)|^2}{\frac{KN}{M^2}\sum\limits_{m=0}^{M-1}|v(m)|^2} = 10\log_{10}\frac{\sum\limits_{n=0}^{N-1}|x(n)|^2}{\frac{K}{M}\sum\limits_{n=0}^{N-1}|v(n)|^2}. \tag{6.100}$$

Hence, we have:

$$SNR_{out} = SNR_i - 10\log_{10}(K/M). \tag{6.101}$$

The resulting SNR is a function of the sparsity level and the number of available samples.

## 6.5 Relationship Between the Robust Estimation Theory and Compressive Sensing

Next, we analyze and discuss a relationship between the robust estimation theory (discussed in Chap. 1) and compressive sensing. This relationship is established through a simple algorithm for signal reconstruction with a flexibility of using different types of minimization norms for different noisy environments.

Let us consider a multicomponent signal that consists of $K$ sinusoidal components:

$$f(n) = \sum_{i=1}^{K} f_i(n) = \sum_{i=1}^{K} a_i e^{j2\pi k_i n/N}, \tag{6.102}$$

where $a_i$ and $k_i$ denote the $i$-th signal amplitude and frequency, while $N$ is the total number of samples. According to the compressive sensing concept, assume that only $M$ ($M < N$) samples are available. Furthermore, we observe the generalized loss function $L$, used in the definition of the robust estimation theory:

$$L\{\varepsilon(n)\} = L\left\{\left|f(n)\psi(n,k) - \widetilde{F}(k)\right|\right\}, \tag{6.103}$$

where $\psi(n, k)$ are basis functions (e.g., $\psi(n,k) = e^{-j2\pi kn/N}$ in the DFT case), while $\widetilde{F}(k)$ is a robust estimate (the DFT of the signal)[6]:

$$\widetilde{F}(k) = \begin{cases} \underset{n}{mean}\{f(n)\psi(n,k)\}, \text{ standard form} - \text{Gaussian noise} \\ \underset{n}{median}\{f(n)\psi(n,k)\}, \text{ robust form} - \text{impulsive noise} \end{cases}. \tag{6.104}$$

The generalized loss function is defined as the $p$-th norm of the estimation error:

$$L\{\varepsilon\} = |\varepsilon|^p. \tag{6.105}$$

Here, we have a flexibility to use any value of $p$ and to adapt the norm to the problem of interest. For example, the standard transforms are obtained by using the $\ell_2$-norm: $L\{\varepsilon\} = |\varepsilon|^2$, while the robust forms based on $\ell_1$-norm ($L\{\varepsilon\} = |\varepsilon|$) are used in the presence of impulse and heavy-tailed noises.

In the case of compressive sampling the number of available samples is much lower than $N$. The samples $f(n_1), f(n_2), \ldots, f(n_M)$ are chosen randomly at the points from the set $\mathbf{N}_M = \{n_1, n_2, \ldots, n_M\}$.

---

[6] Recall that, when the data set is complex-valued, the marginal median is applied independently to the real and imaginary parts of set values.

The errors can be calculated for each available sample:

$$\varepsilon(n_m,k) = \left|f(n_m)\psi(n_m,k) - \widetilde{F}(k)\right|, \tag{6.106}$$

where $n_m \in \mathbf{N}_M$ and $k=0,1,\ldots, N-1$. The total deviations of estimation errors is defined as an average of the loss function $L\{\varepsilon(n_m,k)\}$ calculated for each frequency $k=0, 1,\ldots, N-1$:

$$D(k) = \frac{1}{M}\sum_{n_m \in \mathbf{N}_M} L\{\varepsilon(n_m,k)\}, \tag{6.107}$$

and will be referred as generalized deviations. For the sake of simplicity, let us focus to the case of DFT with $\psi(n,k) = e^{-j2\pi kn/N}$ and the loss function $L\{\varepsilon\} = |\varepsilon|^2$. The following notation will be used:

$$e(n_m,k) = f(n_m)e^{-j2\pi kn_m/N} \Rightarrow \widetilde{F}(k) = mean\{e(n_1,k),\ \ldots, e(n_M,k)\}.$$

According to Eqs. (6.106) and (6.107), the total deviations of estimation errors can be calculated as:

$$D(k) = \frac{1}{M}\sum_{n_m \in \mathbf{N}_M} |e(n_m,k) - mean\{e(n_1,k),\ldots,e(n_M,k)\}|^2, \tag{6.108}$$

or equivalently, it can be written in terms of variance as follows:

$$D(k) = \mathrm{var}\{e(n_1,k),\ e(n_2,k),\ \ldots,\ e(n_M,k)\}. \tag{6.109}$$

Furthermore, since *f*(*n*) is a *K*-components signal defined by Eq. (6.102), the variable $e(n_m, k)$ can be written as:

$$e(n_m,k) = \sum_{i=1}^{K} a_i e^{j2\pi k_i n_m/N} e^{-j2\pi kn_m/N} = \sum_{i=1}^{K} a_i e^{j2\pi(k_i-k)n_m/N}. \tag{6.110}$$

It is important to note that $e(n_m, k)$ has the form of random variables given by Eq. (6.76) in Sect. 6.4. Depending on whether the frequency point *k* corresponds to a position of signal component or not, we observe two characteristic cases:

(a) $k = k_q$ (position of the *q*-th component):

$$e_a(n_m,k) = \sum_{i=1,\ i\neq q}^{K} a_i e^{j2\pi(k_i-k)n_m/N}. \tag{6.111}$$

(b) $k \neq k_i$, for any $i = 1, 2, \ldots, K$:

$$e_b(n_m, k) = \sum_{i=1}^{K} a_i e^{j2\pi(k_i - k)n_m/N}. \tag{6.112}$$

According to the derivations in Sect. 6.4, particularly the expressions Eqs. (6.83) and (6.84), the variances of random variables are given by:

$$\sigma^2\{e_a(n_m, k)\} = M\frac{N-M}{N-1}\sum_{i=1, i\neq q}^{K} a_i^2,$$
$$\sigma^2\{e_b(n_m, k)\} = M\frac{N-M}{N-1}\sum_{i=1}^{K} a_i^2. \tag{6.113}$$

Therefore, the observed cases are asymptotically related as:

$$\frac{\sum_{i=1,\ i\neq q}^{K} a_i^2}{\sum_{i=1}^{K} a_i^2} < 1. \tag{6.114}$$

A similar inequality holds for other norms. It can be used as an indicator whether a considered frequency $k$ belongs to the signal component or not. In other words, the variance (or any deviation in general) will have lower values at $K$ signal frequencies than at the $N-K$ frequencies that do not contain signal components.

An illustration of variance at signal and non-signal positions is given in Fig. 6.12 (signal with three sinusoidal components is used). The Fourier transform of the full set of samples is shown in Fig. 6.12a in order to see the signal components positions. The total deviations are given in Fig. 6.12b (the number of available samples is $M = 25\%N$). It is obvious that the total deviations have smaller values at the positions of signal components, which can be used to detect their frequencies.

Another way to combine compressive sensing with the robust estimation theory is in implementation of the L-estimation approach for noisy signals presented in

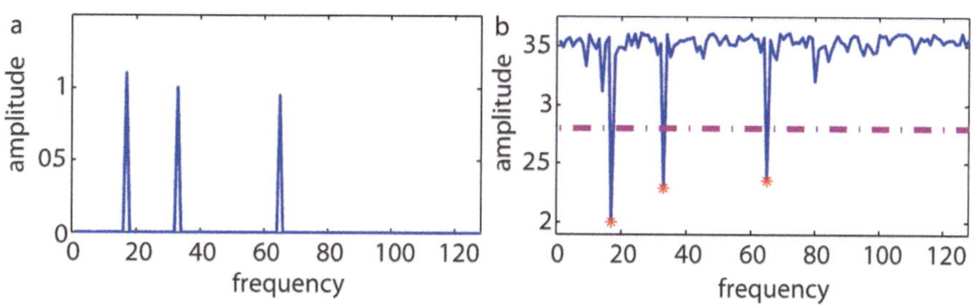


**Fig. 6.12** (**a**) The Fourier transform of original signal, (**b**) variances at signal (marked by *red symbols*) and non-signal positions

Chap. 1. Namely, in the L-estimation, the possibly corrupted signal samples are discarded and the transform domain representation is calculated using the remaining samples. Although the noise will be removed, discarding a certain amount of samples will decrease the quality of transform domain representation. If the discarded samples are treated as the missing samples in the context of CS, then these discarded samples can be accurately recovered using CS reconstruction algorithms. Therefore, the L-estimation transforms can benefit from the CS approach to achieve high quality signal representation.

### 6.5.1 Algorithm Based on Generalized Deviations

In order to summarize the above theoretical considerations, we present one simple algorithm based on the total deviations. The description of the algorithm steps follows.

1. Calculate the Fourier transform estimate $\widetilde{F}(k)$ and the generalized deviation $D(k)$. Choose the norm $p$ according to the assumed noise model.
2. Find the total deviation minima by applying the threshold:

$$\mathbf{k} = \arg\{D(k) < T\}, \tag{6.115}$$

   where T can be calculated, for example, with respect to $median\{D(k)\}$, e.g., $\alpha \cdot median\{D(k)\}$ (where α is a constant close to 1). The vector of positions $\mathbf{k}$ should contain all signal frequencies:
   $k_i \in \mathbf{k}$ for any $i = 1,\ldots,K$.
3. Set $\widetilde{F}(k) = 0$ for frequencies $k \notin \mathbf{k}$.
4. The estimates of the DFT values can be calculated by solving the set of equations, at the localized frequencies from the vector $\mathbf{k}$. Assume that $\mathbf{k}$ contains only $K$ frequencies belonging to the signal components: $\mathbf{k} = \{k_1, k_2,\ldots,k_K\}$. We set a system of equations:

$$\sum\nolimits_{i=1}^{K} F(k_i)e^{j2\pi k_i n_m} = f(n_m) \quad \text{for } n_m \in \mathbf{N}_M. \tag{6.116}$$

The CS matrix is formed as a partial DFT matrix: columns correspond to the positions of available samples, rows correspond to the selected frequencies. The system is solved in the least square sense:

$$\mathbf{F} = (\mathbf{A}_{cs}^T \mathbf{A}_{cs})^{-1} \mathbf{A}_{cs}^T \mathbf{f}(\mathbf{N}_M).$$

The reconstructed components in $\mathbf{F}$ are exact for all frequencies $k_i$.

# 6.6 Applications of Compressive Sensing Approach

## 6.6.1 *Multicomponent One-Dimensional Signal Reconstruction*

In order to illustrate compressive sensing reconstruction, first we consider a sparse signal composed of a few non-zero components, e.g., 5 different sinusoids. The signal length is $N = 500$ samples. The analytic form of signal can be written as:

$$f_x(n) = \sum_{i=1}^{5} \sin(2\pi f(i)n/N), \quad n = 0, \ldots, N-1, \tag{6.117}$$

where the vector of frequencies is: $f = [25 \;\; 45 \;\; 80 \;\; 100 \;\; 176]$.

In the Fourier domain, the signal consists of five components. Consequently, the signal can be considered as sparse in the frequency domain. Thus, the signal reconstruction can be done by using a small set of samples that are chosen randomly from 500 signal samples. The measurements are taken from the time domain, while the sparsifying matrix is obtained by taking the first $M = 150$ rows of the permuted inverse Fourier basis matrix.

The signal is reconstructed by using 30 % of the total number of coefficients. The original and the reconstructed signals in the time domain are shown in Fig. 6.13.

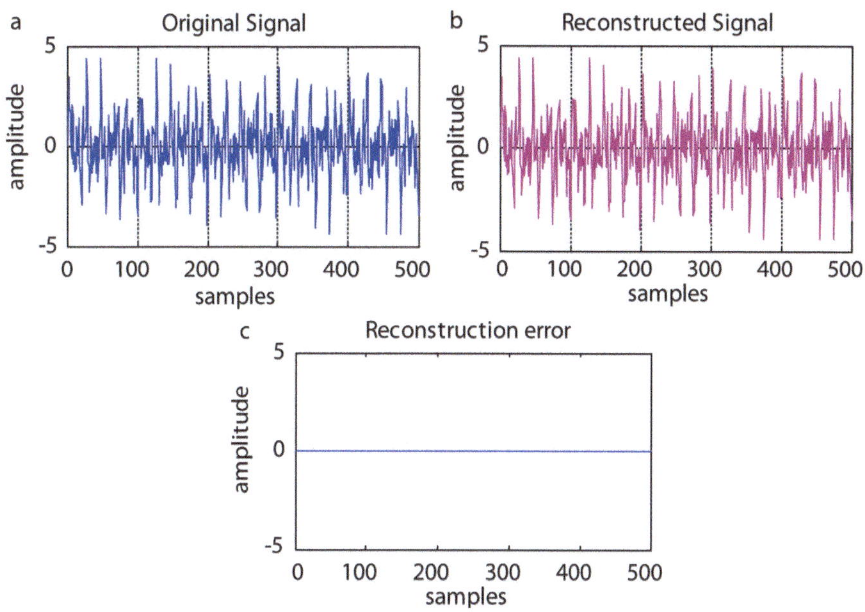


**Fig. 6.13** (**a**) The original signal, (**b**) the reconstructed signal, (**c**) the reconstruction error

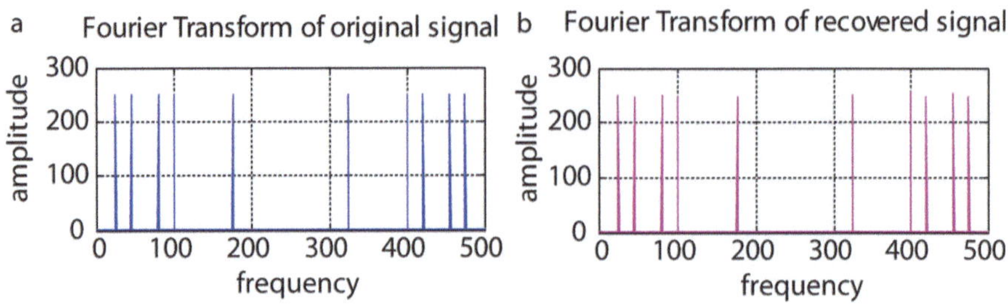


**Fig. 6.14** (**a**) The Fourier transform of the original signal, (**b**) The Fourier transform of the reconstructed signal

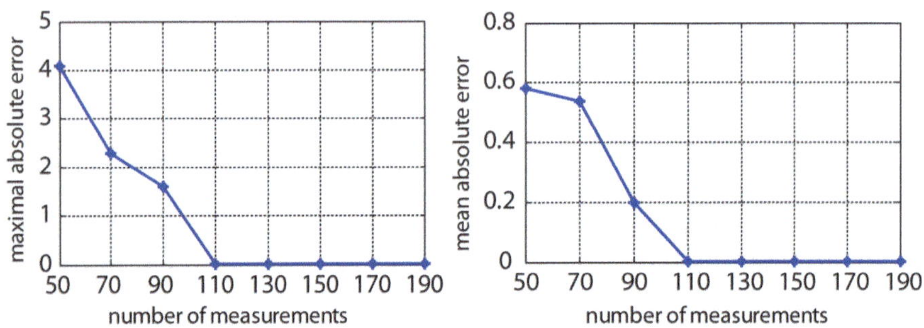


**Fig. 6.15** The maximal and the mean absolute errors of signal reconstruction

The original and the reconstructed Fourier transform of this signal are shown in Fig. 6.14.

The reconstruction error is very small and negligible comparing to the signal amplitudes (an average absolute error is $e \sim 10^{-4}$, while the average signal amplitude is higher than 1). The maximal and mean absolute errors for different numbers of measurements $M$ are plotted in Fig. 6.15.

Next we consider an audio signal representing a flute tone, with a total length of 2000 samples. In this application we will use the DCT transform domain (for the matrix $\mathbf{\Psi}$). The rows of $\mathbf{\Psi}$ are then randomly permuted and the first $M$ rows are used for sensing. The signal is reconstructed by using $M = 700$ random measurements out of $N = 2000$ signal samples. The results are shown in Fig. 6.16. By using the listening test it is proved that the quality of the reconstructed audio file is preserved without introducing any audible distortions.

### *6.6.2 Compressive Sensing and Image Reconstruction*

In this section, we consider the compressive sensing applications to image processing. Hence, let us consider the images Lena and Baboon (of size

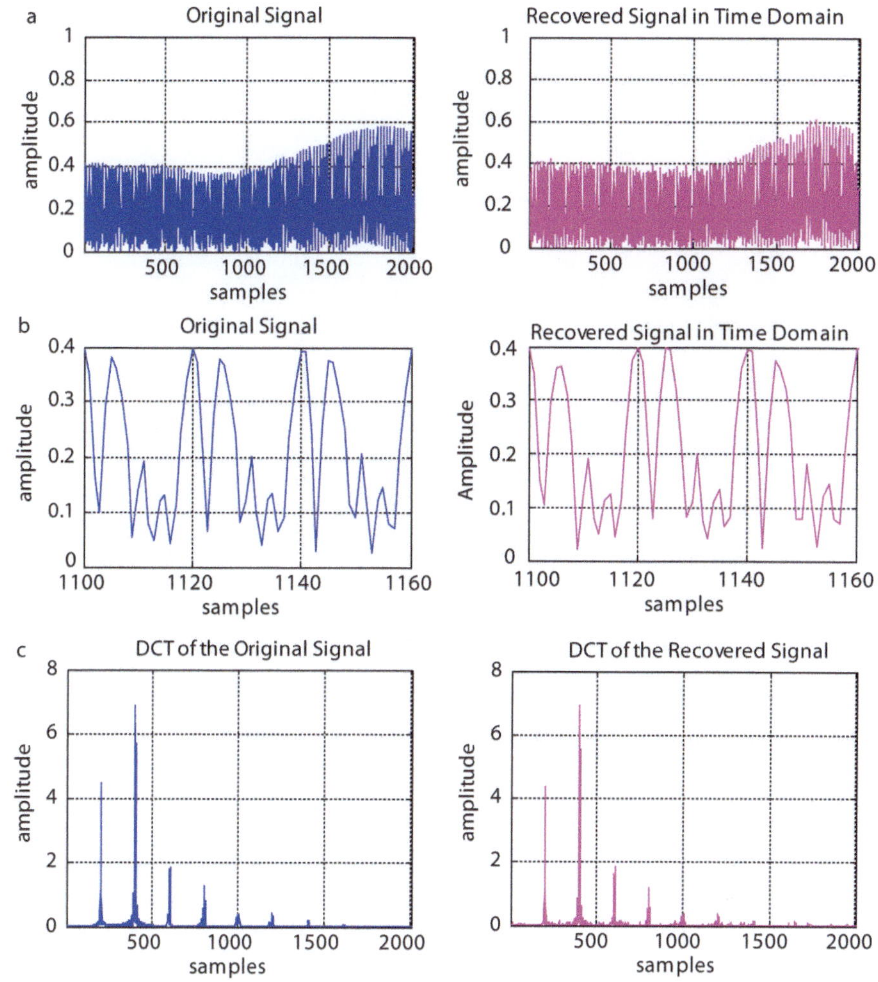


**Fig. 6.16** (**a**) The original (*left*) and reconstructed (*right*) flute signal, (**b**) a zoomed-in region of the original (*left*) and reconstructed (*right*) signal, (**c**) the Fourier transform of the original (*left*) and reconstructed (*right*) signals

$256 \times 256$). First we split the image into block of size $64 \times 64$. The compressive sensing is performed as follows:

- Each block is represented as a vector $\mathbf{f}$ with $N = 4096$ elements;
- As an observation set we select only $M = 1500$ random measurements (within a vector $\mathbf{y}$) from the block elements;
- The DCT (of size $4096 \times 4096$) is used, while $\mathbf{A} = \mathbf{\Phi\Psi}$ is obtained by taking $M$ rows of the randomly permuted transform matrix $\mathbf{\Psi}$.

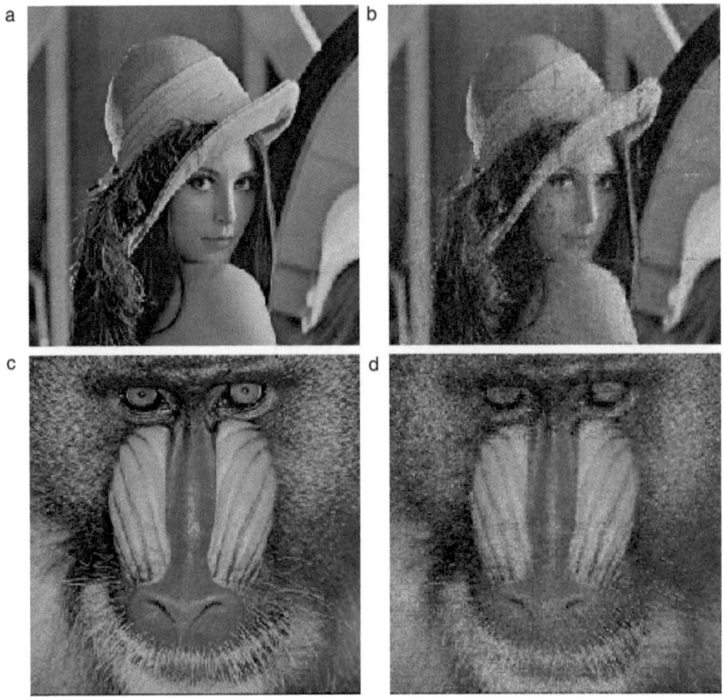

**Fig. 6.17** $\ell_1$-minimization based reconstruction: (**a**) the original "Lena" image, (**b**) the reconstructed "Lena" image, (**c**) the original "Baboon" image, (**d**) the reconstructed "Baboon" image

The original and the reconstructed Lena and Baboon images are shown in Fig. 6.17. Note that the quality of reconstructed images needs to be further improved due to the fact that the images are not strictly sparse in the DCT domain.

#### 6.6.2.1 Total-Variation Method

One of the approaches used in various image processing applications is based on the variational parameters, i.e., the total-variation of an image. An example of using the total-variation method is denoising and restoring of noisy images. Hence, if $x_n = x_0 + e$ is a "noisy" observation of $x_0$, we can restore $x_0$ by solving the following minimization problem:

$$\min_x \ TV(x) \ \ s.t. \ \|x_n - x\|_2^2 < \varepsilon^2, \tag{6.118}$$

where $\varepsilon = \|e\|_2^2$ should holds and TV denotes the total-variation. The total-variation of $x$ represents the sum of the gradient magnitudes at each point and can be approximated as:

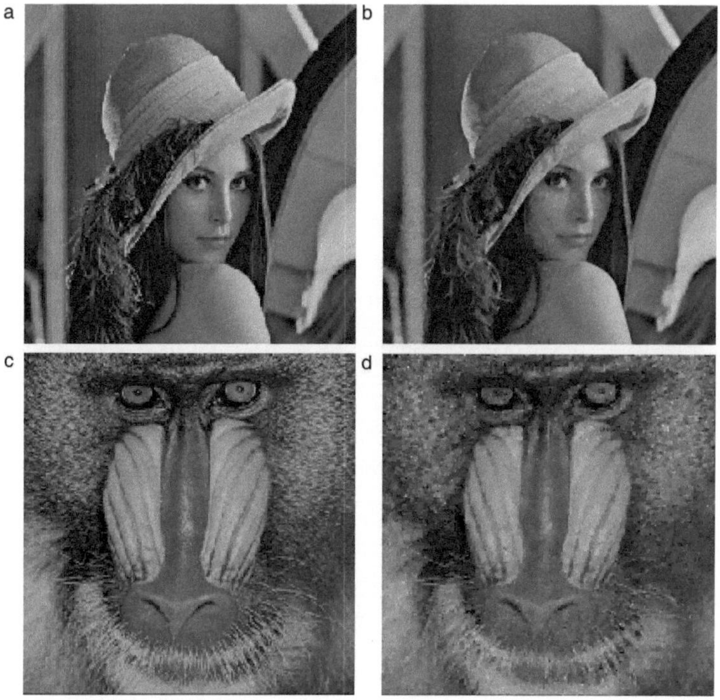


**Fig. 6.18** TV reconstruction: (**a**) the original "Lena" image, (**b**) the reconstructed "Lena" image, (**c**) the original "Baboon" image, (**d**) the reconstructed "Baboon" image

$$TV(x) = \sum_{i,j} \left\| D_{i,j}x \right\|_2, \; D_{i,j}x = \begin{bmatrix} x(i+1,j) - x(i,j) \\ x(i,j+1) - x(i,j) \end{bmatrix}. \tag{6.119}$$

The TV based denoising methods tend to remove the noise while retaining the details and edges. The TV approach could be applied in compressive sensing to define an efficient reconstruction method. Thus, in the light of compressive sensing we may write:

$$\min_x \; TV(x) \;\; s.t. \; \|Ax - y\|_2^2 < \varepsilon^2. \tag{6.120}$$

TV minimization provides a solution whose variations are concentrated on a small number of edges. The results obtained by applying the TV minimization algorithm for image reconstructions are given in Fig. 6.18. The algorithm is applied to $64 \times 64$ blocks. The total number of samples per block is $N = 4096$, while the random $M = 1500$ measurements are used for reconstruction (the DFT matrix is used). Note that the quality of results is significantly improved compared to the images in Fig. 6.17. The $l_1$-magic toolbox (see the link given in the literature) is used for solving the minimization problems.

#### 6.6.2.2 Gradient-Based Image Reconstruction

Next, we consider the application of gradient-based algorithm presented in Sect. 6.3.2 to the image reconstruction. The reconstruction is performed on a block basis. Assume that the $i$-th image block is denoted by $I(n,m)$ with $M$ available pixels out of $N$ elements.

$$y(n,m) = \begin{cases} I(n,m), & \text{for } (n,m) \in \Omega \\ 0, & \text{for } (n,m) \in \mathbf{N}\backslash\Omega \end{cases}, \tag{6.121}$$

where the set of available pixels is denoted by $\Omega$, while the total set of a block pixels is $\mathbf{N}$. Then, only on the position of missing pixels the values of $y$ are changed as follows:

$$\begin{aligned} y^{+}(n,m) &= y_i(n,m) + \Delta, \ \text{for}\,(n,m) \in \mathbf{N}\backslash\Omega, \\ y^{-}(n,m) &= y_i(n,m) - \Delta, \ \text{for}\,(n,m) \in \mathbf{N}\backslash\Omega \end{aligned} \tag{6.122}$$

where $\Delta = \max_{(n,m)\in\Omega}\{|I(n,m)|\}$, while $i$ denotes the $i$-th iteration. Furthermore, the gradient vector is calculated as follows:

$$G^{i}(n,m) = \frac{\eta(\mathbf{DCT}^{+}) - \eta(\mathbf{DCT}^{-})}{N}, \ (n,m) \in \mathbf{N}\backslash\Omega, \tag{6.123}$$

where $\mathbf{DCT}^{+}$ and $\mathbf{DCT}^{-}$ denote 2D DCT of image blocks with changed values $y^{+}(n,m)$ and $y^{-}(n,m)$, and:

$$\begin{aligned} \eta(\mathbf{DCT}^{+}) &= \sum\nolimits_{k_1=0}^{N-1}\sum\nolimits_{k_2=0}^{N-1}|DCT^{+}(k_1,k_2)|, \\ \eta(\mathbf{DCT}^{-}) &= \sum\nolimits_{k_1=0}^{N-1}\sum\nolimits_{k_2=0}^{N-1}|DCT^{-}(k_1,k_2)|. \end{aligned}$$

The gradient vector on the positions of available pixels $(n,m) \in \Omega$ is zero. The pixels values are updated as follows:

$$y^{i+1}(n,m) = y^{i}(n,m) + G^{i}(n,m). \tag{6.124}$$

The result obtained by applying the gradient-based image reconstruction algorithm using $16 \times 16$ blocks is given in Fig. 6.19. The Lena image is firstly preprocessed in the DCT domain: all except the 32 highest coefficients are set to zero in each block.

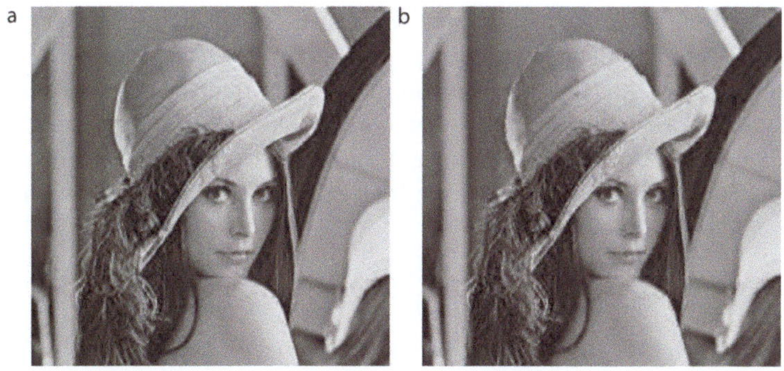


**Fig. 6.19** Gradient-based image reconstruction: (**a**) the original "Lena" sparsified image, (**b**) the reconstructed "Lena" image

**Fig. 6.20** The Fourier transform vector (8 coefficients) of **y**

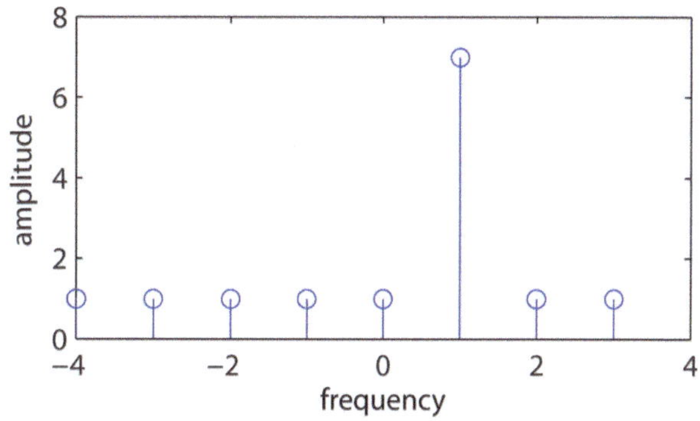


## 6.7 Examples

6.1. Consider a set of samples with one missing sample at the last position:

$\mathbf{f} = [0.7071 + 0.7071j,\ j,\ -0.7071 + 0.7071j,\ -1,\ -0.7071 - 0.7071j,\ -j,\ 0.7071 - 0.7071j,\ \boldsymbol{P}]$.

Determine the sample value $\boldsymbol{P}$ such that the solution is sparse in the Fourier transform domain.

Solution:
The total number of samples is $N = 8$, while the number of available samples is $M = 7$. The missing sample is on the 8th position within the original signal vector **f**. Hence, the measurement vector **y** can be defined as follows:

$$\mathbf{y} = [f(1), f(2), \ldots, f(7)].$$

Now we can calculate $N$ coefficients of the Fourier transform (Fig. 6.20) corresponding to the measurement vector **y**. In other words, it has the same result as if we calculate Fourier transform of **f** assuming that $\boldsymbol{P} = 0$.

Due to the sparsity assumption, we can observe from the Fourier transform that the signal contains one component at $k = 1$ which obviously originates from a single sinusoid. The remaining spectral components appear as a consequence of missing sample. Now we can show that the sparse solution is achieved if $\boldsymbol{P}$ corresponds to the appropriate sample of the sinusoid.

According to the definition, the Fourier transform of signal $\mathbf{f}$ with missing sample $\boldsymbol{P}$ can be written as:

$$\begin{aligned} X(k) &= \sum_{n=1}^{7} f(n)e^{-j2\pi nk/8} + Pe^{-j2\pi 8k/8} \\ &= \sum_{n=1}^{8} f(n)e^{-j2\pi nk/8} + (P - f(8))e^{-j2\pi 8k/8} \\ &= N\delta(k-1) + (P - f(8))e^{-j2\pi 8k/8} \end{aligned}$$

Since we are searching for the sparsest $\mathbf{X}$:

$$\min\|\mathbf{X}\|_0 \ \textit{subject to}\ \mathbf{y} = \mathbf{AX},$$

we can determine the missing value $P$ as: $P = f(8)$, where $f$ represents a sinusoid in the form: $f(n) = e^{j2\pi n/N}$, $n = 1, 2, \ldots, 8$. Therefore, we can conclude $P = f(8) = e^{j2\pi 8/8} = 1$.

6.2. Consider a signal in the form:

$$f(n) = 2.5\exp(2\pi j8n/N) + 4\exp(2\pi j24n/N), \quad n = 1, \ldots, 32.$$

The number of available samples is $M = 16$ out of the total number of samples $N = 32$. The positions of available samples are given in the vector $\mathbf{p}$:

$\mathbf{p} = [7, 22, 23, 14, 8, 27, 2, 32, 19, 15, 26, 5, 31, 29, 10, 20]$;

Using the threshold with probability $P = 0.95$ perform the signal reconstruction from the available set of measurements.

Solution:

The first step is to calculate the variance of spectral noise that appears in the Fourier transform domain due to the missing samples.

$$\begin{aligned} \sigma^2{}_{MS} &= M\frac{N-M}{N-1}\sum_{i=1}^{K} a_i{}^2 = M\frac{N-M}{N-1}\left(\frac{1}{M}\sum_{n\in\mathbf{p}}|f(n)|^2\right) \\ &= 8.2581 \cdot 22.25 = 183.7419 \end{aligned}$$

Further, we calculate the threshold value for $P(T) = 0.95$ using the following expression:

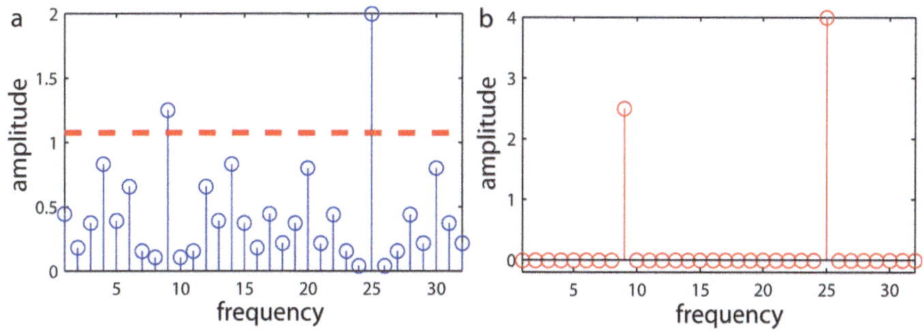


**Fig. 6.21** (**a**) The initial discrete Fourier transform and normalized threshold, (**b**) the reconstructed discrete Fourier transform of signal **f**

$$T = \sqrt{-\sigma_{MS}{}^2 \log\left(1 - P(T)^{\frac{1}{N}}\right)} = 34.39.$$

Since we need the normalized threshold value: $T/N = 1.0747$.

The initial Fourier transform is calculated by assuming zero values at the position of missing samples. The Fourier transform coefficients are plotted in Fig. 6.21a. The thresholding procedure is then applied to the initial Fourier transform vector **X**.

The positions of signal components in the Fourier transform vector **X** are determined as follows:

$$\mathbf{k} = \arg\left\{ |\mathbf{X}| > \frac{T}{N} \right\} = [9,\ 25].$$

The compressive sensing matrix is obtained from the inverse Fourier transform matrix by keeping only the rows that corresponds to available samples (rows defined by vector **p**) and columns that correspond to vector **k**. The inverse DFT matrix for $N = 32$ is given below.

$$\Psi = \begin{bmatrix} 1 & 1 & 1 & 1 & \dots & 1 \\ 1 & e^{j\frac{2\pi}{32}} & e^{2j\frac{2\pi}{32}} & e^{3j\frac{2\pi}{32}} & \dots & e^{31j\frac{2\pi}{32}} \\ 1 & e^{2j\frac{2\pi}{32}} & e^{4j\frac{2\pi}{32}} & e^{6j\frac{2\pi}{32}} & \dots & e^{62j\frac{2\pi}{32}} \\ \dots & \dots & \dots & \dots & \dots & \dots \\ 1 & e^{30j\frac{2\pi}{32}} & e^{60j\frac{2\pi}{32}} & e^{90j\frac{2\pi}{32}} & \dots & e^{930j\frac{2\pi}{32}} \\ 1 & e^{31j\frac{2\pi}{32}} & e^{62j\frac{2\pi}{32}} & e^{93j\frac{2\pi}{32}} & \dots & e^{961j\frac{2\pi}{32}} \end{bmatrix},$$

Then, $\mathbf{A}_{cs} = \boldsymbol{\Psi}(\mathbf{p}, \mathbf{k})$. The least square problem is solved:

$$\widetilde{\mathbf{X}} = \left(\mathbf{A}_{cs}{}^{\mathrm{T}}\mathbf{A}_{cs}\right)^{-1}\mathbf{A}_{cs}{}^{\mathrm{T}}\mathbf{y},$$

resulting in $\widetilde{\mathbf{X}} = [4j, \ -2.5j]$ and thus $\left|\widetilde{\mathbf{X}}\right| = [4, \ 2.5]$. The reconstructed vector of the original Fourier transform is obtained as (Fig. 6.21b):

$$\mathbf{F} = \begin{cases} \widetilde{\mathbf{X}}, & \text{for } k \in [9, \ 25] \\ 0, & \text{otherwise} \end{cases}.$$

6.3. The original signal is defined in the form:

$$f(n) = a_1\psi_5(n) + a_2\psi_{10}(n) + a_3\psi_{13}(n) + a_4\psi_{21}(n) + a_5\psi_{40}(n) + a_6\psi_{45}(n),$$

where $\psi_i(n)$ represents a Hermite function of the $i$-th order, while $a_1 = a_2 = a_3 = a_4 = a_5 = a_6 = 1$. The original signal length is $N = 60$, while the number of available samples is $M = 30$. Define the signal reconstruction problem in the Hermite transform domain using the analogy with the Fourier transform domain and reconstruct the missing samples.

Solution:

Note that the signal is made of a linear combination of 6 Hermite functions. Assume that we use $N = 60$ Hermite functions in the Hermite expansion. Thus, the Hermite transform coefficients are calculated as:

$$\mathbf{C} = \mathbf{H}\mathbf{f},$$

where $\mathbf{C}$ is a column vector of Hermite transform coefficients, $\mathbf{f}$ is a column vector of original time domain signal samples, while the Hermite transform matrix is given by:

$$\mathbf{H} = \frac{1}{N}\begin{bmatrix} \frac{\psi_0(0)}{(\psi_{N-1}(0))^2} & \frac{\psi_0(1)}{(\psi_{N-1}(1))^2} & \cdots & \frac{\psi_0(N-1)}{(\psi_{N-1}(N-1))^2} \\ \frac{\psi_1(0)}{(\psi_{N-1}(0))^2} & \frac{\psi_1(1)}{(\psi_{N-1}(1))^2} & \cdots & \frac{\psi_1(N-1)}{(\psi_{N-1}(N-1))^2} \\ \cdots & \cdots & \cdots & \cdots \\ \frac{\psi_{N-1}(0)}{(\psi_{N-1}(0))^2} & \frac{\psi_{N-1}(1)}{(\psi_{N-1}(1))^2} & \cdots & \frac{\psi_{N-1}(N-1)}{(\psi_{N-1}(N-1))^2} \end{bmatrix}$$

Recall that: $\Psi_0(n) = \frac{1}{\sqrt[4]{\pi}}e^{-n^2/2}$, $\Psi_1(n) = \frac{\sqrt{2}n}{\sqrt[4]{\pi}}e^{-n^2/2}$,

$\Psi_p(n) = n\sqrt{\frac{2}{p}}\Psi_{p-1}(n) - \sqrt{\frac{p-1}{p}}\Psi_{p-2}(n)$, $\forall p \geq 2$, where points $n$ are zeros of the $N$-th order Hermite polynomial.

The original signal and its Hermite transform are illustrated in Fig. 6.22.

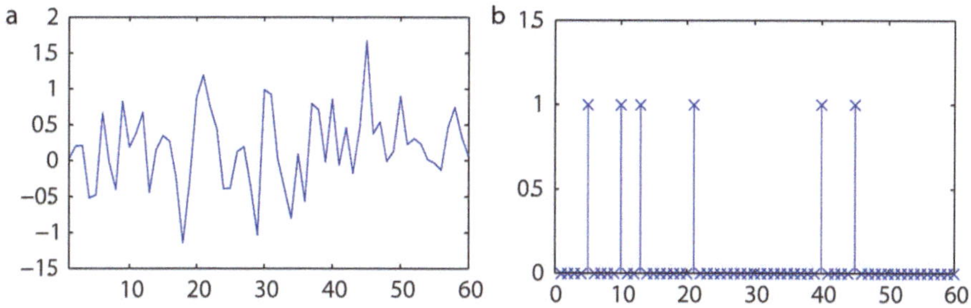


**Fig. 6.22** (**a**) The reconstructed signal in the time domain, (**b**) the reconstructed Hermite transform of the signal

**Fig. 6.23** A vector of Hermite coefficients corresponding to the available measurements **y**

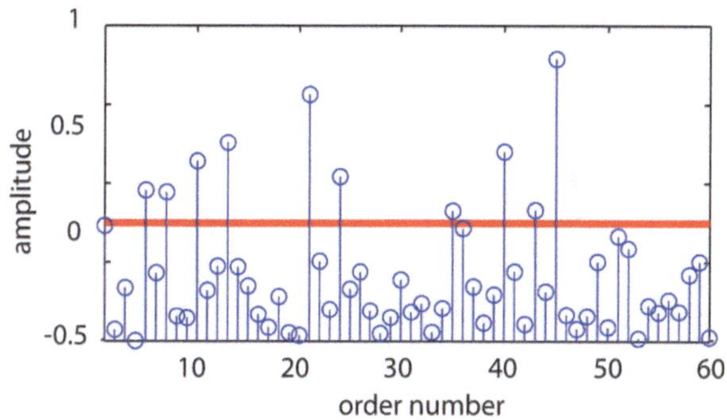


Now, we consider the measurement vector **y** obtained from **f** by randomly selecting $M$ out of $N$ samples, at the positions:

$\mathbf{p}$ = [46, 58, 15, 26, 9, 11, 8, 27, 54, 28, 35, 2, 47, 6, 17, 29, 60, 51, 16, 23, 3, 42, 56, 57, 34, 45, 38, 22, 43, 37].

The initial vector of Hermite coefficients that correspond to measurements **y** is calculated as:

$$\mathbf{C}^y = \mathbf{H}_\mathbf{p}\mathbf{y},$$

where $\mathbf{H_p}$ is obtained from $\mathbf{H}$ by omitting rows that correspond to missing samples and keeping only the rows that correspond to positions of available samples defined by vector of positions **p**. Vector $\mathbf{C}^y$ is plotted in Fig. 6.23.

Now, we can apply a simple threshold on $\mathbf{C}^y$, for example:

$$\mathbf{k} = \arg\{|\mathbf{C}^y| > 0.3\},$$

resulting in a vector of positions **k**:

$$\mathbf{k} = [\boxed{5}, 7, \boxed{10}, \boxed{13}, \boxed{21}, 24, 35, \boxed{40}, 43, \boxed{45}].$$

Within the vector of selected positions **k**, we have 6 positions corresponding to the signal components (marked in red), while remaining 4 positions are false. However, false components can be eliminated by solving the system in the least square sense, which will set zero values for false components and the exact values for the true signal components.

The solution of the equation:

$$\mathbf{y} = \mathbf{A}_{cs}\widetilde{\mathbf{C}}, \tag{6.125}$$

can be found in the least square sense as follows:

$$\widetilde{\mathbf{C}} = \left(\mathbf{A}_{cs}^T\mathbf{A}_{cs}\right)^{-1}\mathbf{A}_{cs}^T\mathbf{y}. \tag{6.126}$$

The compressive sensing matrix $\mathbf{A}_{cs}$ is obtained from the inverse Hermite transform matrix $\boldsymbol{\Psi}$ by keeping only rows on the positions defined by $\mathbf{p}$ and columns on the positions defined by $\mathbf{k}$:

$$\mathbf{A}_{cs} = \boldsymbol{\Psi}(\mathbf{p}, \mathbf{k}), \tag{6.127}$$

$$\boldsymbol{\Psi} = \begin{bmatrix} \psi_0(0) & \psi_1(0) & \ldots & \psi_{N-1}(0) \\ \psi_0(1) & \psi_1(1) & \ldots & \psi_{N-1}(1) \\ \ldots & \ldots & \ldots & \ldots \\ \psi_0(N-1) & \psi_1(N-1) & \ldots & \psi_{N-1}(N-1) \end{bmatrix}$$

The resulting vector $\widetilde{\mathbf{C}}$ has only 10 entries (the number of entries corresponds to the length of vector $\mathbf{k}$):

$$\widetilde{\mathbf{C}} = [1,\ 0,\ 1,\ 1,\ 1,\ 0,\ 0,\ 1,\ 0,\ 1].$$

Note that the non-zero coefficients are obtained only at the positions that correspond to true signal components.

The reconstructed sparse Hermite transform vector is given by:

$$\mathbf{C}_R = \begin{cases} \widetilde{\mathbf{C}}, & \text{for } k \in \mathbf{k} \\ 0, & \text{otherwise} \end{cases} \tag{6.128}$$

and it is illustrated in Fig. 6.24, as well as the reconstructed signal in the time domain.

6.4. For a signal in the form:

$$f(n) = \sum_{i=1}^{K} a_i e^{j2\pi k_i n/N}, \quad \text{for } N = 128 \text{ and } n = 1, 2, \ldots, N \tag{6.129}$$

perform the reconstruction using single iteration threshold base algorithm for a different number of components $K = \{1, \ldots, 30\}$, and a different number of

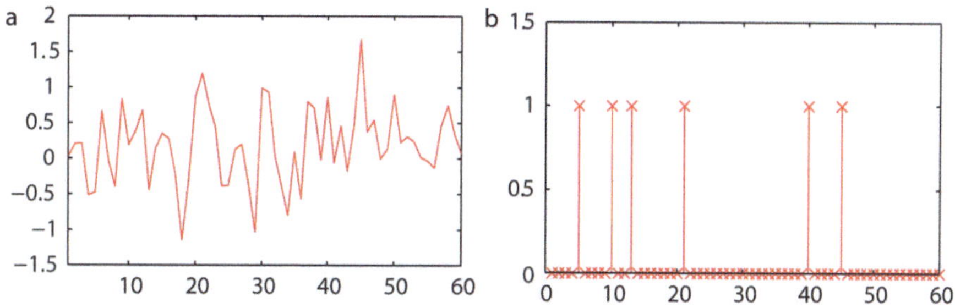

**Fig. 6.24** (**a**) The reconstructed signal in the time domain, (**b**) the reconstructed Hermite transform of the signal

available samples $M = \{1,2,\ldots,120\}$. The amplitudes of components should be chosen randomly from the set:

$a_i = 2 : 0.05 : 5$, while the signal frequencies are chosen randomly from the integer set $k_i = 1{:}128$. For each chosen par of ($M$, $K$) perform the reconstruction 10 times and find the mean values of a signal to reconstruction-error ratio (SRR).

Solution:

The SRR is calculated as follows:

$$SRR(m,k) = 10\log_{10}\frac{\sum_{n=1}^{N}|f(n)|^2}{\sum_{n=1}^{N}\left|f(n) - f_{rec}^{m,k}(n)\right|^2}, \tag{6.130}$$

where $f_{rec}^{m,k}(n)$ represents the reconstructed signal for a number of available samples $M = m$ and a number of components (sparsity level) $K = k$. The threshold is set as follows:

$$T = \sqrt{-\sigma_{MS}{}^2\log\left(1 - P(T)^{\frac{1}{N}}\right)} = \sqrt{-\sigma_{MS}{}^2\log\left(1 - 0.9^{\frac{1}{N}}\right)},$$

or in other words the probability is set to 0.9 (90 %).

The SRR is calculated for different values of $M$ and $K$, namely: $M = 1,2,..,120$ and $K = 1,2,\ldots,30$. The results are depicted in Fig. 6.25. The values of $K$ are on the horizontal axis, while the values of $M$ are on the vertical axis. It can be observed that as the number of components increases, we need to increase the number of measurements in order to achieve successful reconstruction with SRR > 50 dB.

6.5. Sparsify the grayscale image in the DCT domain such that in each $16 \times 16$ block there is only 1/8 of nonzero coefficients. From each block discard randomly 60 % of pixels. Perform image reconstruction using the adaptive gradient-based algorithm in the $16 \times 16$ block-based DCT domain.

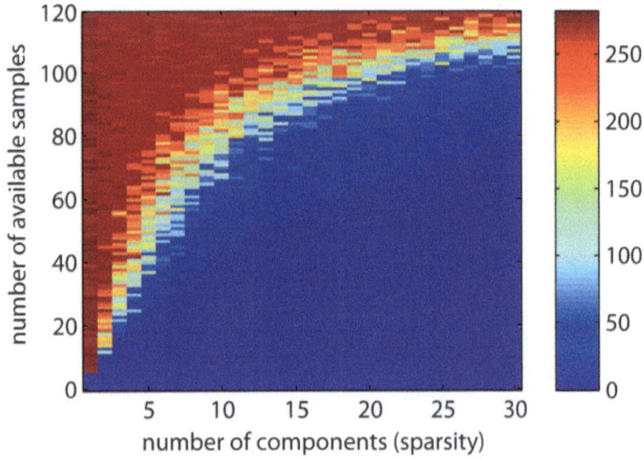


**Fig. 6.25** The signal to reconstruction-error ratio in [dB]

Solution:

The image is firstly divided into $16 \times 16$ blocks and each block is then sparsified as follows:

1: **Block** $\leftarrow$ *Image* $(i{:}i+16, j{:}j+16)$;
2: **X** $\leftarrow$ DCT{**Block**}; (Calculate DCT of each block)
3: **V** $\leftarrow$ **X** (Rearrange block to vector)
4: **Ind** $\leftarrow$ arg{sort(abs(**V**)}; (Sort in descending order)
5: **X**(ind($K+1$:end)) $\leftarrow$ **0**; (Set all except $K = 32$ values to 0)
6: **Block** $\leftarrow$ IDCT(**Block**);

For each $16 \times 16$ Block, select 60 % of samples that will be declared as missing and put a referent value 0 on the positions of missing samples:

7: **Pos** $\leftarrow$ *randperm*(256);
8: **Pos** $\leftarrow$ **Pos**(1: *round*(0.6*256));
9: **Block (Pos)** $\leftarrow$ 0;

Then set the number of iteration to $Iter = 150$ and perform the gradient based reconstruction method on each Block using the DCT transform to calculate the concentration measure and the gradient vector. Note that $\Delta = 128$ is used in the case of images. The set of available samples is denoted by $(n,m) \in \{(n_1,m_1),(n_2,m_2),\ldots,(n_M,m_M)\}$.

10: ***For*** $q = 1{:}Iter$
11: ***Set*** $i \notin \{n_1,n_2,\ldots,n_M\}$ and $j \notin \{m_1,m_2,\ldots,m_M\}$
12: $y^+(n,m) = y_q(n,m) + \Delta\delta(n-i, m-j)$
$y^-(n,m) = y_q(n,m) - \Delta\delta(n-i, m-j)$
13: $X^+(k_1,k_2) \leftarrow DCT2\{y^+(n,m)\}$
14: $X^-(k_1,k_2) \leftarrow DCT2\{y^-(n,m)\}$
15: $G_q(n,m) \leftarrow \frac{1}{N}\sum_{k_1=0}^{N-1}\sum_{k_2=0}^{N-1}|X^+(k_1,k_2)| - |X^-(k_1,k_2)|$

(Gradient corresponding to the missing pixel at $(m,n)$ position)

16: $y_{q+1}(n,m) = y_q(n,m) - G_q(n,m)$

**Fig. 6.26** (**a**) The original image, (**b**) the sparsified image, (**c**) the image with 60 % missing samples, (**d**) the reconstructed image

(Each value at the position of missing pixel is changed in the direction of the gradient)

Note that the change on the positions of available pixels is 0, and consequently the gradient on the position of available signal samples is 0 as well.

$$\beta_q = \operatorname{arcos} \frac{\sum_{n=0}^{N-1}\sum_{m=0}^{N-1} G_{q-1}(n,m)G_q(n,m)}{\sqrt{\sum_{n=0}^{N-1}\sum_{m=0}^{N-1}\left(G_{q-1}(n,m)\right)^2}\sqrt{\sum_{n=0}^{N-1}\sum_{m=0}^{N-1}\left(G_q(n,m)\right)^2}}$$

**if** $\beta_q > 170^\circ$

$\Delta \leftarrow \Delta/\sqrt{10}$

**End**

**End**

The procedure should be repeated in the same way for each image block (Fig. 6.26).

## References


1. Abolghasemi V, Ferdowsi S, Makkiabadi B, Sanei S (2010) On optimization of the measurement matrix for compressive sensing. 18th European Signal Processing Conference (EUSIPCO-2010), Aalborg
2. Ahmad F, Amin MG (2013) Through-the-wall human motion indication using sparsity-driven change detection. IEEE Trans Geosci Remote Sens 51(2):881–890
3. Ailon N, Rauhut H (2013) Fast and RIP-optimal transforms. Discrete Comput Geom 52:780
4. Bandeira AS, Fickus M, Mixon DG, Wong P (2013) The road to deterministic matrices with the restricted isometry property. J Fourier Anal Appl 19(6):1123–1149
5. Baraniuk R (2007) Compressive sensing. IEEE Signal Process Mag 24(4):118–121
6. Baraniuk RG, Davenport M, De Vore RA, Wakin M (2008) A simple proof of the restricted isometry property for random matrices. Constr Approx 28(3):253–263
7. Blumensath T, Davies M (2009) Iterative hard thresholding for compressed sensing. Appl Comput Harmon Anal 27(3):265–274
8. Boyd S, Vandenberghe L (2004) Convex optimization. Cambridge University Press, Cambridge
9. Candès E (2006) Compressive sampling. Int Congress Math 3:1433–1452
10. Candès E (2011) A probabilistic and RIPless theory of compressed sensing. IEEE Trans Inf Theory 57(11):7235–7254
11. Candès E, Romberg J (2007) Sparsity and incoherence in compressive sampling. Inverse Probl 23(3):969–985
12. Candès E, Romberg J, Tao T (2006) Robust uncertainty principles: exact signal reconstruction from highly incomplete frequency information. IEEE Trans Inf Theory 52(2):489–509
13. Candès E, Wakin M (2008) An introduction to compressive sampling. IEEE Signal Process Mag 25(2):21–30
14. Chartrand R (2007) Exact reconstructions of sparse signals via nonconvex minimization. IEEE Signal Process Lett 14(10):707–710
15. Chen SS, Donoho DL, Saunders MA (1999) Atomic decomposition by Basis pursuit. SIAM J Sci Comput 20(1):33–61
16. Daković M, Stanković L, Orović I (2014) Adaptive gradient based algorithm for complex sparse signal reconstruction. 22nd Telecommunications Forum TELFOR 2014, Belgrade
17. Davenport MA, Boufounos PT, Wakin MB, Baraniuk RG (2010) Signal processing with compressive measurements. IEEE J Select Top Signal Process 4(2):445–460
18. Donoho D (2006) Compressed sensing. IEEE Trans Inf Theory 52(4):1289–1306
19. Donoho DL, Tsaig Y, Drori I, Starck JL (2007) Sparse solution of underdetermined linear equations by stagewise orthogonal matching pursuit. IEEE Trans Inf Theory 58(2):1094–1121
20. Duarte M, Wakin M, Baraniuk R (2005) Fast reconstruction of piecewise smooth signals from random projections. SPARS Workshop, Rennes
21. Duarte M, Davenport M, Takhar D, Laska J, Sun T, Kelly K, Baraniuk R (2008) Single-pixel imaging via compressive sampling. IEEE Signal Process Mag 25(2):83–91
22. Flandrin P, Borgnat P (2010) Time-frequency energy distributions meet compressed sensing. IEEE Trans Signal Process 8(6):2974–2982
23. Fornasier M, Rauhut H (2011) Compressive sensing. In: Scherzer O (ed) Handbook of mathematical methods in imaging. Springer, New York, NY, Chapter in Part 2
24. Foucart S, Rauhut H (2013) A mathematical introduction to compressive sensing. Springer, New York, NY
25. Gurbuz AC, McClellan JH, Scott WR Jr (2009) A compressive sensing data acquisition and imaging method for stepped frequency GPRs. IEEE Trans Geosci Remote Sens 57 (7):2640–2650
26. Jokar S, Pfetsch ME (2007) Exact and approximate sparse solutions of underdetermined linear equations. SIAM J Sci Comput 31(1):23–44

27. Laska J, Davenport M, Baraniuk R (2009) Exact signal recovery from sparsely corrupted measurements through the pursuit of justice. Asilomar Conf. on Signals, Systems, and Computers, Pacific Grove, CA
28. L1-MAGIC: http://users.ece.gatech.edu/~justin/l1magic/
29. Orovic I, Stankovic S, Thayaparan T (2014) Time-frequency based instantaneous frequency estimation of sparse signals from an incomplete set of samples. IET Signal Process 8 (3):239–245
30. Orovic I, Stankovic S (2014) Improved higher order robust distributions based on compressive sensing reconstruction. IET Signal Process 8(7):738–748
31. Orovic I, Stankovic S, Stankovic L (2014) Compressive sensing based separation of lfm signals. 56th International Symposium ELMAR, Zadar, Croatia
32. Peyré G (2010) Best basis compressed sensing. IEEE Trans Signal Process 58(5):2613–2622
33. Romberg J (2008) Imaging via compressive sampling. IEEE Signal Process Mag 25(2):14–20
34. Saab R, Chartrand R, Yilmaz Ö (2008) Stable sparse approximation via nonconvex optimization. IEEE Int. Conf. on Acoustics, Speech, and Signal Processing (ICASSP), Las Vegas, NV
35. Saligrama V, Zhao M (2008) Thresholded basis pursuit: quantizing linear programming solutions for optimal support recovery and approximation in compressed sensing, arXiv:0809.4883
36. Stankovic I, Orovic I, Stankovic S (2014) Image reconstruction from a reduced set of pixels using a simplified gradient algorithm. 22nd Telecommunications Forum TELFOR 2014, Belgrade, Serbia
37. Stankovic L, Orovic I, Stankovic S, Amin M (2013) Compressive sensing based separation of non-stationary and stationary signals overlapping in time-frequency. IEEE Trans Signal Process 61(18):4562–4572
38. Stanković L, Stanković S, Thayaparan T, Daković M, Orović I (2015) Separation and reconstruction of the rigid body and micro-doppler signal in ISAR Part I – Theory. IET Radar, Sonar & Navigation, December 2015
39. Stanković L, Stanković S, Thayaparan T, Daković M, Orović I (2015) Separation and reconstruction of the rigid body and micro-doppler signal in ISAR Part II – Statistical Analysis. IET Radar, Sonar & Navigation, December 2015.
40. Stanković L, Stanković S, Orović I, Amin M (2013) Robust time-frequency analysis based on the L-estimation and compressive sensing. IEEE Signal Process Lett 20(5):499–502
41. Stanković L, Stanković S, Amin M (2014) Missing samples analysis in signals for applications to L-estimation and compressive sensing. Signal Process 94:401–408
42. Stankovic L (2015) Digital signal processing with selected topics: adaptive systems, sparse signal processing, time-frequency analysis, draft 2015
43. Stanković L, Daković M, Vujović S (2014) Adaptive variable step algorithm for missing samples recovery in sparse signals. IET Signal Process 8(3):246–256
44. Stankovic L, Dakovic M (2015) Reconstruction of Randomly Sampled Sparse Signals Using an Adaptive Gradient Algorithm, arXiv:1412.0624
45. Stankovic L, Stanković S, Orović I, Zhang Y (2014) Time-frequency analysis of micro-doppler signals based on compressive sensing. Compressive Sensing for Urban Radar. CRC-Press, Boca Raton, FL
46. Stanković S, Orović I, Stanković L (2014) An automated signal reconstruction method based on analysis of compressive sensed signals in noisy environment. Signal Process 104(Nov 2014):43–50
47. Stankovic S, Stankovic L, Orovic I (2013) L-statistics combined with compressive sensing. SPIE Defense, Security and Sensing, Baltimore, MD
48. Stanković S, Stanković L, Orović I (2014) Relationship between the robust statistics theory and Sparse compressive sensed signals reconstruction. IET Signal Process 8:223
49. Stankovic S, Orovic I, Amin M (2013) L-statistics based modification of reconstruction algorithms for compressive sensing in the presence of impulse noise. Signal Process 93 (11):2927–2931

50. Stanković S, Orović I, Stanković L, Draganić A (2014) Single-iteration algorithm for compressive sensing reconstruction. Telfor J 6(1):36–41
51. Tropp J, Gilbert A (2007) Signal recovery from random measurements via orthogonal matching pursuit. IEEE Trans Inf Theory 53(12):4655–4666
52. Tropp J, Needell D (2009) CoSaMP: iterative signal recovery from incomplete and inaccurate samples. Appl Comput Harmon Anal 26:301
53. Vujović S, Daković M, Stanković L (2014) Comparison of the L1-magic and the gradient algorithm for sparse signals reconstruction. 22nd Telecommunications Forum, TELFOR, 2014, Belgrade
54. Yoon Y, Amin MG (2008) Compressed sensing technique for high-resolution radar imaging. Proc. SPIE, 6968, Article ID: 69681A, 10 pp

# Chapter 7
# Digital Watermarking

Advances in the development of digital data and the Internet have resulted in changes in the modern way of communication. A digital multimedia content, as opposed to an analog one, does not lose quality due to multiple copying processes. However, this advantage of digital media is also their major disadvantage in terms of copyright and the unauthorized use of data.

Cryptographic methods and digital watermarking techniques have been introduced in order to protect the digital multimedia content. Cryptography is used to protect the content during transmission from sender to recipient. On the other hand, digital watermarking techniques embed permanent information into a multimedia content. The digital signal embedded in the multimedia data is called digital watermark. A watermarking procedure can be used for the following purposes: ownership protection, protection and proof of copyrights, data authenticity protection, tracking of digital copies, copy and access controls.

The general scheme of watermarking is shown in Fig. 7.1. In general, a watermarking procedure consists of watermark embedding and watermark detection. Although the low watermark strength is preferable in order to meet the imperceptibility requirement, one must ensure that such a watermark is detectable, as well. This can be achieved by using an appropriate watermark detector.

Watermark embedding can be based on additive or multiplicative procedures. In multiplicative procedures, the watermark is multiplied by the original content. Watermark detection can be blind (without using the original content) or non-blind (in the presence of the original content).

## 7.1 Classification of Digital Watermarking Techniques

A number of different watermarking techniques have been developed. Most of them can be classified into one of the categories given in Fig. 7.2. From the perceptual aspect, the watermark can be classified as either perceptible or imperceptible.



S. Stanković et al., *Multimedia Signals and Systems*,
DOI 10.1007/978-3-319-23950-7_7

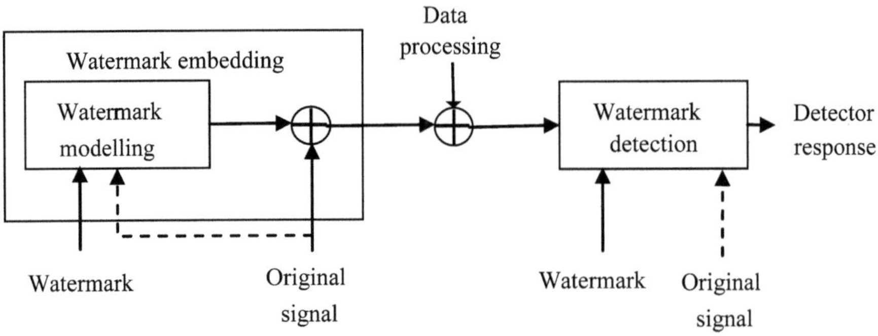


**Fig. 7.1** A block scheme of a watermarking procedure

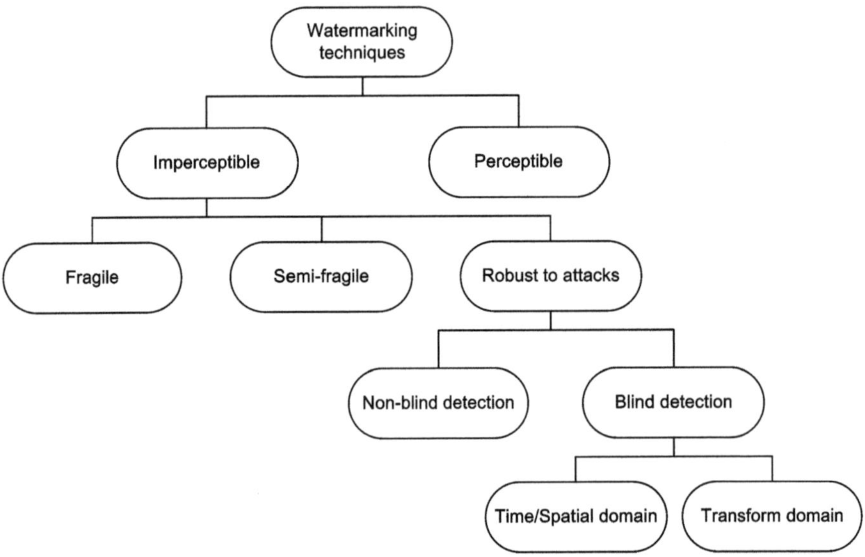


**Fig. 7.2** Classification of digital watermarking techniques

Noticeable watermark visibly changes the original content. It is sometimes used to protect images and videos, but generally it is not very popular nowadays. This technique involves embedding characters that uniquely identify the owners of the content and appear as a background image, or as a visible sign. However, watermark of this type can be removed. Almost all currently used techniques fall into the class of imperceptible techniques.

Imperceptible techniques are further divided into robust techniques, semi-fragile and fragile watermarking techniques. Fragile watermarking assumes embedding of certain watermark that will be significantly damaged or removed in an attempt to modify the content. These techniques are useful in proving the data authenticity. In semi-fragile watermarking, the watermark should be resistant to certain signal processing techniques (e.g., compression), while it is fragile under any other attack. However, the most commonly used techniques are based on the robust watermarking and are considered in the next sections.

Robust techniques involve embedding a watermark in the original signal, such that the watermark removal causes serious degradation of the signal quality. Watermark should be designed to be robust to the standard signal processing approaches (compression, filtering, etc.), as well as to intentional attempts to remove the watermark.

*Classification in terms of the embedding domain*

Watermarking techniques are further divided by the domains in which the watermark is embedded. Namely, the watermark can be embedded directly in the signal domain, or in one of the transform domains. The choice of the watermarking domain depends on the type of multimedia data and the watermarking application. The most frequently used transform domains are based on the DFT, DCT, and DWT transforms. The transform domain watermarking is more convenient for modeling the spectral characteristics of watermark according to the human perceptual model. For highly nonstationary signals, the modeling can be achieved by using time-frequency transforms.

## 7.2 Common Requirements Considered in Watermarking

Depending on the application and the type of data to be watermarked, the watermarking procedure should fulfill a number of requirements. In the sequel, we discuss some general and very common watermarking requirements.

1. The watermark should be accessible only to the authorized users. This issue is referred as security of the watermarking procedure and it is generally achieved by using cryptographic keys.
2. The watermark detectability should be assured regardless of the conventional signal processing or malicious attacks that may be applied.
3. Generally, although one should provide an unremovable watermark, it should be imperceptible within the host data.
4. The watermark should convey a sufficient amount of information.

As stated above, the first requirement is related to the security of the watermark and watermarking procedure in general. In some applications, the specific security keys (which can be encrypted) are used during the watermark embedding and extraction. If the watermark is created as a pseudo-random sequence, then the key used to generate a sequence can be considered as a watermarking key.

The next requirement is watermark robustness, which is one of the main challenges when designing the watermarking procedure. The watermark should be robust not only to the standard signal processing techniques, but also to the malicious attacks aiming to remove the watermark. All algorithms that may lead to

Table 7.1 Common attacks in audio and image watermarking procedures

| Attacks | |
|---|---|
| Audio watermarking | Image watermarking |
| Resampling | Requantization |
| Wow and flutter | JPEG compression |
| Requantization | Darkening |
| mp3 with constant bit rate | Lightening |
| mp3 with variable bit rates | Mean filter (of size $3 \times 3$, $5 \times 5$, $7 \times 7$) |
| Pitch scaling | Median filter (of size $3 \times 3$, $5 \times 5$, $7 \times 7$) |
| Audio samples cropping | Image cropping |
| Echo and time-scale modifications | Image resize |
| Filtering | Rotation |
| Amplitude normalization | Adding noise Gaussian or impulse |

the loss of the watermark information are simply called attacks. Some of the common examples are compression algorithms, filtering, change of the data format, noise, cropping signal samples, resampling, etc. The list of commonly present attacks for audio signals and images is given in Table 7.1.

Perceptual transparency is one of the most important requirements. Watermark should be adapted to the host content, and should not introduce any perceptible artifacts or signal quality degradations. However, the imperceptibility is usually confronted with the watermark robustness requirement. In order to be imperceptible, the watermark strength should be low, which directly affects its robustness. Hence, an efficient watermarking procedure should always provide a trade-off between the imperceptibility and robustness. In order to perform the watermark embedding just below the threshold of perception, various masking procedures can be employed.

In some applications it is desirable that the watermark convey a significant number of bits, which will be extracted by detector. Hence, it is sometimes required that the watermark data rate (payload) is high. The property that describes the ability to embed a certain amount of information is known as a capacity of the watermarking algorithm.

Besides the general watermarking requirements discussed above, there could be some specific requirements, as well, related to the following issues:

- Real-time implementation,
- Complete extraction/reconstruction of the watermark at the decoder,
- The absence of the original data during the watermark extraction (blind extraction), etc.

## 7.3 Watermark Embedding

This section considers the additive and multiplicative watermark embedding techniques.

Additive embedding techniques can be defined as:

$$I_w = I + \alpha w, \tag{7.1}$$

where $I$ represents the vector of signal samples or transform domain coefficients used for watermarking, $I_w$ is the vector of watermarked coefficients, $w$ is the watermark, while the parameter $\alpha$ controls the watermark strength. If the parameter $\alpha$ should be adjusted to the signal coefficients then the watermark embedding can be written as:

$$I_w = I + \alpha(I)w. \tag{7.2}$$

Another frequently used approach is multiplicative embedding, given by the relation:

$$I_w = I + \alpha w I. \tag{7.3}$$

In order to provide that the watermark does not depend on the sign of selected watermarking coefficients, a modified version of Eq. (7.3) can be used:

$$I_w = I + \alpha w|I|. \tag{7.4}$$

Multiplicative watermarking is often used in the frequency domain, to ensure that the watermark energy at a particular frequency is proportional to the image energy at that frequency. An additional advantage of multiplicative watermark embedding is that it is difficult to estimate and remove watermark by averaging a set of watermarked signals, which is one of the common attacks.

Let us consider an example of robust image watermarking in the transform domain. The two-dimensional DFT of an image is shown in Fig. 7.3a, while its two-dimensional DCT is illustrated in Fig. 7.3b. Note that the DCT is real and has only positive part of the spectrum, making it suitable for applications in watermarking.

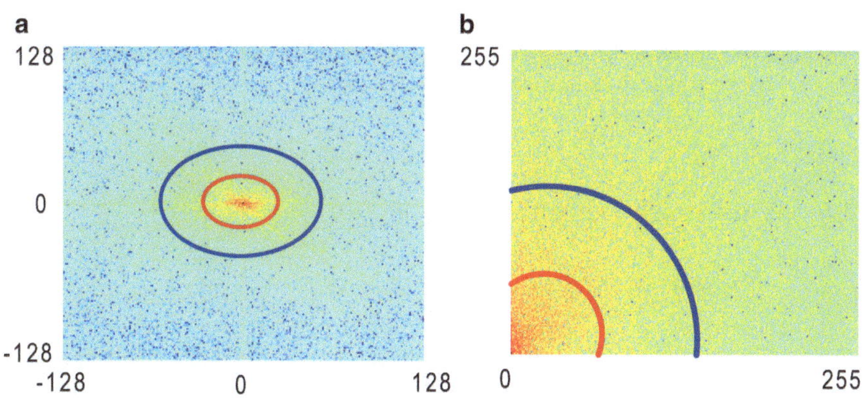


**Fig. 7.3** (**a**) The DFT of "Baboon" image, (**b**) The DCT of "Baboon" image

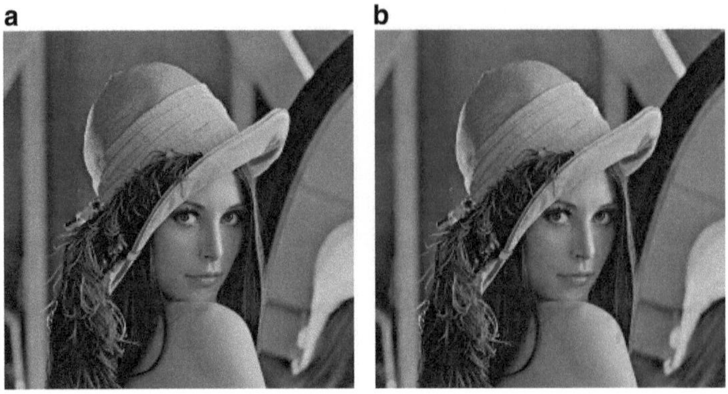

**Fig. 7.4** (**a**) Original "Lena" image, (**b**) watermarked "Lena" image

The region marked by the red line corresponds to low-frequency coefficients, which contain most of the image energy. Consequently, the modification of these coefficients can cause significant image quality degradation. Therefore, the low-frequency coefficients are usually avoided in watermarking. Outside the blue circle, we have high-frequency components. These components carry certain image details and can be filtered out, without significant image degradation. Therefore, the high-frequency components are also often omitted in watermarking. It follows that the watermarking should be done in the middle frequency part (between the blue and red circles in Fig. 7.3).

Consider the sorted DCT coefficients of an image. Given the nature of the DCT transform, it is necessary to omit the first $L$ coefficients (which are mostly the low-frequency components) and choose the next $M$ coefficients (mostly belonging to the middle frequencies). Watermarking is then performed as:

$$I_w(i) = I(i) + \alpha|I(i)|w(i) \quad \text{for } i = L+1, L+2, \ldots, L+M, \tag{7.5}$$

where $I(i)$ denotes a DCT coefficient of an image. The watermark $w$ can be created as a pseudo-random sequence. The inverse DCT is then applied to obtain the watermarked image. The original and watermarked "Lena" images are shown in Fig. 7.4 (peak signal to noise ratio PSNR = 47 dB).

## 7.4 Watermark Detection

### 7.4.1 Hypothesis Testing Approach

The goal of each algorithm for watermark detection is to provide a reliable proof of the watermark presence within the signal. Denote by $I_x$ a set of coefficients on which the watermark detection is performed ($I_x$ can be either $I_w$ or $I$ depending whether the watermark is present or not), and the watermark is $w$. A general approach for watermark detection is based on *a hypothesis testing problem*. The assumptions are:

$H_0$: $I_x$ does not contain watermark $w$,
$H_1$: $I_x$ contains watermark $w$.

The problem of watermark detection is based on a reliable threshold, used to decide whether a watermark is present or not. The threshold is determined by defining a criterion that ensures a minimum probability of detection error. Since the watermark detection can be viewed as a detection of signal in noise, the *likelihood ratio* is used to minimize the error. Detection errors can occur in two cases: $G_{10}$—when the assumption of $H_0$ is accepted as true, although the correct hypothesis is $H_1$, $G_{01}$—when the assumption $H_1$ is accepted as true, but the correct hypothesis is $H_0$.

The criterion that determines the presence of the watermark is defined as follows:

$$\Phi(I_x) = \begin{cases} 1, & I_x \in R_1, \\ 0, & I_x \in R_0, \end{cases} \tag{7.6}$$

where $R_1$ and $R_0$ are regions in which the assumptions $\mathrm{H}_1$ and $\mathrm{H}_0$ are tested. In order to minimize error during detection, a likelihood ratio $l$ is defined by using the conditional probability density functions $p(I_x|H_1)$ and $p(I_x|H_0)$:

$$l(I_x) = \frac{p(I_x|H_1)}{p(I_x|H_0)}. \tag{7.7}$$

The minimum probability of error will be achieved when the region $R_1$ is determined as:

$$R_1 = \left\{ I_x : l(I_x) > \frac{p_0 P_{01}}{p_1 P_{10}} \right\}, \tag{7.8}$$

where $p_0$ and $p_1$ are a priori known probabilities of the assumptions $H_0$ and $H_1$ occurrence, while $P_{01}$ and $P_{10}$ are decision weights associated with $G_{01}$ and $G_{10}$, respectively. The criterion for the detection can be written as:

$$\Phi(I_x) = \begin{cases} 1, & l(I_x) > \dfrac{p_0 P_{01}}{p_1 P_{10}} \\ 0, & \text{otherwise.} \end{cases} \tag{7.9}$$

Therefore, the detection is done by comparing the likelihood ratio with:

$$\lambda = \frac{p_0 P_{01}}{p_1 P_{10}}. \tag{7.10}$$

The threshold λ can be set to minimize the total probability of error that occurs during detection:

$$P_e = p_0 P_f + p_1(1 - P_d), \tag{7.11}$$

where $P_f$ is the probability that the watermark is detected, when in fact it is not present (false alarm), and $(1 - P_d)$ is the probability of watermark misdetection. The error minimization procedure is commonly performed under the assumption that $P_{01} = P_{10}$ and $p_0 = p_1$, or in other words for $\lambda = 1$. It means that the probabilities of false alarm $P_f$ and misdetection $P_m = (1 - P_d)$ are the same.

In practice, we usually have a predefined maximum false alarm probability from which the threshold λ is calculated as follows:

$$\int_{\lambda}^{\infty} p(l|H_0)dl = P_f, \tag{7.12}$$

where $p(l|H_0)$ is the pdf of $l$ under $H_0$. After the threshold λ is determined, the probability of misdetection is calculated as:

$$P_m = \int_{-\infty}^{\lambda} p(l|H_1)dl. \tag{7.13}$$

#### 7.4.1.1 Additive White Gaussian Model

Let us consider the procedure to minimize the detection error in the case of additive white Gaussian noise (AWGN) model, which is the simplest one encountered in practice. This model assumes that the coefficients are uncorrelated and have a Gaussian distribution. Note that the watermark is considered as a noisy signal:

$$I_x = I + w + n, \tag{7.14}$$

where $I_x$, $I$ and $w$ are the coefficients of the watermarked content, the original content and the watermark, respectively. The watermarked content can be modified in the presence of attack, which is modeled by noise $n$ (white Gaussian noise). Under the assumption that the original coefficients, as well as the noise coefficients, are uncorrelated and follow the Gaussian distribution, Eq. (7.14) can be written as follows:

$$I_x = I_n + w. \tag{7.15}$$

$I_n$ also has the Gaussian distribution with the modified mean value and the variance compared to the original content $I$. Now, the previously defined hypothesis can be written as:

$$H_0 : I_x = I_n$$
$$H_1 : I_x = I_n + w.$$

In order to minimize the similarity measure $l(I_x) = \frac{p(I_x|H_1)}{p(I_x|H_0)}$, it is necessary to know the conditional probability density function, which in the case of the Gaussian distribution is defined as:

$$p(I_x|H_1) = \prod_{i=1}^{M} \frac{1}{\sqrt{2\pi\sigma_x^2}} e^{-\frac{(I_x(i)-\mu_x-w(i))^2}{2\sigma_x^2}}$$
$$p(I_x|H_0) = \prod_{i=1}^{M} \frac{1}{\sqrt{2\pi\sigma_x^2}} e^{-\frac{(I_x(i)-\mu_x)^2}{2\sigma_x^2}}, \tag{7.16}$$

where $\mu_x$ is the mean value of signal coefficients used in watermark detection, while $M$ is the length of watermark. Now the measure of similarity is calculated as:

$$l(I_x) = \frac{p(I_x|H_1)}{p(I_x|H_0)} = \frac{\prod_{i=1}^{M} e^{-\frac{(I_x(i)-\mu_x-w(i))^2}{2\sigma_x^2}}}{\prod_{i=1}^{M} e^{-\frac{(I_x(i)-\mu_x)^2}{2\sigma_x^2}}}. \tag{7.17}$$

Equation (7.17) can be written in a simplified form by applying the logarithmic function:

$$\ell(I_x) = \sum_{i=1}^{n} \frac{1}{2\sigma_x^2}\left[(I_x(i) - \mu_x)^2 - (I_x(i) - \mu_x - w(i))^2\right]$$
$$= \frac{1}{2\sigma_x^2}\left[\sum_{i=1}^{n} 2I_x(i)w(i) - \sum_{i=1}^{n} 2\mu_x w(i) - \sum_{i=1}^{n} w^2(i)\right], \tag{7.18}$$

where $\ell(I_x)$ indicates the natural logarithm function of $l(I_x)$. Note that the last two terms within the brackets do not depend on $I_x$. Therefore, the term representing linear correlation of $I_x$ and $w$ is used as a watermark detector:

$$D = \sum_{i=1}^{M} I_x(i)w(i), \tag{7.19}$$

which is optimal under the considered assumptions and is called the standard correlator. In the case when the signal statistics is not distributed according to the Gaussian distribution, other detector forms can be used.

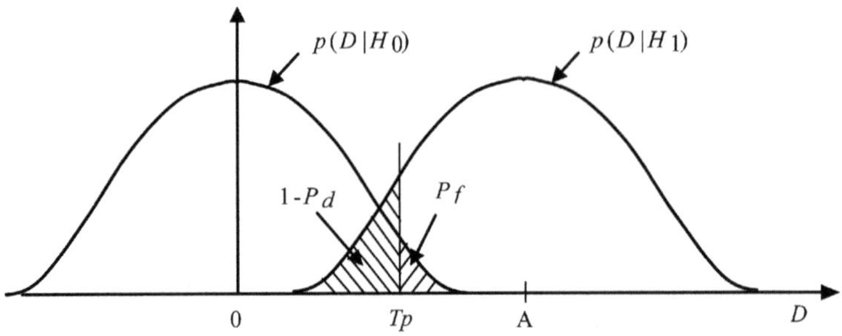


**Fig. 7.5** Illustration of the errors that may occur in watermark detection

According to the procedure for determining the general detection threshold λ, we can now determine the threshold for the standard correlator as:

$$\int_{T_p}^{\infty} p(D|H_0)dD = P_f, \tag{7.20}$$

where $p(D|H_0)$ is the pdf of detector responses $D$ under $H_0$. The pdf of $D$ under $H_0$ and $H_1$ are illustrated in Fig. 7.5. If the response of the detector is $D < T_p$, we conclude that the watermark is not present, and vice versa. In the case of equal probabilities $P_f = 1 - P_d$, the optimum threshold is $A/2$ (Fig. 7.5).

In order to determine the threshold and the probability of error, we need to know how the watermark is embedded, the statistical characteristics of the image coefficients, as well as the characteristics of attacks.

### 7.4.2 A Class of Locally Optimal Detectors

According to the signal detection theory, it is difficult to define a general test that maximizes the signal detection probability. Also, it is known that for detection of weak signals a locally optimal detector can be created (in our case a watermark signal is weak in comparison to the host signal). It is defined as follows:

$$D = g_{LO}(I_x) \cdot w, \tag{7.21}$$

where $g_{LO}$ is the local optimum nonlinearity, defined by:

$$g_{LO}(I_x) = -\frac{p'(I_x)}{p(I_x)}, \tag{7.22}$$

with $p(I_x)$ and $p'(I_x)$ indicating the coefficients probability density function and its derivative, respectively. Note that, the detector contains the nonlinear part $g_{LO}$, which is correlated with the watermark signal. If the coefficients have the Gaussian distribution, the proposed detector corresponds to the standard correlator.

#### 7.4.2.1 The Most Commonly Used Distribution Functions and the Corresponding Detector Forms

The coefficients distribution for most images can be modeled by the Gaussian, Laplace, generalized Gaussian or Cauchy distribution functions. For example, recall that the generalized Gaussian function can be defined as:

$$GGF = \frac{\alpha\beta}{2\Gamma(1/\alpha)} e^{(-\beta|x-\mu|)^{\alpha}}, \ \alpha > 0, \ \beta = \frac{1}{\sigma}\left[\frac{\Gamma(3/\alpha)}{\Gamma(1/\alpha)}\right]^{1/2}. \tag{7.23}$$

For $\alpha = 1$, this function is equal to the Laplace distribution, and for $\alpha = 2$ it is equal to the Gaussian distribution. Figure 7.6 shows the coefficients distribution of an image. The form of the detector, which corresponds to the generalized Gaussian distribution, is given by:

$$D_1 = \sum_{i=1}^{M} sign(I_x(i))|I_x(i)|^{\alpha-1} w(i), \tag{7.24}$$

while the detector form for Cauchy distribution, $CF = \frac{\gamma}{\pi\left(\gamma^2+(x-\delta)^2\right)}$, is equal to:

$$D_2 = \sum_{i=1}^{M} \frac{2(I_x(i) - \delta)}{(I_x(i) - \delta)^2 + \gamma^2} w(i). \tag{7.25}$$

Note that $x$ (in the pdf) corresponds to the watermarked coefficients $I_x$ in the detector form ($M$ is the length of watermarked sequence and watermark). It is

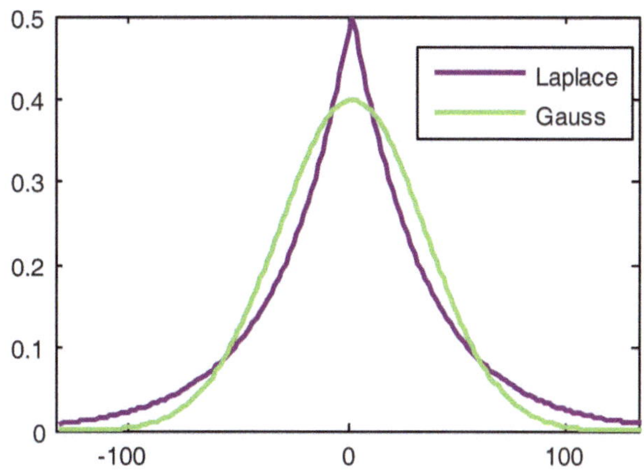


**Fig. 7.6** The distribution of coefficients: Gaussian (*green line*) and Laplace distribution (*purple line*)

important to emphasize that the locally optimum detector form can be quite sensitive to the pdf variations.

A simple measure of detection quality can be defined as:

$$R = \frac{\overline{D}_{wr} - \overline{D}_{ww}}{\sqrt{{\sigma_{wr}}^2 + {\sigma_{ww}}^2}}, \tag{7.26}$$

where $D$ and $\sigma$ are mean values and standard deviations of detector responses, while the indices $wr$ and $ww$ are used for right keys (watermarks) and wrong keys (wrong trials), respectively. The wrong trial is any sequence which is not the watermark, but is generated in the same way.

### *7.4.3 Correlation Coefficient and Similarity Measure*

In order to determine the similarity between the original watermark $w$ and the watermark $w^*$ extracted from the protected data at the detection side, we can use the similarity measure defined as follows:

$$Sim(w, w^*) = \frac{\mathbf{w} \cdot \mathbf{w}^*}{\sqrt{\mathbf{w} \cdot \mathbf{w}^*}}. \tag{7.27}$$

The similarity measure is usually given in the form of the correlation coefficient, which can be calculated as:

$$\rho(w, w^*) = \frac{\sum_{i=1}^{M} w(i) w^*(i)}{\sqrt{\sum_{i=1}^{M} (w(i))^2} \sqrt{\sum_{i=1}^{M} (w^*(i))^2}}, \tag{7.28}$$

where $M$ is the length of watermark.

## 7.5 Examples of Watermarking Procedures

### *7.5.1 Audio Watermarking Techniques*

Audio watermarking procedures are mainly based on the specific audio signal characteristics and psychoacoustics. In the next subsections, a brief description of audio watermarking approaches such as the spread-spectrum audio watermarking, two-sets method and echo embedding, is provided.

#### 7.5.1.1 Spread-Spectrum Watermarking

Spread-spectrum watermarking is an example of correlation based method that assumes a pseudo-random sequence embedding, where the standard correlator is used for detection. This is a commonly used watermarking approach. The pseudo-random sequence, *r*(*n*), i.e., the wide-band noise sequence, can be embedded in the time or in the transform domain. This sequence is used to modulate the binary message $\upsilon = \{0, 1\}$, or equivalently $b = \{-1, 1\}$. The watermarked sequence $w(n) = br(n)$ obtained in this way, is scaled according to the energy of the host signal $s(n)$, to provide a compromise between the watermark imperceptibility and robustness. The watermark embedding can be done, for example, by an additive procedure: $s_w(n) = s(n) + \alpha w(n)$. A suitable pseudo-random sequence should have good correlation properties, in the way that it should be orthogonal to the other pseudo-random sequences. The commonly used sequence is called the *m-sequence* (maximum length sequence), and its autocorrelation is given by:

$$\frac{1}{M}\sum_{i=0}^{M-1} w(i)w(i-k) = \begin{cases} 1, & \text{for } k = 0, \\ -\frac{1}{M}, & \text{for } k \neq 0. \end{cases} \tag{7.29}$$

#### 7.5.1.2 Two Sets Method

This blind audio watermarking procedure is based on the two sets A and B of audio samples. A value *d* (watermark) is added to the samples within the set A, while it is subtracted from the samples in B:

$$a_i^* = a_i + d, \quad b_i^* = b_i - d,$$

where $a_i$ and $b_i$ are samples from A and B, respectively. When making decision about watermark presence, the expected value $E\left[\bar{a}^* - \bar{b}^*\right]$ is employed, where $\bar{a}^*$ and $\bar{b}^*$ are mean values of samples $a_i^*$ and $b_i^*$. This method is based on the assumption that the mean values of the samples from different signal blocks are the same, i.e., that $E\left[\bar{a} - \bar{b}\right] = 0$ holds (which may not be always the case in the practice). Only in this case, the watermark can be detected as:

$$E\left[\bar{a}^* - \bar{b}^*\right] = E\left[(\bar{a} + d) - (\bar{b} - d)\right] = E(\bar{a} - \bar{b}) + 2d = 2d. \tag{7.30}$$

#### 7.5.1.3 Echo Embedding

The echo embedding procedure can be realized according to:

$$x(n) = s(n) + \alpha s(n - d), \tag{7.31}$$

where $d$ represents a certain delay of the echo signal. The extraction of the embedded echo requires the detection of delay $d$. The signal copy is usually delayed for approximately 1 ms. The echo amplitude is significantly lower than the original signal amplitude, and hence, the signal quality is not degraded. On the contrary, the sound is enriched. There is also a variant of this procedure, where two delays are considered: one is related to the logical value "1", while the other is related to "0". The double echo embedding operation can be written as:

$$x(n) = s(n) + \alpha s(n-d) - \alpha s(n-d-\Delta), \tag{7.32}$$

where the difference between delays corresponding to "1" and "0" is denoted by $\Delta$, and its value does not exceed four samples. The delay detection is done by using the cepstrum autocorrelation, which is the inverse Fourier transform of the - log-magnitude spectrum. The complexity of cepstrum calculation is one of the main disadvantages of this method.

#### 7.5.1.4 Watermarking Based on the Time-Scale Modifications

Time-scale modifications are related to compressing and expanding of the time axis. The basic idea of time-scale watermarking is to change the time scale between two successive extremes (maximum and minimum). The interval between two extremes is divided into $N$ segments with equal amplitudes. The signal slope is changed within a certain amplitudes range according to the bits that should be embedded. Namely, the steep slope corresponds to bit "0", while the mild slope corresponds to bit "1".

### 7.5.2 *Image Watermarking Techniques*

A simple watermarking algorithm for digital image protection is based on the additive watermark embedding procedure in the $8 \times 8$ DCT domain. First, an image is divided into $8 \times 8$ blocks of pixels as in the case of JPEG algorithm. Then, the 2D DCT transform is applied to each block separately. The watermark is embedded into the set of selected coefficients. In order to provide a good compromise between the watermark imperceptibility and robustness, the coefficients are selected from the middle frequency region, as illustrated in Fig. 7.7.

Watermark embedding is based on the standard additive procedure: $I_w = I + \alpha w$, where $I$ denotes the original middle-frequency DCT coefficients (from $8 \times 8$ block), while $I_w$ are the watermarked DCT coefficients. Next, we perform the inverse DCT transform that results in watermarked $8 \times 8$ block. This is repeated for each block.

Watermark detection can be performed by using a standard correlation detector (assuming that the distribution of selected coefficients can be modeled by the Gaussian function). However, more accurate modeling can be obtained by using

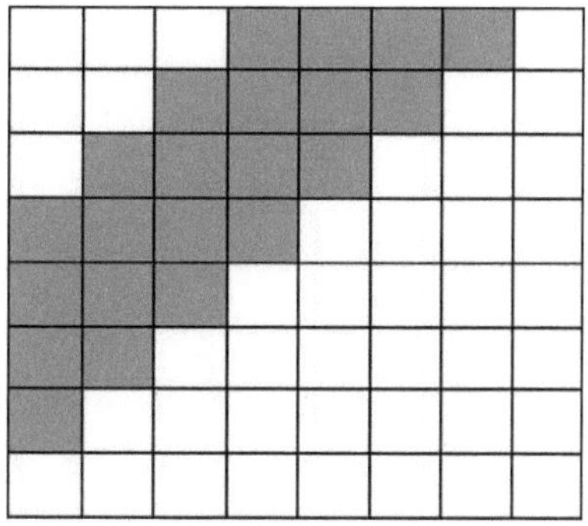

**Fig. 7.7** A region of middle frequency DCT coefficients within 8 × 8 block (shaded in *gray*)

generalized Gaussian and Cauchy function, where the corresponding detectors $D_1$ and $D_2$ (defined by Eqs. (7.24) and (7.25)) are used for detection.

### 7.5.3 *The Procedure for Watermarking of Color Images*

Unlike the previous procedure, where the block-based DCT is performed, here we use the 2D DCT transform of the entire image. The procedure is described in the sequel.

(a) The selection of coefficients for watermark embedding is done through the following steps:

1. The color channels are separated (e.g., R, G, B), Fig. 7.8.
2. 2D DCT is computed for each color matrix
3. The matrices of DCT coefficients are transformed into vectors and sorting operation is performed
4. The largest $L$ coefficients are omitted and the next $M$ coefficients are selected for watermarking.

(b) Watermark embedding
Let us denote the sorted DCT coefficients by $I$, while $w$ is the watermark created as a pseudo-random sequence. The watermarked DCT coefficients are calculated as:

$$I_w(i) = I(i) + \alpha \cdot |I(i)| \cdot w(i), \quad i = L+1, \ldots, M+L,$$

where $i$ denotes the coefficient position in the sorted sequence.

(c) Reorder the sequence into matrix form.

(d) Calculate the 2D inverse DCT (with rounding to integer values).

**Fig. 7.8** Color image "Lena" and the separated color channels

### 7.5.4 An Overview of Some Time-Frequency Based Watermarking Techniques

The time-frequency based watermarking can be used for different types of multimedia data: audio signals, images and video signals. The time-frequency domain can be efficient regarding the watermark imperceptibility and robustness. Namely, the watermark with specific time-frequency characteristics can be designed and adapted to the host signal components, which enhances the efficiency of the watermarking procedure. Note that the time-frequency representations defined for one-dimensional signals can be extended to two-dimensional case in order to be applied to images. In this case, they are usually referred as the space/spatial-frequency representations.

1. The watermark can be created with specific space/spatial-frequency characteristics, while its embedding can be done even in the space domain. This approach is among the first space/spatial-frequency based image watermarking procedures. Namely, a two dimensional chirp signal is used as watermark:

$$W(x, y) = 2A\cos\left(ax^2 + by^2\right) = A\left(e^{j\left(ax^2+by^2\right)} + e^{-j\left(ax^2+by^2\right)}\right). \tag{7.33}$$

The watermark is embedded within the entire image:

$$I_w(x, y) = I(x, y) + W(x, y). \tag{7.34}$$

It is interesting to observe that multiple different chirps with small amplitudes can be used for watermarking. The parameters of the chirp signals and the random sequence that defines the amplitudes of chirps, serve as the watermark key. Since the watermark is embedded within the entire image in the spatial domain, a proper masking that provides imperceptibility should be applied. Note that the Wigner distribution provides an ideal representation for the chirp signal. Hence, the watermark detection is performed by using a form of the Radon–Wigner distribution:

$$P(\omega_x, \omega_y; W_v) = |FT_{2D}(I_w(x, y)W_v(x, y))|^2 = \left| \int_{-\infty}^{\infty} \int_{-\infty}^{\infty} I_w(x, y)W_v(x, y)e^{-j(x\omega_x + y\omega_y)}dxdy \right|^2, \tag{7.35}$$

where $FT_{2D}$ denotes the 2D Fourier transform, while:

$$W_v(x, y) = e^{-j(a_v x^2 + b_v y^2 + c_v xy)}. \tag{7.36}$$

Different values of parameters $a_v$, $b_v$, and $c_v$ define a set of projection planes. The additional term $c_v xy$ is used to detect some geometrical transformations, as well. In order to make a decision about the watermark presence within the image, the maxima of the Radon–Wigner distribution are calculated:

$$M(a_v, b_v, c_v) = \max_{\omega_x, \omega_y} P(\omega_x, \omega_y; W_v), \tag{7.37}$$

and compared with a reference threshold. This procedure provides robustness to various attacks, some being a median filter, geometrical transformations (translation, rotation and cropping simultaneously applied), a high-pass filter, local notch filter and Gaussian noise.

2. Digital audio watermarking can be done by using time-frequency expansion and compression. The audio signal is firstly divided into frames of size 1024 samples, where the successive frames have 512 samples overlapping. If the original frame is lengthened or shortened, the logical value 1 is assigned, otherwise the "normal frames" corresponds to the logical value 0. The watermark is a sequence obtained as a binary code of the alphabet letters, converted to the ASCII code. The frames with signal energy level above a certain threshold are selected. The signal is transformed to frequency domain and a psychoacoustic model is used to determine the masking threshold for each selected frame. The frames length is changed in frequency domain by adding or removing four samples with amplitudes that do not exceed the masking threshold. It prevents a perceptual distortion. In order to preserve the total signal length, the same number of expanded

and compressed frames is used (usually an expanded frame is followed by a compressed frame). The detection procedure is non-blind, i.e., the original signal is required. The difference between the original and watermarked samples in time domain will have diamond shape for the pair expanded-compressed frame (Diamond frames), while the difference is flat and close to zero for unaltered frames. The pair of Diamond frames is used to represent the binary 1, while the logical values 0 are assigned to the unaltered frames. Hence, it is possible to detect binary values, and consequently the corresponding alphabetical letters.

3. *A spread spectrum based watermarking in the time-frequency domain*

The watermark is created as:

$$w_i(n) = a(n)m_i(n)\,p_i(n)\cos(\omega_0(n)n), \tag{7.38}$$

where $m_i(n)$ is the watermark before spreading, $p_i(n)$ is the spreading code or the pseudo-noise sequence (bipolar sequence taking the values $+1$ and $-1$ with equal probabilities), while $\omega_0$ is the time-varying carrier frequency which represents the instantaneous mean frequency of the signal. The parameter $a(n)$ controls the watermark strength. The masking properties of the human auditory system are used to shape an imperceptible watermark. The pseudo-noise sequence is low-pass filtered according to the signal characteristics. Two different scenarios of masking have been considered. The tone- or noise-like characteristics are determined by using the entropy:

$$H = -\sum_{i=1}^{\omega_{\max}} P(x_i)\log_2 P(x_i). \tag{7.39}$$

The probability of energy for each frequency (within a window used for the spectrogram calculation) is denoted by $P(x_i)$, while $\omega_{max}$ is the maximum frequency. A half of the maximum entropy $H_{max} = log_2\omega_{max}$ is taken as a threshold between noise-like and tone-like characteristics. If the entropy is lower than $H_{max}$, it is considered as a tone-like, otherwise it is a noise-like characteristic.

The time-varying carrier frequency is obtained as the instantaneous mean frequency of the host signal, calculated by:

$$\omega_i(n) = \frac{\sum_{\omega=0}^{\omega_{\max}} \omega TFD(n,\omega)}{\sum_{\omega=0}^{\omega_{\max}} TFD(n,\omega)}. \tag{7.40}$$

The instantaneous mean frequency is computed over each time window of the STFT, and the $TFD(n,\omega)$ is the energy of the signal at a given time and frequency.

Finally, after the watermark is modulated and shaped, it is embedded in the time domain as: $s_{w_i}(n) = s_i(n) + w_i(n)$. During the detection, the demodulation is done by using the time-varying carrier and then the watermark is detected by using the standard correlation procedure.

4. *Watermarking approach based on the time-frequency shaped watermark*
In order to ensure imperceptibility constraints, the watermark can be modeled according to the time-frequency characteristics of the signal components. For this purpose the concept of nonstationary filtering is adapted and used to create a watermark with specific time-frequency characteristics. The algorithm includes the following steps:

   1. Selection of signal regions suitable for watermark embedding;
   2. Watermark modeling according to the time-frequency characteristics of the host signal;
   3. Watermark embedding and watermark detection in the time-frequency domain.

   Due to the multicomponent nature of multimedia signals (e.g., speech signals), the cross-terms free time-frequency distributions (*TFD*) should be used, such as the spectrogram and the S-method. If a region selected from the *TFD* is:

$$D = \{(t, \omega) : t \in (t_1, t_2), \omega \in (\omega_1, \omega_2)\}, \tag{7.41}$$

we can define a time-frequency mask as follows:

$$L_M(t, \omega) = \begin{cases} 1 \text{ for } (t, \omega) \in D \text{ and } |TFD(t, \omega)| > \xi, \\ 0 \text{ for } (t, \omega) \notin D \text{ or } |TFD(t, \omega)| < \xi. \end{cases} \tag{7.42}$$

   The parameter $\xi$ is a threshold which can be calculated as a portion of the *TFD* maximum: $\xi = \lambda 10^{\lambda \log_{10}(\max(|TFD(t,\omega)|))}$ ($\lambda$ is a constant). The mask $L_M$ contains the information about the significant components within the region *D*. Hence, if we start with an arbitrary random sequence *p*, the modeled watermark is obtained at the output of the nonstationary (time-varying) filter:

$$w(t) = \sum_{\omega} L_M(t, \omega) STFT_p(t, \omega), \tag{7.43}$$

where $STFT_p$ stands for the short-time Fourier transform of *p*. The watermark embedding is done according to:

$$STFT_{I_w}(t, \omega) = STFT_I(t, \omega) + STFT_{w_{key}}(t, \omega), \tag{7.44}$$

where $I_w$, *I* and *w* are related to the watermarked signal, original signal and watermark, respectively.

The watermark detector can be made by using the correlation in the time-frequency domain:

$$D = \sum_{i=1}^{M} STFT^{i}_{wkey} STFT^{i}_{I_w}. \tag{7.45}$$

Note that the time-frequency domain provides a larger number of coefficients for correlation (compared to time or frequency domains), which enhances the detection performance.

## 7.6 Examples

7.1. Consider a vector with a few image DFT coefficients chosen for watermarking.
DFT = [117 120 112 145 136 115].
The watermarking procedure should be done in the following way:

(a) Sort the vector of DFT coefficients.
(b) Add a watermark given by $w = [-3.5\ -2\ 4\ 5\ 9\ -7]$.
(c) Assume that the sequence $wrong = [3\ 2\ -5\ -7\ 2\ 4]$ provides the highest response of the correlation based detector among large number of wrong trials (wrong keys) used for testing.
(d) Prove that the watermark can be successfully detected by using the standard correlator.

Solution:

$$DFT_{sort} = [112 \quad 115 \quad 117 \quad 120 \quad 136 \quad 145],$$
$$DFT_w = DFT_{sort} + w = [108.5 \quad 113 \quad 121 \quad 125 \quad 145 \quad 138].$$

In order to ensure a reliable watermark detection using the standard correlator, the detector response for the watermark should be higher than the maximal detector response when using wrong trials:

$$\Sigma DFT_w \cdot w > \Sigma DFT_w \cdot wrong,$$
$$\Sigma DFT_w \cdot w = 842.25,$$
$$\Sigma DFT_w \cdot wrong = -86.5.$$

Having in mind the results, we may conclude that the watermark detection is successful.

7.2. Write a program in Matlab which perform the image watermarking as follows:

(a) Calculate and sort the DCT coefficients of the considered image;

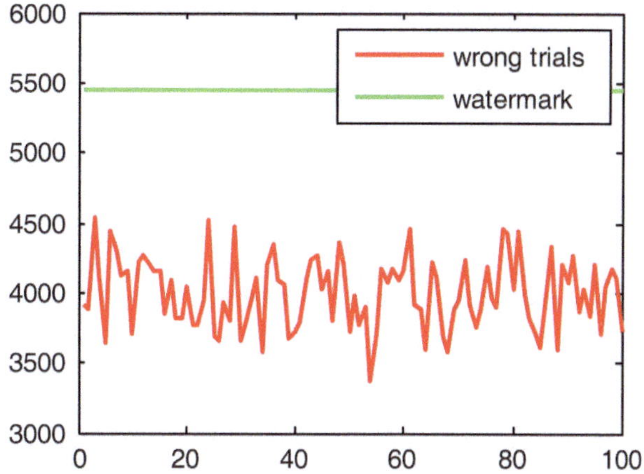


**Fig. 7.9** Results of watermark detection

(b) Create a watermark as a pseudo-random sequence, e.g., watermark = 0.5·rand(1500,1);
(c) After omitting the strongest 1500 coefficients, embed the watermark into the next 1500 coefficients by using the multiplicative procedure with $\alpha = 0.8$;
(d) Check if the watermark is imperceptible within the protected image;
(e) Perform the watermark detection in the DCT domain by using the standard correlator. It is necessary to demonstrate that the detector response for the watermark is higher than the detector response for any of the 100 wrong trials (Fig. 7.9).

Solution:

```
alfa=0.8;
Det=zeros(2,100);
image=imread('lena512.bmp');
image=image(1:2:512,1:2:512);
N=256;
DCT1=dct2(image);
Vector=DCT1(:);
[g,v]=sort(abs(Vector));
watermark=0.5*rand(1500,1);
Vectorwat=Vector;
Vectorwat(v(N*N-1500-1500+1:N*N-1500))=Vector(v(N*N-1500-
1500+1:N*N-1500))+alfa*abs(Vector(v(N*N-1500-1500+1:N*N-
1500))).*watermark;
DCTwat=DCT1;
DCTwat(:)=Vectorwat;
```

```
imagewat=idct2(DCTwat);
figure,imshow(uint8(imagewat))
DCTwat1=dct2(imagewat);
DCTwat1=DCTwat1(:);
x=DCTwat1(v(N*N-1500-1500+1:N*N-1500));
for k=1:100
wrong=0.5*rand(1500,1);
Det(1,k)=sum(x.*watermark);
Det(2,k)=sum(x.*wrong);
end
figure,
plot(1:100,Det(2,1:100),'r',1:100,Det(1,1:100),'g')
```

7.3. Consider the watermarking procedure described in the sequel. A block of the $8 \times 8$ DCT coefficients is selected. The watermark is added to the block coefficients: $I_w = I + w$. The watermarked image is exposed to the quantization attack defined by the quantization matrix $Q$. Determine which watermark samples will contribute to the difference between the watermarked and the original coefficient after quantization attack.

$$DCT = \begin{matrix} 45 & 20 & 54 & 81 & 0 & 0 & 0 & 0 \\ 15 & 77 & 0 & 11 & 0 & 0 & 0 & 0 \\ 21 & 0 & 0 & 39 & 0 & 0 & 0 & 0 \\ 27 & 44 & 52 & 75 & 0 & 0 & 0 & 0 \\ 0 & 0 & 0 & 0 & 0 & 0 & 0 & 0 \\ 0 & 0 & 0 & 0 & 0 & 0 & 0 & 0 \\ 0 & 0 & 0 & 0 & 0 & 0 & 0 & 0 \\ 0 & 0 & 0 & 0 & 0 & 0 & 0 & 0 \end{matrix} \qquad w = \begin{matrix} 5 & 4 & 1 & -3 & 0 & 0 & 0 & 0 \\ 3.5 & 5 & 0 & 5 & 0 & 0 & 0 & 0 \\ 3 & 3 & 2 & -5 & 0 & 0 & 0 & 0 \\ 0 & -2 & 0 & 6.5 & 0 & 0 & 0 & 0 \\ 0 & 0 & 0 & 0 & 0 & 0 & 0 & 0 \\ 0 & 0 & 0 & 0 & 0 & 0 & 0 & 0 \\ 0 & 0 & 0 & 0 & 0 & 0 & 0 & 0 \\ 0 & 0 & 0 & 0 & 0 & 0 & 0 & 0 \end{matrix}$$

$$Q = \begin{matrix} 3 & 5 & 7 & 9 & 11 & 13 & 15 & 17 \\ 5 & 7 & 9 & 11 & 13 & 15 & 17 & 19 \\ 7 & 9 & 11 & 13 & 15 & 17 & 19 & 21 \\ 9 & 11 & 13 & 15 & 17 & 19 & 21 & 23 \\ 11 & 13 & 15 & 17 & 19 & 21 & 23 & 25 \\ 13 & 15 & 17 & 19 & 21 & 23 & 25 & 27 \\ 15 & 17 & 19 & 21 & 23 & 25 & 27 & 29 \\ 17 & 19 & 21 & 23 & 25 & 27 & 29 & 31 \end{matrix}$$

Solution:
Approach I: It is possible to perform the quantization of the original and the watermarked coefficients, to compare them and to select the coefficients that are different after quantization.

$$DCT_q = \begin{matrix} 15 & 4 & 8 & 9 & 0 & 0 & 0 & 0 \\ 3 & 11 & 0 & 1 & 0 & 0 & 0 & 0 \\ 3 & 0 & 0 & 3 & 0 & 0 & 0 & 0 \\ 3 & 4 & 4 & 5 & 0 & 0 & 0 & 0 \\ 0 & 0 & 0 & 0 & 0 & 0 & 0 & 0 \\ 0 & 0 & 0 & 0 & 0 & 0 & 0 & 0 \\ 0 & 0 & 0 & 0 & 0 & 0 & 0 & 0 \\ 0 & 0 & 0 & 0 & 0 & 0 & 0 & 0 \end{matrix} \qquad DCT_{wq} = \begin{matrix} \boxed{17} & \boxed{5} & 8 & 9 & 0 & 0 & 0 & 0 \\ \boxed{4} & \boxed{12} & 0 & 1 & 0 & 0 & 0 & 0 \\ 3 & 0 & 0 & 3 & 0 & 0 & 0 & 0 \\ 3 & 4 & 4 & 5 & 0 & 0 & 0 & 0 \\ 0 & 0 & 0 & 0 & 0 & 0 & 0 & 0 \\ 0 & 0 & 0 & 0 & 0 & 0 & 0 & 0 \\ 0 & 0 & 0 & 0 & 0 & 0 & 0 & 0 \\ 0 & 0 & 0 & 0 & 0 & 0 & 0 & 0 \end{matrix}$$

Hence, the positions of the selected coefficients are: (1,1), (1,2), (2,1), (2,2).
Approach II: Select the watermark samples higher than $Q/2$.

$$\frac{Q}{2} = \begin{matrix} 1.5 & 2.5 & 3.5 & 4.5 & 5.5 & 6.5 & 7.5 & 8.5 \\ 2.5 & 3.5 & 4.5 & 5.5 & 6.5 & 7.5 & 8.5 & 9.5 \\ 3.5 & 4.5 & 5.5 & 6.5 & 7.5 & 8.5 & 9.5 & 10.5 \\ 4.5 & 5.5 & 6.5 & 7.5 & 8.5 & 9.5 & 10.5 & 11.5 \\ 5.5 & 6.5 & 7.5 & 8.5 & 9.5 & 10.5 & 11.5 & 12.5 \\ 6.5 & 7.5 & 8.5 & 9.5 & 10.5 & 11.5 & 12.5 & 13.5 \\ 7.5 & 8.5 & 9.5 & 10.5 & 11.5 & 12.5 & 13.5 & 14.5 \\ 8.5 & 9.5 & 10.5 & 11.5 & 12.5 & 13.5 & 14.5 & 15.5 \end{matrix} \qquad w = \begin{matrix} \boxed{5} & \boxed{4} & 1 & -3 & 0 & 0 & 0 & 0 \\ \boxed{3.5} & \boxed{5} & 0 & 5 & 0 & 0 & 0 & 0 \\ 3 & 3 & 2 & -5 & 0 & 0 & 0 & 0 \\ 0 & -2 & 0 & 6.5 & 0 & 0 & 0 & 0 \\ 0 & 0 & 0 & 0 & 0 & 0 & 0 & 0 \\ 0 & 0 & 0 & 0 & 0 & 0 & 0 & 0 \\ 0 & 0 & 0 & 0 & 0 & 0 & 0 & 0 \\ 0 & 0 & 0 & 0 & 0 & 0 & 0 & 0 \end{matrix}$$

The selected watermark samples will produce a difference between the quantized original and watermarked image coefficients.

7.4. Based on the principles introduced in the previous example, the image watermarking procedure is implemented as follows:

(a) DCT is calculated for the $8 \times 8$ image blocks
(b) The watermark $w$ is added to the quantized coefficients:
$I_q(i,j) = K(i,j)Q(i,j)$, ($K(i,j)$ are integers), but the coefficients quantized to zero value are immediately omitted
(c) The selection of coefficients suitable for watermarking is done according to the constraints:

- The watermarked DCT coefficients after quantization with $Q$ should have non-zero values;
- The watermarked coefficients should not be rounded to the same value as the original coefficients.

Analyze and define the values of $K(i,j)$ and watermark $w$, that satisfy the above constraints.

Solution and explanation:
To ensure that the watermarked DCT coefficients after quantization with $Q$, have non-zero values, the following relation should hold:

$$|K(i,j)Q(i,j)| - |w| \geq \frac{Q(i,j)}{2}, \tag{7.46}$$

or equivalently, the watermark should satisfy the condition:

*Condition 1*

$$|w| \leq |K(i,j)Q(i,j)| - \frac{Q(i,j)}{2}. \tag{7.47}$$

The watermarked coefficients will not be quantized to the same value as the original coefficients if the following condition is satisfied:

*Condition 2*

$$\begin{array}{l} K(i,j)Q(i,j) + w < K(i,j)Q(i,j) - Q/2 \\ or \\ K(i,j)Q(i,j) + w \geq K(i,j)Q(i,j) + Q/2 \end{array}. \tag{7.48}$$

From Eq. (7.48) we have: $|w| > Q(i,j)/2$. Combining *Condition 1* and *Condition 2*, we get:

$$w \subset \left\{ -[|K(i,j)| - 1/2]Q(i,j), \ -\frac{Q(i,j)}{2}) \cup (\frac{Q(i,j)}{2}, [|K(i,j)| - 1/2]Q(i,j) \right\},$$

Note that $|K(i,j)| \geq 2$ should hold.

7.5. Define the form of a locally optimal watermark detector, that corresponds to the watermarked coefficients pdf, assuming that they are selected by using the criterion $|K(i,j)| \geq 2$. The coefficients pdf can be modeled by a function illustrated in Fig. 7.10.

Solution:
In the considered case, the coefficients pdf from Fig. 7.10 can be approximately described by using the following function:

$$p(I_x) = \frac{\left(\frac{I_x}{a}\right)^{2n}}{1 + \left(\frac{I_x}{a}\right)^{2n}} e^{-\left|\frac{I_x}{a}\right|^{2\gamma}}, \tag{7.49}$$

where parameter $a$ defines the positions of the pdf maxima, while $n$ controls the pdf decay between the maximum and the origin. The parameter $\gamma$ is usually equal to 1/2, 1 or 2.

A locally optimal detector can be defined as:

$$D_{opt} = -\frac{p'(I_x)}{p(I_x)} \cdot w, \tag{7.50}$$

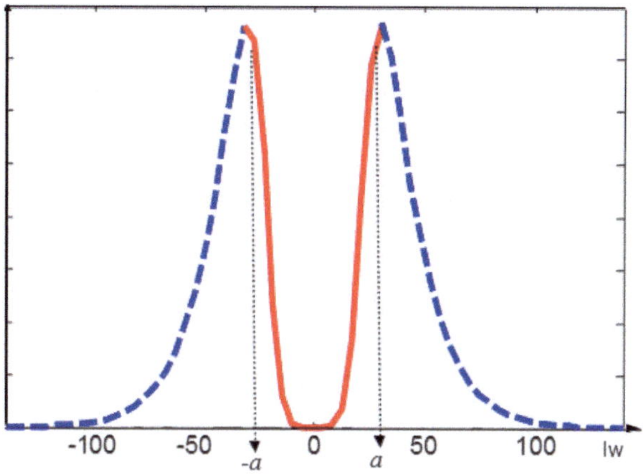


**Fig. 7.10** The distribution of coefficients after omitting low-frequency components

which in the case of the specified function $p$ becomes:

$$D_{opt} = \sum_{i=1}^{M} w_i \left( \frac{\gamma}{a^{2\gamma}} I_{x_i}^{2\gamma-1} \operatorname{sgn}\left(\frac{I_{x_i}}{a}\right)^{2\gamma} - \frac{n}{I_{x_i}\left(1+\left(\frac{I_{x_i}}{a}\right)^{2n}\right)} \right), \tag{7.51}$$

where $M$ is the number of watermarked coefficients.

7.6. By using the results obtained in the Example 7.4, derive the condition for watermarked coefficients selection which would provide the robustness to a certain JPEG quantization degree defined by the matrix $Q'$. By robustness we assume that the coefficients pdf is preserved even after quantization with matrix $Q'$, in order to provide successful detection by using locally optimal detector. Assume that the watermark embedding is done according to:

$$I_w(i,j) = \operatorname{round}\left(\frac{I(i,j)}{Q(i,j)}\right) Q(i,j) + Q(i,j) w, \tag{7.52}$$

where $Q$ is the quantization matrix with high quality factor $QF$ (i.e., a low compression ratio).

Solution:
In order to provide robustness to the quantization defined by the matrix $Q'$ (quality factor $QF'$), the criterion for coefficients selection should be modified. The watermarked coefficients will be robust after applying $Q'$ if they are not rounded to zero, i.e., if the following condition is satisfied:

$$|K(i,j)Q(i,j)| - |Q(i,j)w| > \frac{Q'(i,j)}{2}. \tag{7.53}$$

Note that the worst case is assumed in Eq. (7.53): the coefficient and the watermark have opposite signs. Hence, we may observe that for efficient watermark detection, the coefficients should be selected for watermarking if:

$$|K(i,j)| \geq w + \frac{Q'(i,j)}{2Q(i,j)}. \tag{7.54}$$

Therefore, if $Q$ with an arbitrary high $QF$ is used for watermark embedding, the robustness is satisfied even for matrix $Q'$ as long as the criterion Eq. (7.54) is satisfied. In this way the procedure provides the full control over the robustness to any JPEG quantization degree.

Note that if the criterion is satisfied for $QF' < QF$, then the watermark detection will certainly be successful for any quantization $Q_x$ defined by $QF_x$ for which $QF_x > QF'$ holds.

7.7. A speech watermarking procedure in the time-frequency domain can be designed according to the following instructions:

1. Voiced speech regions are used for watermarking.
2. Watermark is modeled to follow the time-frequency characteristics of speech components in the selected region.
3. Watermark embedding and detection is done in the time-frequency domain by using the S-method and time-varying filtering procedure.

(a) Design a time-frequency mask for watermark modeling and define the modeled watermark form.

(b) Define a watermark detection procedure in the time-frequency domain which includes the cross-terms.

Solution:

(a) By using the S-method (with $L=3$), a voiced speech region is selected:

$$D = \{(t,\omega) : t \in (t_1,t_2), \omega \in (\omega_1,\omega_2)\},$$

where $t_1$ and $t_2$ are the start and end points in the time domain, while frequency range is $\omega \in (\omega_1,\omega_2)$. According to Eq. (7.42), the time-frequency mask can be defined as:

$$L_M(t,\omega) = \begin{cases} 1 \text{ for } (t,\omega) \in D \text{ and } |SM(t,\omega)| > \xi \\ 0 \text{ for } (t,\omega) \notin D \text{ or } |SM(t,\omega)| < \xi \end{cases},$$

where parameter $\lambda$ within the energy floor $\xi$ can be set to 0.7. The illustration of speech region is given in Fig. 7.11a, the corresponding mask is shown in Fig. 7.11b, while the time-frequency representation of the modeled watermark is shown in Fig. 7.11c. The modeled version of the watermark is obtained by using Eq. (7.43).

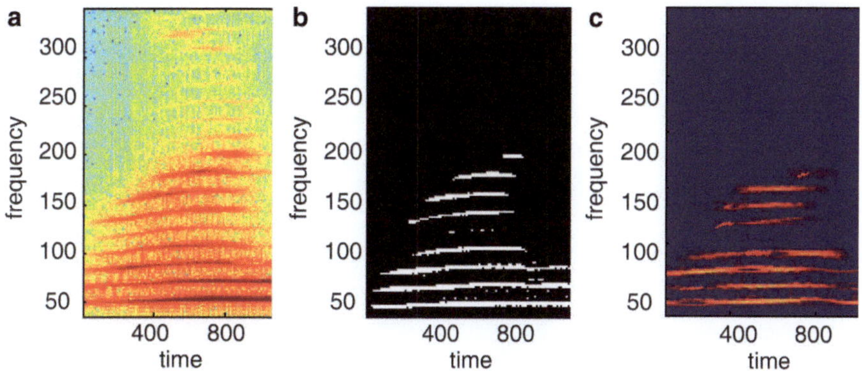


**Fig. 7.11** (**a**) Speech region selected for watermarking, (**b**) mask function, (**c**) time-frequency representation of modeled watermark

Note that the time-frequency characteristics of watermark correspond to the speech components. Hence, it would be difficult to remove the watermark without introducing serious signal quality degradation.

(b) The watermark detection can be performed by using the S-method with $L = 32$ to intentionally produce the cross-terms:

$$D = \sum_{i=1}^{N} SM_{wkey}^{i} SM_{x_w}^{i} + \sum_{\substack{i,\, j=1 \\ i \neq j}}^{N} SM_{wkey}^{i,j} SM_{x_w}^{i,j}, \tag{7.55}$$

where the index $w$ is related to watermark and $x_w$ to watermarked coefficients. Although the cross-terms are usually undesirable in the time-frequency analysis, they may increase performance of watermark detector.

7.8. In analogy with one-dimensional case described in the previous example, design a space/spatial-frequency based image watermarking procedure.

Note: Space/spatial-frequency representation is calculated for each pixel and it reflects the two-dimensional local frequency content around the pixel. The 2D form of the STFT for the window of size $N \times N$ is extended from the 1D version as:

$$STFT(n_1, n_2, k_1, k_2) = \sum_{i_1=-N/2}^{N/2-1} \sum_{i_2=-N/2}^{N/2-1} I(n_1 + i_1, n_2 + i_2) w(i_1, i_2) e^{-j\frac{2\pi}{N}(k_1 i_1 + k_2 i_2)}.$$

Solution:

Space/spatial-frequency representation can be used for classification between the flat and busy image regions. Namely, busy image regions are preferred in watermarking, because it is easier to provide watermark imperceptibility.

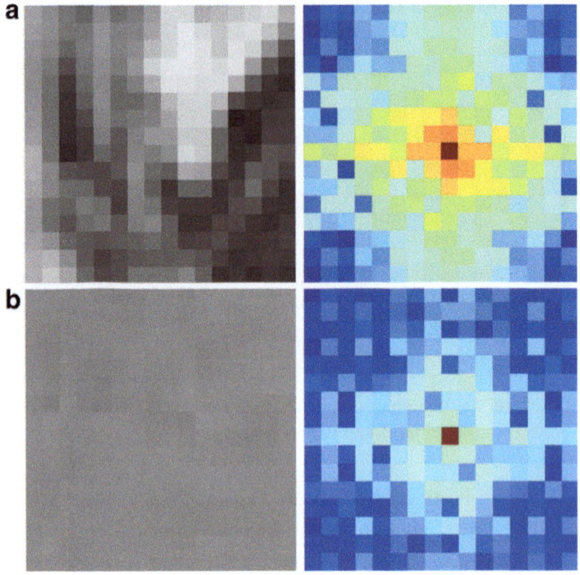

**Fig. 7.12** (**a**) Busy region and its spectrogram, (**b**) flat region and its spectrogram

The examples of busy and flat image regions are shown in Fig. 7.12a, b, respectively. Note that, unlike the busy regions, the flat regions contain small number of significant components in the space/spatial-frequency domain, which can be used as a criterion for regions classification.

Following analogy with the procedure for speech signals, watermark can be modeled according to the local frequency characteristics defined by the mask $L$:

$$w_{key}(n_1, n_2) = \sum_{\omega_1} \sum_{\omega_2} STFT_p(n_1, n_2, \omega_1, \omega_2) L(n_1, n_2, \omega_1, \omega_2), \tag{7.56}$$

where $STFT_p$ is a short-time Fourier transform of the two-dimensional pseudo-random sequence. The mask is obtained as:

$$L(n_1, n_2, \omega_1, \omega_2) = \begin{cases} 1 \text{ for } (\omega_1, \omega_2) : & |STFT(n_1, n_2, \omega_1, \omega_2)|^2 > \xi, \\ 0 \text{ for } (\omega_1, \omega_2) : & |STFT(n_1, n_2, \omega_1, \omega_2)|^2 \leq \xi. \end{cases}$$

Watermark embedding and detection can be done in the space/spatial-frequency domain in the same way as in the case of speech signals.

# References


1. Al-khassaweneh M, Aviyente S (2005) A time-frequency based perceptual and robust watermarking scheme. Proc. of EUSIPCO 2005, Antalya
2. Barni M, Bartolini F (2004) Watermarking systems engineering. Marcel Dekker, Inc., New York, NY
3. Barkat B, Sattar F (2010) Time-frequency and time-scale-based fragile watermarking methods for image authentication. EURASIP J Adv Signal Process 2010, 408109. doi:10.1155/2010/408109
4. Battiato S, Emanuel S, Ulges A, Worring M (2012) Multimedia in forensics, security, and intelligence. IEEE Multimedia Mag 19(1):17–19
5. Briassouli A, Strintzis MG (2004) Locally optimum nonlinearities for DCT watermark detection. IEEE Trans Image Process 13(12):1604–1618
6. Cox IJ, Miller ML, Bloom JA (2002) Digital watermarking. Academic, London
7. Dittmann J, Megías D, Lang A, Herrera-Joancomartí J (2006) Theoretical framework for a practical evaluation and comparison of audio watermarking schemes in the triangle of robustness, transparency and capacity. In: Transactions on Data Hiding and Multimedia Security I, vol 4300, Lecture Notes in Computer Science. Springer, Berlin, pp 1–40
8. Djurović I, Stanković S, Pitas I (2001) Digital watermarking in the fractional Fourier transformation domain. J Netw Comput Appl 24(2):167–173
9. Esmaili S, Krishnan S, Raahemifar K (2003) Audio watermarking time-frequency characteristics. Can J Elect Comput Eng 28(2):57–61
10. Foo SW, Ho SM, Ng LM (2004) Audio watermarking using time-frequency compression expansion. Proc. of the Int. Symp. on Circuits and Systems, ISCAS 04, 3: III - 201-4
11. Hannigan BT, Reed A, Bradley B (2001) Digital watermarking using improved human visual system model. Proc SPIE 4314:468–474
12. Heeger D (1997) Signal detection theory, Teaching Handout. Department of Psychology, Stanford University, Stanford, CA
13. Hernandez JR, Amado M, Perez Gonzales F (2000) DCT-domain watermarking techniques for still images: detector performance analysis and a new structure. IEEE Trans Image Process 9:55–68
14. Kang X, Huang J, Zeng W (2008) Improving robustness of quantization-based image watermarking via adaptive receiver. IEEE Trans Multimedia 10(6):953–959
15. Katzenbeisser S, Petitcolas F (2000) Information hiding: techniques for steganography and digital watermarking. Artech House, Norwood, MA
16. Kirovski D, Malvar HS (2003) Spread-spectrum watermarking of audio signals. IEEE Trans Signal Process 51(4):1020–1033
17. Mobasseri BG, Zhang Y, Amin MG, Dogahe BM (2005) Designing robust watermarks using polynomial phase exponentials. Proc. of Acoustics, Speech, and Signal Processing (ICASSP'05), Philadelphia, PA, vol 2, pp ii/833–ii/836
18. Muharemagić E, Furht B (2006) Survey of watermarking techniques and applications, Chapter 3. In: Furht B, Kirovski D (eds) Multimedia watermarking techniques and applications. Auerbach Publication, Boca Raton, FL, pp 91–130
19. Nikolaidis A, Pitas I (2003) Asymptotically optimal detection for additive watermarking in the DCT and DWT domains. IEEE Trans Image Process 12(5):563–571
20. Proceedings of the IEEE: Special Issue on Identification and Protection of Multimedia Information, vol 87, July 1999
21. Podilchuk CI, Zheng W (1998) Image adaptive watermarking using visual models. IEEE J Select Areas Commun 16:525–539
22. Stankovic L, Stankovic S, Djurovic I (2000) Space/spatial-frequency based filtering. IEEE Trans Signal Process 48(8):2343–2352
23. Stanković S, Djurović I, Herpers R, Stanković LJ (2003) An approach to the optimal watermark detection. AEUE Int J Elect Commun 57(5):355–357

24. Stanković S, Djurović I, Pitas I (2001) Watermarking in the space/spatial-frequency domain using two-dimensional Radon-Wigner distribution. IEEE Trans Image Process 10:650–658
25. Stankovic S, Orovic I, Chabert M, Mobasseri B (2013) Image watermarking based on the space/spatial-frequency analysis and Hermite functions expansion. J Elect Imaging 22 (1):013014
26. Stankovic S, Orovic I, Mobasseri B, Chabert M (2012) A Robust procedure for image watermarking based on the Hermite projection method. Automatika 53(4):335
27. Stanković S, Orović I, Žarić N (2008) Robust watermarking procedure based on JPEG-DCT image compression. J Elect Imaging 17(4):043001
28. Stanković S, Orović I, Žarić N (2010) An application of multidimensional time-frequency analysis as a base for the unified watermarking approach. IEEE Trans Image Process 19 (2):736–745
29. Stanković S (2000) About time-variant filtering of speech signals with time-frequency distributions for hands-free telephone systems. Signal Process 80(9):1777–1785
30. Stanković S, Orović I, Žarić N (2008) Robust speech watermarking in the time-frequency domain. EURASIP J Adv Signal Process 2008:519206
31. Steinebach M, Dittmann J (2003) Watermarking-based digital audio data authentication. EURASIP J Appl Signal Process 2003(10):1001–1015
32. Wang FH, Pan JS, Jain LC (2009) Innovations in digital watermarking techniques. Studies in computational intelligence. Springer, New York, NY
33. Wickens TD (2002) Elementary signal detection theory. Oxford University Press, Oxford

# Chapter 8
# Multimedia Signals and Systems in Telemedicine

Due to the advances in technology and medicine, humans tend to live longer. This has increased the pressure on health care systems worldwide to provide higher quality health care for a greater number of patients. A greater demand for health care services prompted researchers to seek new ways of organizing and delivering health care services. Telemedicine, as a new research area, promotes the use of multimedia systems as a way of increasing the availability of care for patients in addition to cost and time-saving strategies. In other words, telemedicine provides a way for patients to be examined and treated, while the health care provider and the patient are at different physical locations. Using telemedicine technologies, future hospitals will provide health care services to patients all over the world using multimedia systems and signals that can be acquired over distances. Signal and image transmission, storage, and processing are the major components of telemedicine.

## 8.1 General Health Care

### *8.1.1 Telenursing*

Telenursing requires the use of multimedia systems and signals to provide nursing practice over the distance. It was developed as a need to alter the current nursing practices and provide home care to older adults and/or other patient groups, which preferred to stay in the comfort of their own homes. Multimedia technologies (e.g., video-telephony) allow patients to maintain their autonomy by enhancing their emotional, relational, and social abilities. Generally, the patients welcome the use of multimedia systems to communicate with a nurse about their physical and psychological conditions. So far the use of advanced technology did not have any



S. Stanković et al., *Multimedia Signals and Systems*,
DOI 10.1007/978-3-319-23950-7_8

significant effects on health care providers and patients, as well as on their abilities to communicate.

Multimedia systems, when used for patient care, provide various advantages. As an example, let us mention that the multimedia signals and systems have been used to significantly reduce congestive heart failure readmission charges. Also, these systems can reduce the amount of needed time to care for a patient, while providing the same level of health care as in-patient visits. Similarly, an analysis of telenursing in the case of metered dose inhalers in a geriatric population have shown that multimedia systems can provide most of services, and only small percentage require on-site visits. It should be mentioned that some of the telenursing systems can reduce the number of visits to emergency departments or doctors in private practice.

It should be mentioned that other potential applications of telenursing also include, but are not limited to:

- training nurses remotely;
- caring for patients in war zones;
- global collaboration between nurses.

### 8.1.2 Telepharmacy

Particularly, in rural and remote areas the pharmacy services to patients are often limited. This has led to creation of service called telepharmacy, which assumes providing pharmaceutical care to patients and medication dispensing from distance. In this way, the Multimedia systems and technology preserve pharmacy services in remote rural communities. The telepharmacy services adhere to all official regulations and services as traditional pharmacies, including verification of drugs before dispensing and patient counseling. In other words, telepharmacy services maintain the same services as the traditional pharmacies and provide additional value-added features. Additional services provided by telepharmacies can also include point-of-care refill authorization and medication assistance referrals. Specifically, in a recent study analyzing the utility of telepharmacy services for education on a metered-dose inhaler technique, it has been shown that patient education provided by pharmacists via video was superior to education provided via written instructions on an inhaler package insert.

### 8.1.3 Telerehabilitation

Rehabilitation is based on the idea that therapeutic interventions can enhance patient outcomes, since human physiological system can dynamically alter as a function of inputs (e.g., exercise). Therefore, telerehabilitation tools enable us to

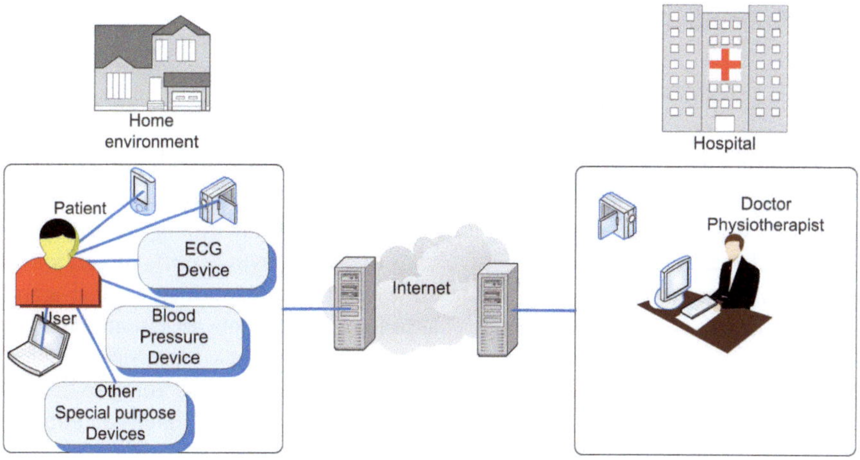


**Fig. 8.1** A typical telerehabilitation system

decrease the distance between patients and clinicians/researchers, which opens up new possibilities for discovering and implementing optimized intervention strategies.

Telerehabilitation was established in 1997 when the National Institute on Disability and Rehabilitation Research (US Department of Education) ranked it as one of the top priorities for a newly established Rehabilitation Engineering Research Center (RERC). While telerehabilitation covers diverse fields of investigations (e.g., intelligent therapeutic robots and other health gadgets), it also addresses societal challenges in the delivery of rehabilitative services. The main efforts have been made to provide telecommunication techniques that are capable to support rehabilitation services at a distance, then to provide technology for monitoring and evaluating the rehabilitation progress, and finally to provide technology for therapeutic intervention at a distance, Fig 8.1.

Having these comprehensive objectives, telerehabilitation may have far-reaching effects on patients. One such example is based on using a videoconsulting system in a community-based post-stroke program that involves educational talks, exercise, and psychosocial support, proving significant improvements in the health status of the patients after the intervention.

Furthermore, the feasibility of telerehabilitation has been applied for functional electrical stimulation of affected arm after stroke or for evaluating the speech and swallowing status of laryngectomy patients following discharge from acute care. Telerehabilitation tools have been also used to address the fall risks. Nevertheless, we still need to acquire strong evidence regarding the impact of telerehabilitation on resources and associated costs to support clinical and policy decision-making.

## 8.2 Specialist Health Care

### *8.2.1 Telecardiology*

Telecardiology encompasses merging technology with cardiology in order to provide a patient with a proper medical care without disturbing the patient's daily routines. Due to the advances in multimedia systems, telecardiology is currently a well-developed medical discipline involving many different aspects of cardiology (e.g., acute coronary syndromes, congestive heart failure, sudden cardiac arrest, arrhythmias). It is safe to state that telecardiology has become an essential tool for cardiologists in either a hospital-based or community-based practices. Patient consultations with cardiologists via multimedia systems are becoming extremely common. For example, a consulting cardiologist receives many signals and images in real time to assess the patient condition (Fig. 8.2).

Further technological advances will be fueled by the development of novel sensors and multimedia systems. This will result in a move from device-centered to patient-oriented telemonitoring. By focusing on patient-oriented monitoring, a comprehensive approach of disease management, based on coordinating health care interventions, is provided. Such a possibility will not only help us with early diagnosis and quick interventions, but will also prove to be cost-effective.

Therefore, telecardiology has the two major aims. The first aim is to reduce the health care cost. The second aim is to evaluate the efficiency of telecardiac tools (e.g., wireless ECG) at variable distances. By accomplishing these two aims, telecardiology will enhance the psychological well-being of patients in addition to bridging the gap between rural areas and hospitals. Note that, the search for new telecardiac technologies is a big challenge. However, various constraints such as institutional and financial factors may play a significant role in the further development of these multimedia systems needed in telecardiology before we see an increase of these tools in clinical practices.

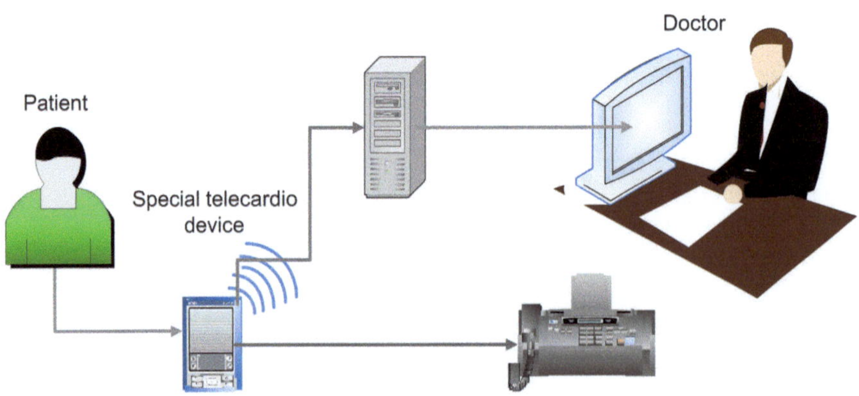


**Fig. 8.2** An example of multimedia systems in telecardiology

In order to fulfill the aims of telecardiology, multimedia based technologies have been increasingly applied to patients in small rural communities needing distance monitoring of their chronic heart conditions. These multimedia systems provided an alternative modality for effective cardiac care from a variable distances by utilizing information and communication technology.

### 8.2.2 Teleradiology

Teleradiology defines a field in which multimedia systems are used to acquire and interpret multimedia signals (e.g., radiological images) at different geographical locations. Teleradiology has especially flourished in recent years due to the increased development of digital imaging systems and the Internet. Nevertheless, it should be mentioned that teleradiology is not only acquisition and transmission of digital images between different locations but also involves sharing of expertise between radiologists across the globe, providing radiological services to remote areas, around the clock coverage, etc.

Initially, teleradiology was developed to provide health care to wounded soldiers. In 1980s, the first commercial teleradiology system was developed with ability to capture and transfer radiological videos. However, further development of teleradiological systems was slowed down to a lack of systems for inexpensive transfer of radiological videos. Due to the advances in multimedia systems for acquisition and transfer of video data (e.g., wavelet compression algorithms for images) and the development of the Internet telecommunication systems, we have witnessed a significant growth in teleradiological services. An illustration of teleradiological communication system is shown in Fig. 8.3. A Picture Archive Communication System is used to store, transfer, and display digital images acquired in Diagnostic Imaging. The archive server is used to provide a long-term backup of the data. The Radiology Information Server is used to connect different aspects of the radiology information system.

Given the current state of the art when it comes to multimedia systems, most of the current efforts in teleradiology are geared towards medicolegal issues. Teleradiology is one of the first fields where the development of technologies in the recent years has sparked intense professional and even legal debates regarding the role of radiologists in the patient care. For example, teleradiology can provide great benefits in emergency departments, when used correctly. However, poorly implemented teleradiological services can degrade the quality of patient care. Hence, it has been urged that during the design, management, and performance of teleradiology services radiologists should play a significant role.

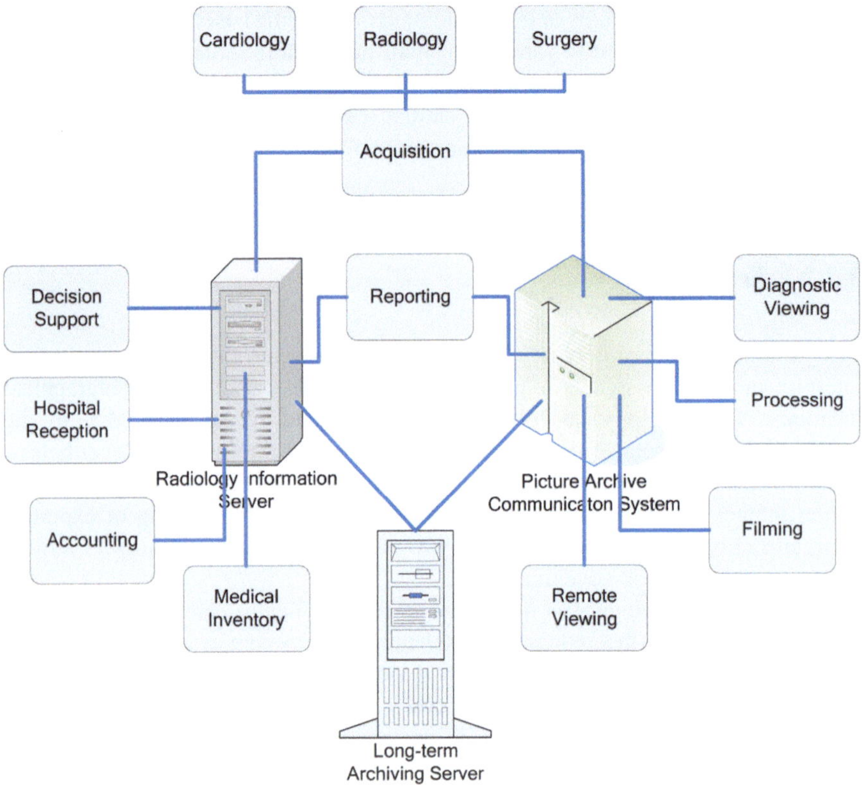


**Fig. 8.3** A teleradiology system

## 8.2.3 *Telesurgery*

Dissemination of new surgical skills and techniques across the wide spectrum of practicing surgeons is often difficult and time consuming, especially because the practicing surgeons can be located very far from large teaching centers. Therefore, telesurgery provides multiple advantages to practicing surgeons, including but not limited to dissemination of expertise, widespread patient care, cost savings, and improved community care (Fig. 8.4). It is expected that more widespread multimedia systems and technologies may exist to launch everyday telesurgery procedures within a few years.

Specifically, telesurgery has already shown to be a powerful method for performing minimally invasive surgery (MIS) because patients recover more rapidly when small MIS incisions are made in comparison to conventional methods. To examine the practicality of telesurgery over long distances, a recent study showed that operators using a telesurgery platform can complete maneuvers with

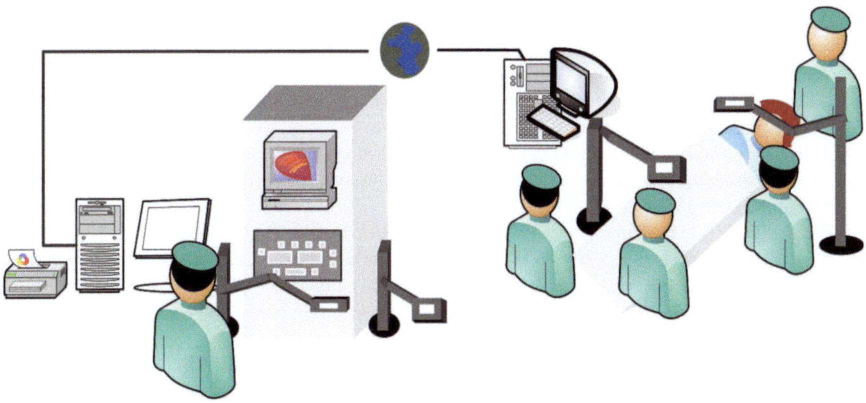

**Fig. 8.4** A typical use of a telesurgery system

delays up to 500 ms and no additional increase in error rates. Also, the emulated surgery in animals can be effectively executed using either ground or satellite, while keeping the satellite bandwidth above 5 Mb/s.

## References


1. Anvari M (2007) Telesurgery: remote knowledge translation in clinical surgery. World J Surg 31(8):1545–1550
2. Arnaert A, Delesie L (2001) Telenursing for the elderly. The case for care via video-telephony. J Telemed Telecare 7(6):311–316
3. Barneveld Binkhuysena FH, Ranschaert ER (2011) Teleradiology: evolution and concepts. Eur J Radiol 78(2):205–209
4. Brunetti ND, Amodio G, De Gennaro L, Dellegrottaglie G, Pellegrino PL, Di Biase M, Antonelli G (2009) Telecardiology applied to a region-wide public emergency health-care service. J Thromb Thrombolysis 28(1):23–30
5. Bynum A, Hopkins D, Thomas A, Copeland N, Irwin C (2001) The effect of telepharmacy counseling on metered-dose inhaler technique among adolescents with asthma in rural Arkansas. Telemed J E-Health 7(3):207–217
6. Chan WM, Hjelm NM (2001) The role of telenursing in the provision of geriatric outreach services to residential homes in Hong Kong. J Telemed Telecare 7(1):38–46
7. Friesner DL, Scott DM, Rathke AM, Peterson CD, Anderson HC (2011) Do remote community telepharmacies have higher medication error rates than traditional community pharmacies? Evidence from the North Dakota Telepharmacy Project. J Am Pharm Assoc 51(5):580–590
8. Garrelts JC, Gagnon M, Eisenberg C, Moerer J, Carrithers J (2010) Impact of telepharmacy in a multihospital health system. Am J Health Syst Pharm 67(17):1456–1462
9. Giansanti D, Morelli S, Maccioni G, Costantini G (2009) Toward the design of a wearable system for fall-risk detection in telerehabilitation. Telemed J E-Health 15(3):296–299
10. Hagan L, Morin D, Lepine R (2000) Evaluation of telenursing outcomes: satisfaction, self-care practices, and cost savings. Public Health Nurs 17(4):305–313
11. Hermann VH, Herzog M, Jordan R, Hofherr K, Levine P, Page SJ (2010) Telerehabilitation and electrical stimulation: an occupation-based, client-centered stroke intervention. Am J Occup Ther 64(1):73–81

12. Jerant AF, Azari R, Martinez C, Nesbitt TS (2003) A randomized trial of telenursing to reduce hospitalization for heart failure: patient-centered outcomes and nursing indicators. Home Health Care Serv Q 22(1):1–20
13. Jönsson AM, Willman A (2009) Telenursing in home care services: experiences of registered nurses. Electron J Health Inf 4(1):e9.1–7
14. Kairy D, Lehoux P, Vincent C, Visintin M (2009) A systematic review of clinical outcomes, clinical process, healthcare utilization and costs associated with telerehabilitation. Disabil Rehabil 31(6):427–447
15. Katz ME (2010) Pediatric teleradiology: the benefits. Pediatr Radiol 40(8):1345–1348
16. Lai JCK, Woo J, Hui E, Chan WM (2004) Telerehabilitation - a new model for community-based stroke rehabilitation. J Telemed Telecare 10(4):199–205
17. Lam AY, Rose D (2009) Telepharmacy services in an urban community health clinic system. J Am Pharm Assoc 49(5):652–659
18. Mitchell JR, Sharma P, Modi J, Simpson M, Thomas M, Hill MD, Goyal M (2011) A smartphone client-server teleradiology system for primary diagnosis of acute stroke. J Med Internet Res 2:e31
19. Nikus K, Lähteenmäkib J, Lehto P, Eskola M (2009) The role of continuous monitoring in a 24/7 telecardiology consultation service - a feasibility study. J Electrocardiol 42(6):473–480
20. Pappasa Y, Sealeb C (2010) The physical examination in telecardiology and televascular consultations: a study using conversation analysis. Patient Educ Couns 81(1):113–118
21. Peterson CD, Anderson HC (2004) The North Dakota telepharmacy project: restoring and retaining pharmacy services in rural communities. J Pharm Technol 20(1):28–39
22. Wakefield DS, Ward MM, Loes JL, O'Brien J, Sperry L (2010) Implementation of a telepharmacy service to provide round-the-clock medication order review by pharmacists. Am J Health Syst Pharm 67(23):2052–2057
23. Ward E, Crombie J, Trickey M, Hill A, Theodoros D, Russell T (2009) Assessment of communication and swallowing post-laryngectomy: a telerehabilitation trial. J Telemed Telecare 15(5):232–237
24. Whitten P, Mair F, Collins B (2001) Home telenursing in Kansas: patients' perceptions of uses and benefits. J Telemed Telecare 3(1):67–69
25. Winters JM (2002) Telerehabilitation research: emerging opportunities. Annu Rev Biomed Eng 4:287–320

# Index




S. Stanković et al., *Multimedia Signals and Systems*,
DOI 10.1007/978-3-319-23950-7

## J



## L



## M



## N



## O



## P



## Q

MIX
Papier aus verantwortungsvollen Quellen
Paper from responsible sources
FSC® C105338

If you have any concerns about our products,
you can contact us on
**ProductSafety@springernature.com**

In case Publisher is established outside the EU,
the EU authorized representative is:
**Springer Nature Customer Service Center GmbH**
**Europaplatz 3, 69115 Heidelberg, Germany**

Printed by Libri Plureos GmbH
in Hamburg, Germany